Modern Statistical, Systems, and GPSS Simulation

Second Edition

Modern Statistical, Systems, and GPSS Simulation

Second Edition

Zaven A. Karian
Denison University
Granville, Ohio

Edward J. Dudewicz
Syracuse University
Syracuse, New York

CRC Press
Boca Raton London New York Washington, D.C.

Library of Congress Cataloging-in-Publication Data

Karian, Zaven A.
    Modern statistical, systems, and GPSS simulation : the first course
/ Zaven A. Karian, Edward J. Dudewicz. -- 2nd ed.
        p.   cm.
    Includes bibliographical references and index.
    ISBN 0-8493-3922-7 (alk. paper)
    1. Digital computer simulation. 2. GPSS (Computer program
language)   I. Dudewicz, Zaven A. II. Title.
QA76.9.C65K37 1998
003'.35133—dc21   98-35396
                                                                                                            CIP

Preface to the First Edition

The primary intent of this book is to provide a modern text that introduces the theory and implementation of discrete-event simulation. Most texts on this subject fall into one of two categories. Some concentrate on theoretical and statistical concepts and issues. With such a text, a student may learn the best ways to design a simulation and analyze its output, but will lack appreciation for the complexities of implementation, since such texts treat implementation in a particular programming language in only a cursory way. Other texts describe a specific simulation language in great detail, almost like a language manual, but give scant attention to conceptual issues. With such a text, a student may learn how to program, but not what to program, how long to run the program, or how to analyze the output.

In our opinion, a modern approach requires more than either of these options offers. A text today should provide at least three elements. It should establish a theoretical basis for simulation methodology, so that students will know which random number generators are good, how long to run a simulation, and how to analyze the output from simulations run for some of the most important experimental goals. It should also give enough details of an important simulation language (GPSS—General Purpose Simulation System—is used here) to enable the reader to develop reasonably complex models, and to include in them good random number generators, rational methods of determining how long to run a simulation, and valid statistical output analysis. A third element in a modern text is to integrate the previous two elements and bring them together at the end in a systems simulation case study. We do this in Chapter 8, where we analyze a case study of a transportation problem and provide details including an efficient experimental design and a GPSS simulation. Our goal has been to strike a balance between extremes so that no one element is studied to the exclusion of the others, but to do so in a way that allows instructors to concentrate on any one or two elements if this seems appropriate.

Chapters 1, 3, 4, and 6 provide the conceptual basis for simulation in any language and include the elements that are most important in the design and analysis of a simulation. In a **course on the statistics of simulation**, one could concentrate on these chapters, covering them all in virtually full detail, and be assured that students would have learned the topics needed for future simulation work. The other chapters might not be covered formally in the course, but would still indicate how to use this material in combination with a specific simulation language. Such courses might be offered in departments of mathematics, with a prerequisite of an undergraduate course in statistics.

Chapters 1, 2, 5, and 7 provide the basis for a **course specifically on the GPSS simulation language**. In such a course, one could cover all these chapters and be assured that students would see how to incorporate into their simulations the best modern random number generators, choice of run length, and methods of analyzing output. The consideration of these aspects of the other chapters in Chapters 1, 2, 5, and 7 is a unique feature of this book.

Chapter 8 can be included with either of the two scenarios as a capstone to the course. This approach might be desirable for courses in systems simulation such as in a department of industrial and systems engineering or for operations research or quantitative analysis courses given in schools of management or business administration. The inclusion of Chapter 8 assumes that students have some familiarity with the rudiments of experimental design from Chapter 6.

The GPSS programming language was chosen as the vehicle for implementing simulations because of its popularity and its ease of use. There may be other languages that are more popular—it seems that more simulations are programmed in FORTRAN than in any other language—or more versatile or more powerful. But no other language seems to have the semantic features, ease of use, and general availability that combine to make GPSS so appropriate for our purposes.

We develop GPSS in three stages, following the general introduction in Chapter 1. Chapter 2 presents the GPSS syntax and introduces features that enable the reader to write programs for simulating simple queueing systems. Chapter 5 develops a deeper understanding of GPSS through a discussion of the internal workings of the language and its use of chain structures. As illustrated by the examples there, it is possible to develop models of reasonable complexity, using the content of Chapters 2 and 5. Chapter 7 starts with a discussion of the GPSS Standard Numerical Attributes, which add considerable flexibility to model development. It also includes many of the GPSS features not covered in Chapters 2 and 5.

The GPSS used throughout this book is generic in the sense that almost all GPSS implementations have all the features that we describe. We have intentionally excluded features that are specific only to a particular version of GPSS (such as GPSS/PC, GPSS/H, or GPSS/VX), to make the text appropriate for use with whatever GPSS implementation is available to the reader. This approach also ensures that students will not have to relearn features in order to program with another version in their workplace. Such a course requires programming experience in some high-level language as a prerequisite and might be given in computer science departments.

GPSS/PC is an interactive implementation of GPSS for the IBM-PC compatible family of microcomputers. The disk accompanying this book contains the limited educational version of GPSS/PC along with many of the illustrative examples discussed in the text. With this disk, students who have access to an IBM-PC compatible microcomputer with the DOS operating system will be able to develop GPSS models as they progress through the book. Appendix A provides an introduction to using GPSS/PC, describes some of its interactive features, and indicates which examples are included on the disk.

This text has been tested in classes not only at our own universities, but also at the University of Pittsburgh at Bradford (by Richard Melka), at Ithaca College (by Diane Schwartz), and at John Brown University (by Calvin Piston). Professor Schwartz's course was similar to the one we described on the statistics of simulation, while Professor Piston's included extensive use of GPSS/PC. We gratefully acknowledge the suggestions of these faculty and their students, as well as those of our own students at Denison University and Syracuse University, from which we and the text have benefited greatly. We have also received valuable suggestions from participants in short courses on this material that we have offered at the Center for Statistics, Quality Control, and Design at the State University of New York at Binghamton, at the Ohio State University, and at annual meetings of the American Statistical Association and of the Society for Computer Simulation.

Susan Streiff typed much of the manuscript. We particularly appreciate the efforts of Susan Karian in typesetting the entire manuscript in TEX, which was made available through the Denison University Research Foundation. We also wish to acknowledge the assistance of the staff at W. H. Freeman and Company, particularly that of Diana Siemens, throughout the editorial process.

The **numbering system** used in this book is designed to enable the reader to locate theorems, algorithms, figures, etc., quickly. Theorems, lemmas, algorithms, definitions, remarks, equations, and expressions are numbered sequentially (e.g., equation (4.8.7) is in Chapter 4, Section 8 and is followed by algorithm 4.8.8). This is simpler to use than the common system wherein theorems, algorithms, definitions, etc., are numbered sequentially *separately* from equations and expressions; such a system may have an algorithm 4.8.8 as well as an equation (4.8.8), possibly several pages apart.

Figures and tables in the text (except for tables appearing in the appendices), have their own combined numbering sequence. For example, Figure 4.6–2 is the second such item in Section 6 of Chapter 4 and we have Table 4.6–1 followed by Figure 4.6–2 in Section 6 of Chapter 4. This is simpler than the more common system where one may have both a Table 4.6–1 and a Figure 4.6–1; in our scheme, a person searching for x.y–z knows that when x.y–(z–1) is found, the item of interest is ahead and x.y–(z–2) is behind.

Zaven A. Karian
Denison University
Granville, Ohio

Edward J. Dudewicz
Syracuse University
Syracuse, New York

August 1990

Preface to the Second Edition

Our experience using this book in teaching for most of the 1990s, and comments from our students and instructors at a variety of institutions, have convinced us of the desirability of making a number of additions in this new edition. Similarly, comments from users in industry and government have led to the inclusion of new materials, as have those of researchers in many fields (such as computer and information sciences, operations research, statistical sciences, environmental sciences and forestry, mathematical sciences, earth sciences, economics, industrial and systems engineering, and other fields). The comments of our readers indicate that, as the field continues its meteoric ascent and we enter the 21st century with speed of computation undreamed of only a few years ago, the following would be valuable additions.

- Coverage of random number generators with astronomic period (such as 10^{30}), which are very attractive to many now that simulations run so fast that the traditional generators risk exhaustion of their number stream (now in Chapter 3).

- Coverage of the new entropy-based tests of uniformity (i.e., of random number generators' goodness), which are currently of high interest (now in Chapter 3).

- Coverage of the goodness of additional random number generators; these continue to proliferate, yet as we observed in the original version of the book, at best only about one in four proposed generators is of suitable quality for serious simulation studies (in Chapter 3).

- Coverage of gamma variate generation, plus additional coverage of beta variate, Student's t variate, and normal variate generation (now in Chapter 4).

- Enhancements to fitting distributions to data, in particular,

  - new results on the GLD (namely, the Extended GLD) which allow it to be fitted to any vector of (mean, variance, skewness, kurtosis) values—previously some regions that occurred in applied studies were excluded

  - new results which extend the GLD to bivariate distribution fitting (and also allow for fitting of distributions with non-convex contours of constant probability density function).

  These are now in Chapter 4.

- Coverage of additional statistical design and analysis aspects, in particular,

  - efficient estimation of a percentile point
  - estimation of $\text{Var}(\bar{X})$
  - variance reduction techniques.

  These are now in Chapter 6.

Of course, there have been a number of other less substantial additions, corrections, and modifications. However, all revisions have been guided by comments which indicated that the existing material was well-received and should, therefore, not be subjected to unnecessary surgery. As one reviewer put it,

> I am very impressed by the text. It is well-written and covers all of the important topics that one would want to teach in a simulation course: event and process world views, data structures, a little queueing theory, a lot of modeling, random number and variate generation, output analysis, experimental design, and a thorough treatment of a simulation language (in this case, GPSS). I like the organization, mathematical level, and technical presentation of the book. The book also seems to be technically up-to-date...students will benefit a great deal from the presentation and exercises.

We look forward to receiving comments from users of this new edition (as we have benefited greatly from comments from users of its predecessor and of the Arabic Edition published by King Saud University Press in Saudi

Arabia). These may be sent to us by what is now called "snail mail" (which we do not use in a pejorative sense) or by e-mail (electronic mail), for which our current addresses are provided below. In cases of electronic communication, whenever possible, please also send messages by snail mail (one of our systems deletes messages after either 40 days, or if the system is "full," and thus messages may still be lost due to insufficient storage capacity at the university's mail server).

We wish to thank our students and our colleagues, from academic institutions, industry, and government, whose comments and suggestions have helped shape this edition. We also wish to express our gratitude to Susan Karian for her considerable typesetting assistance, to Patricia A. Dudewicz for office support, to Kevin L. Stultz for programming assistance, and to Mr. Robert B. Stern, Executive Editor at the CRC Press LLC, for his insightful comments and reviews. It has been a pleasure working with him and his staff. Their genuine concern for both users and authors is exhilarating.

Zaven A. Karian
Denison University
Granville, Ohio
Karian@Denison.edu

Edward J. Dudewicz
Syracuse University
Syracuse, New York
Dudewicz@syr.edu

August 1998

About the Authors

Dr. Zaven A. Karian holds the Benjamin Barney Chair of Mathematics, and is Professor of Mathematics and Computer Science at Denison University in Ohio. He has been active as instructor, researcher and consultant in mathematics, computer science, statistics, and simulation for over thirty years. He has taught workshops in these areas for a dozen educational institutions and national and international conferences (International Conference on Teaching Mathematics, Greece; Asian Technology Conference in Mathematics, Japan; Joint Meetings of the American Mathematical Society/Mathematical Association of America).

Dr. Karian has taught short courses of varying lengths for colleges and universities (Howard University, Washington, D.C.; The Ohio State University; State University of New York; and Lyndon State College, Vermont), for professional societies (Society for Computer Simulation, American Statistical Association, Mathematical Association of America (MAA), the Ohio Section of the MAA), private and public foundations (Alfred P. Sloane Foundation, National Science Foundation). His consulting activities include Cooper Tire and Rubber Company, Computer Task Group, and Edward Kelcey and Associates (New Jersey), as well as over forty colleges and universities.

Dr. Karian is the author and co-author of eight texts, reference works, and book chapters and he has published over thirty articles. He serves as Editor for computer simulation of the *Journal of Applied Mathematics and Stochastic Analysis*. Dr. Karian holds the bachelor's degree from American International College in Massachusetts, master's degrees from the University of Illinois (Urbana-Champaign) and The Ohio State University, and his doctoral degree from The Ohio State University. He has been a Plenary Speaker, on two occasions, at the Asian Conference on Technology in Mathematics (Singapore and Penang, Malaysia).

Dr. Karian has served on the International Program Committees of conferences in Greece and Japan, the Board of Governors of the MAA, and the governing board of the Consortium of Mathematics and its Applications. He was a member of the Joint MAA/Association for Computing Machinery (ACM) Committee on Retraining in Computer Science and he chaired the Task Force (of the MAA, ACM and IEEE Computer Society) on Teaching Computer Science, the Subcommittee on Symbolic Computation of the MAA, and the Committee on Computing of the Ohio Section of the MAA.

Dr. Karian has been the Acting Director of the Computer Center, Denison University; Visiting Professor of Statistics, Ohio State University; Chair of the Department of Mathematical Sciences, Denison University. He has been the recipient of the R. C. Good Fellowship of Denison University on three occasions and has been cited in *Who's Who in America* (Marquis Who's Who, Inc.).

Dr. Edward J. Dudewicz is Professor of Mathematics at Syracuse University, New York. He has been active as instructor, researcher and consultant in digital simulation for over thirty years. He has taught statistics and digital simulation at Syracuse University, The Ohio State University, University of Rochester, University of Leuven (Belgium), and National University of Comahue (Argentina) and served as a staff member of the Instruction and Research Computer Center at Ohio State University, and as Head Statistician of New Methods Research, Inc. His consulting activities include O. M. Scott and Sons Company, Ohio Bureau of Fiscal Review, Mead Paper Corporation, and Blasland, Bouck, & Lee, Engineers & Geoscientists. Dr. Dudewicz is author, co-author, and editor of eighteen texts and reference works in statistics, simulation, computation, and modeling; he has published over one hundred and fifteen articles, as well as handbook chapters on statistical methods (*Quality Control Handbook* and *Magnetic Resonance Imaging*). He serves as editor of the series *Modern Digital Simulation: Advances in Theory, Application, & Design*, part of the American Series in Mathematical and Management Sciences.

Dr. Dudewicz holds the bachelor's degree from the Massachusetts Institute of Technology and master's and doctoral degrees from Cornell University. He has been Visiting Scholar and Associate Professor at Stanford University, Visiting Professor at the University of Leuven (Belgium) and at the Science University of Tokyo (Japan), Visiting Distinguished Professor at Clemson University, and Titular Professor at the National University of Comahue (Argentina) while Fulbright Scholar to Argentina. His Editorial posts have included *Technometrics* (Management Committee), *Journal of Quality Technology* (Editorial Review Board), *Statistical Theory and Method Abstracts* (Editor, U.S.A.), *Statistics & Decisions* (Germany) (Editor), and *American Journal of Mathematical and Management Sciences* (Founding Editor and Editor-in-Chief), and he also serves the journal *Information Systems Frontiers* (Executive Editorial Board).

Dr. Dudewicz has served as President, Syracuse Chapter, American Statistical Association; Graduate Committee Chairman, Department of Statistics, Ohio State University; Chairman, University Statistics Council, Syracuse University; External Director, Advanced Simulation Project, National University of Comahue, Argentina; Awards Chairman, Chemical Division, American Society for Quality; and Founding Editor, *Basic References in Quality Control: Statistical Techniques*, American Society for Quality (the "How To" series).

Recognitions of Dr. Dudewicz include Research Award, Ohio State Chapter, Society of the Sigma Xi; Chancellor's Citation of Recognition, Syracuse University; Jacob Wolfowitz Prize for Theoretical Advances; Thomas L. Saaty Prize for Applied Advances; Co-author, Shewell Award paper; Jack Youden Prize for the best expository paper in *Technometrics*. He has received the Seal of Banares Hindu University (India), where he was Chief Guest and Special Honoree, presenting an Inaugural Address and Keynote Address. He is an elected Fellow of the New York Academy of Sciences, the American Society for Quality, the Institute of Mathematical Statistics, the American Statistical Association, the International Statistical Institute, and the American Association for the Advancement of Science. He is a subject of biographical record in *Who's Who in America* (Marquis Who's Who, Inc.).

To Susan and Patricia

Contents

Chapter 1

Discrete Event Computer Simulation

Monte Carlo Method describes a technique of solving stochastic problems through experimentation with random numbers. This method can be traced back to physical experiments the French naturalist G. L. L. Buffon used in 1773 to estimate π. However, the American statistician E. L. De Forest may have been the first to use this technique in 1876 with random numbers (see Gentle (1985), p. 612). An early and well-known use of the Monte Carlo Method was by W. S. Gosset who, publishing under the pseudonym "Student," used the method to bolster his faith in the t-distribution in 1908; prior to this the t-distribution had been developed by "theory" that was at best not rigorous. Although the Monte Carlo Method may have originated in 1876, it was not until about 75 years later that S. Ulam and J. von Neumann gave it the name Monte Carlo Method (see Ulam (1976) for an account). The reason for the time lapse was the inapplicability of the method in many important problems until the advent of the digital computer, which was developed between 1946 and 1952 at such institutions as the University of Pennsylvania, Massachusetts Institute of Technology, National Bureau of Standards, and International Business Machines Corporation. **The modern stored-program computer made feasible the voluminous calculations required by the Monte Carlo Method.**

In comparison to today's computers, early computers were slow and had limited memory. For example, the number of arithmetic operations per second (often called floating point operations per second, or flops) that could be performed was below 10,000 in the early 1960s, about 500,000 in the mid-1960s, 20,000,000 in the early 1970s, a billion in the early 1990s, and on the supercomputers of today it exceeds a trillion. Indeed, Moore's Law, articulated by Gordon Moore of the Intel Corporation in 1965, asserts that computing power doubles every 1.5 years (see U.S. News & World Report (1997), pp. 64–65). While growth in certain areas may have been faster in the past and physical limits may slow growth in the future, Moore's Law implies a ten-fold increase in computing power every five years, to perhaps 100 trillion flops by the year 2008. Some estimate an even faster growth: 10 trillion flops in 2000 and 100 trillon flops in 2004!

In addition, programming was done in machine or assembly language until about 1955 because higher level languages such as FORTRAN were not yet available (the first integrated circuit was invented at Texas Instruments in 1958); special-purpose simulation languages did not become available until about a decade later. The early uses of the Monte Carlo Method, which is now known as **simulation**, concentrated on programming techniques, since debugging and running a program was the most arduous task of developing a simulation. The limitations of the early computers often forced oversimplifications of the problem; without such simplifications, programs would not run in feasible computer time or at feasible cost. Often, important issues such as what program runs to make and how to analyze program output were ignored.

We are entering the 21st century with a half-century of development of computer simulation, it is now possible for a text to provide **a theoretical basis for simulation methodology, details of an important simulation language, and the integration of these elements as they are brought to bear on a meaningful case study**. In this book Chapters 1, 3, 4, and 6 provide the conceptual basis for any simulation; Chapters 1, 2, 5, and 7 provide a course on the GPSS (General Purpose Simulation System) language; and Chapter 8 provides a valuable capstone illustrating the gains that modern statistical methods can bring to simulation. With this perspective in mind, we now turn to a simple motivating problem to introduce the basics of simulation, which are the primary concern of this chapter.

If a bank had an automated teller and wanted to study the **waiting time of its customers**, perhaps with a view to expanding the facility by including more tellers if the average waiting time was considered too long, the bank might hire a consultant to observe the system. That person would then note the waiting times of the customers who came to the system; for example, waiting times $W_1, W_2, \ldots, W_{50}$ for the first 50 customers. Then simple **statistics** such as the **average** $\bar{W}$, and more complex statistics such as the **percentile rank curve** of the data, would be computed and, ultimately, a decision might be made. This is a simple, yet indicative, **problem of decision making** that requires minimal (if only $\bar{W}$ is considered) or more substantial (if $\bar{W}(t)$, the average waiting time as a function of time-of-day t, is considered) **statistical analysis for its solution**.

In the situation above, one would gather the data $W_1, W_2, \ldots, W_{50}$ in the **real world** (including, of course, the scenario where "teller" is a unit of

Internet connection capability). Then, various statistical and mathematical tools would be used to analyze the data and reach a decision. In computer simulation, one proceeds just as above, except that instead of observing data in the real world, one generates the data inside a computer. The reasons for this are many. For example, if $\bar{W}(t)$ of the current system is too high at the peak time t, one may wish to expand the system. But should it be expanded to two tellers, or three, or four, or some other number? With simulation, we can "observe," via data generated with a computer, how $\bar{W}$ will vary as the number k of tellers is increased, and choose the most **cost-effective** k (i.e., the smallest k that keeps $\bar{W}$ below some threshold such as 30 seconds). This allows us to **pretest a decision** before implementing it, by observing how well the computer model reflecting the decision would work. In many cases, **real-world experimentation is too complex or expensive;** one does not want to build a teller system with a varying number of tellers, and observe in the real world how $\bar{W}$ varies with k, since this would be both costly and possibly detrimental to customer satisfaction.

Thus, **the fundamental distinction between computer simulation and other statistical problems** is that in computer simulation the data are gathered from a computer model as opposed to being gathered in the real world. In what follows, we will investigate:

1. How such a model is constructed.

2. How random variables $W_1, W_2, \ldots, W_{50}$ can be produced by the computer.

3. How the great control the analyst has over the model can be used to optimize the system.

4. How the simulation process can be used to solve complex mathematical and statistical problems that defy analytical solution.

Although the primary attention in this chapter is devoted to issues of computer implementation of a simulation, some **caveats** are needed:

1. **Simulation results are statistical in nature,** and statistical analysis is needed for their proper use (see Chapter 6 for details).

2. **Replication is essential** for proper use, and one simulation run to "show us how the system operates" will not achieve goals (see Chapter 6 for details of how long to simulate).

3. **Models need to be validated** by, for example, comparing results with those known for existing systems.

1.1   Computer Implementation of Simulations

The difficulty associated with the computer implementation of simulation models using general-purpose programming languages such as FORTRAN, BASIC, or Pascal has led to the development of a variety of special-purpose simulation languages. Many of these languages, in fact, are so "special purpose" that they are not appropriate for the development of a wide class of computer simulations. DYNAMO and CSMP, for example, are particularly suitable for continuous system simulations, but ill suited for the implementation of discrete-event systems. Conversely, GPSS is well suited for discrete-event simulations, but inappropriate for continuous system modeling.

The primary purpose of all specialized programming languages is to make the modeling and programming of specific types of problems conceptually simpler and less tedious to implement. Accordingly, a special-purpose simulation language should provide structures for modeling system entities (such as bank tellers) and their attributes (such as customer waiting times). By using such structures, it is possible to describe the evolution of the model over time and, through control structures embedded within the language (preferably transparent to the user), the user should be able to manage the progress of a simulation over time.

There are, in a broad sense, two distinct types of systems: continuous and discrete. While in both cases model changes occur with respect to time, **continuous systems** reassess the internal configuration of the system at fixed Δt time intervals, as shown in Figure 1.1–1. If time were initialized at t_0, this would involve the computation of all system variables and parameters at each time increment, $t_1 = t_0 + \Delta t$, $t_2 = t_0 + 2\Delta t, \ldots$. By contrast, **discrete systems** are visualized as systems that remain fixed for some, perhaps lengthy and generally unpredictable, length of time (see Figure 1.1–2). Following such a time lapse, they undergo changes and again stay fixed for some, probably different, length of time.

Since this text deals with discrete-event models, we will concentrate on discrete systems and their implementation on modern computers. The development of computer simulations is greatly simplified by the use of special simulation languages such as GPSS, GASP, etc. GPSS is a discrete-event simulation language that is widely used and commonly available on modern computers, including microcomputers. GPSS has a direct methodology for modeling discrete systems and powerful features that simplify the development of complex models. For these reasons, GPSS will be used as the vehicle for the development of simulations throughout this text.

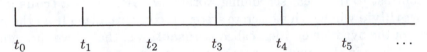

Figure 1.1–1. Time increments for a continuous model with $t_i = i\Delta t_i$.

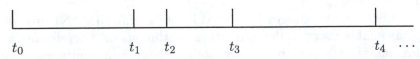

Figure 1.1–2. Time increments for a discrete model.

A discrete system can be visualized as a set of events that occur at specified times. This view requires the development of algorithms (subroutines in actual implementation) that describe the events (changes) that could occur within the system. The entire simulation can then be driven by a scheduling routine that invokes a sequence of subroutines in a specific order. To model events as nonoverlapping activities, all events must occur in zero time. General-purpose programming languages such as FORTRAN or Pascal can be used to write the event subroutines, which can then be called through directives from the simulation language. In this **event-scheduling worldview**, the simulation of a simple, single-line, single-server queueing system (where customers arrive, wait if necessary, obtain service, and leave) can be developed as a scheduling routine and two subroutines. Depending on the internal state of the simulated system, the scheduling routine may advance the simulation clock and call on one of the arrival or service completion subroutines.

The arrival subroutine takes appropriate action, depending on whether the service facility is idle or busy, and schedules the next arrival. If the facility is idle, it changes the facility status to busy and designates the departure time from the facility. If the facility is busy, it makes an entry into a FIFO (First-In, First-Out) linked list. The actions taken by the service completion subroutine are determined by the state of the FIFO list. If the list is empty, the facility status is changed from busy to idle; if it is not empty, the first entry of the list is removed and treated as a new arrival. The logic of the arrival and service completion subroutines is discussed in detail in Section 1.2.

In contrast to the event-scheduling worldview, GPSS uses a **transaction flow worldview** in which a discrete system is conceptualized from the point of view of the dynamic entities, called **transactions**, that move through the system. Thus, in the simple, single-server queueing system, the transaction flow worldview model describes the movement of the dynamic entities (the customers) through the system (waiting line and service facility). Certain activities associated with the customer movement, such as waiting in line or obtaining service, will require that the simulated time be updated before starting another activity.

Some simulation languages (e.g., SIMSCRIPT II.5, SIMAN) provide features that enable users to develop either continuous or discrete models. In such languages, it is possible to use both discrete and continuous features within a simple model.

1.2   The Single-Server Queue

Today, **simulation is used in many areas**. For example, there are recent simulations in such diverse and numerous areas as acoustics, aeronautics, agriculture, food and nutrition, air quality, astronomy and astrophysics, automata, ballistics and military applications, biology, Brownian motion, bus systems, chemical engineering, chemistry, ciphers, climatology, meteorology and solar energy, communications, computer devices, computer networks, correlation, crystallography, dosimetry, electronics, energy, entropy, fermentation, finance, fire science, fisheries, forestry, gaming, health systems, herd management, holography, information theory, insurance, inventory management and policies, irrigation, job shops, queueing, maintenance, management, planning and decision making, manufacturing, production and distribution systems, medical curriculum, medicine, microcomputers, migration, mining, modeling, molecular science, Monte Carlo methods, natural resource planning, navigation, nuclear physics, optics, optometry, paper and pulp, textiles, parasitology, pharmacokinetics, photographic science, physics, police patrols, politics and elections, polymers, population, population ecology and wildlife management, power systems and apparatus, psychiatry, random figures, random numbers and psychical research, random number generation, reliability, robotics and automation, scientific discovery, social systems and public policy, space flight, statistics, traffic engineering, transportation, tumor growth, vehicle design, and water systems, to name but a few. For each of these areas, **references** to recent simulations are given on pp. 342–360 of

Dudewicz and Karian (1985). Other references, especially for complex systems (e.g., traffic in urban transport networks, prices on financial markets, systems arising in social, biological, and behavioral sciences) are provided in the excellent book by science writer John L. Casti (Casti (1997)). Recent advances in applications are detailed in Dudewicz (1996, 1997).

In this section and several that follow, we have chosen a very simple system, **the single-server queue**, to illustrate **the construction, GPSS coding, use, and applications of a simulation program**. With such a program, we have a numerical and logical model of the system under study; the ultimate description of the model is in the form of a computer program. The model is "run" in the same sense that the real-world system runs in real time, except that the simulated system runs in simulated time. While there are many reasons to program a simulation in a simulation language (such as GPSS, SIMSCRIPT, GASP, or others in specific application areas), our initial treatment will be in terms of a basic algorithmic language. Our goal is to illuminate **program construction** (including the problems peculiar to simulation programs, as well as typical program organization), **program use** (including data input, run time, and output analysis), and **applications** (both range and evaluation of quality). Thus, we have chosen an example, the single-server queueing system, that is complex enough to raise the important issues of simulation, but simple enough not to obscure them in details. The single-server queueing system is in fact so simple that one might wonder if it would be complex enough for these purposes; we will see that it is, in fact, complex enough that virtually all the important issues of simulation arise in this model.

In a typical simulation, we have a model that consists of **entities** that have **attributes**. The total collection of entities and their attributes at any point in time is called the **status** of the system, and rules called **events** govern changes in status. Often, we will need to keep track of several events to ensure that they occur in proper order. Thus, time **synchronization** is a vital concern, as are **data input** and **output**.

1.2.1 A Single-Server Queueing Model

In the single-server queue example, assuming a first-come, first-served queue discipline, customers arrive at the facility, join the queue (if the server is busy; otherwise they go directly into service), enter into service when their turn comes, and leave once their service is finished. As we develop a model for this system, we will use entities and attributes with the following definitions:

Entity SERVER: Has the attribute of being busy or idle, which will be denoted by variable BUSY (which = 0 if the SERVER is idle, and = 1 if
the SERVER is busy); and has attribute CTIME (the time the service of
the customer now being served will be finished; if BUSY = 0, so that
there is no customer being served, it will be convenient to set CTIME
to a number larger than any time the simulation will ever reach, and
for convenience we call that number $+\infty$).

Entity QUEUE: Has the attribute of the number of customers who are waiting
(not including the one being served), say QSIZE.

Entity AGEN: Is the mechanism that specifies the times of arrival of the
customers; we will let the time of arrival of the next customer be stored
in variable ATIME (we will think of ATIME as being specified by reading
the next value from a list, in ascending order, of the arrival times of
customers to the system; how one finds such a list is a complexity that
will be considered later).

Entity SGEN: Is the mechanism that specifies the service times of the customers; we will store the service time of the next customer in variable
STIME and, in our first simple system, we will consider this to be the
same for all customers.

A flowchart can now be used to describe the operation of the system. If
the system is operating, then its status will next change when either of the
two events "arrival of a customer" and "completion of service of a customer"
occurs. These happen, respectively, at times ATIME and CTIME, the latter
being $+\infty$ if the server is idle. Thus, we need to next process an arrival
(the arrival event AEVENT) if ATIME $\leq$ CTIME; otherwise, we need to next
process a service completion (the completion event CEVENT). The manner
of choosing the next event to be performed is shown in Figure 1.2–1.

If the next event to occur is an arrival of a customer (AEVENT), then that
customer joins the queue if the server is busy (if BUSY $\neq$ 0), in which case
we increase the QSIZE by 1. We then reset ATIME to the time of the next
arrival by reading the next value from the list of arrival times, and then
continue the processing of the next event. On the other hand, if BUSY = 0
then the customer immediately goes to service (so we set BUSY = 1), and
that will end at CTIME = ATIME + STIME (since the customer has arrived
and begun service at time ATIME, and will take length of time STIME to be
served). We then reset ATIME and continue to process the next event. This
model is shown in Figure 1.2–2.

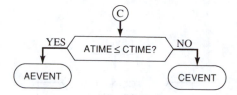

Figure 1.2–1. Choice of AEVENT or CEVENT happening next.

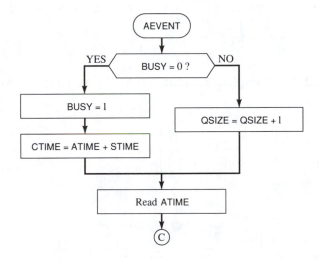

Figure 1.2–2. Arrival event (AEVENT) for single-server queue.

Suppose that the next event to occur is completion of service of a customer (CEVENT). If the queue is empty (QSIZE = 0), the server goes idle (BUSY = 0) and we set CTIME to note that no completion will be scheduled to occur (CTIME = +∞). Otherwise (QSIZE ≠ 0), we take a customer from the queue (decrease QSIZE by 1), and set the time of that customer's service completion to current time (CTIME) plus the service time (STIME). We then continue with the next event. This process is shown in Figure 1.2–3.

1.2.2 An Improved Model for the Single-Server Queue

If we add an initialization block at the front, then the combination of Figures 1.2–1, 1.2–2, and 1.2–3 yields a complete flowchart for the simulation of the single-server queue. When executed, this program runs and functions just as the real-world system does. This model is shown in Figure 1.2–4

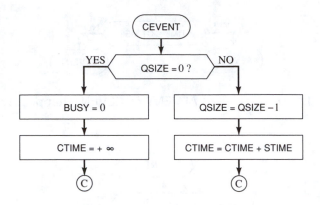

Figure 1.2–3. Completion event (CEVENT) for single-server queue.

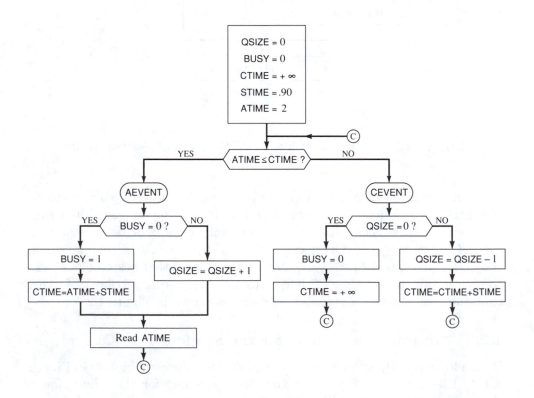

Figure 1.2–4. Single-server queue with initialization.

with initialization for a system that starts empty at time zero, has its first customer arrive at time 2, and has a service time of 0.90 unit per customer.

There are some problems with the flowchart of Figure 1.2–4. First, **the simulation of the queue never ends.** We have written an infinite loop. So one modification needed is the addition of a stopping mechanism. For example, we might specify a time HTIME at which the simulation is to terminate. This is done in Figure 1.2–5 with HTIME 480 minutes.

A second problem is that **"no one is looking."** Our simulation runs just as does the real-world system with no observer. To add an observer, we need to keep track of quantities that will allow us to **answer the questions of interest about the system.** However, we (perhaps amazingly!) do **not** yet know **why** we are simulating this system; we have made one of the cardinal sins of simulation: we have jumped into flowcharting (and perhaps coding) without knowing why the simulation is to be done. This "why" has important consequences regarding the level of detail needed in the simulation, structure of the program, etc.

To allow us to proceed, we will now assume that the question of interest to the experimenter is **"What will the average queue size be after 8 hours (480 minutes) of simulated time?"** To develop a method of accumulating the information needed to answer this question at the end of the simulation, let us examine what the queue size looks like as a function of time. Since the queue size is the number of customers in line at any time, QSIZE can take on only values that are nonnegative integers $0, 1, 2, 3, 4, \ldots$. A possible plot of this variable as a function of time, for a system with few arrivals and relatively long service times, is shown in Figure 1.2–6. There we see that the QSIZE was 0 from 0 to 100 minutes, and also from 450 to 480 minutes, for a total of

$$(100 - 0) + (480 - 450) = 130 \text{ minutes.}$$

Similarly, QSIZE was 1 for

$$(150 - 100) + (370 - 300) + (450 - 400) = 170 \text{ minutes;}$$

QSIZE was 2 for

$$(200 - 150) + (300 - 220) + (400 - 370) = 160 \text{ minutes;}$$

and QSIZE was 3 (the maximum in this simulation) for

$$220 - 200 = 20 \text{ minutes.}$$

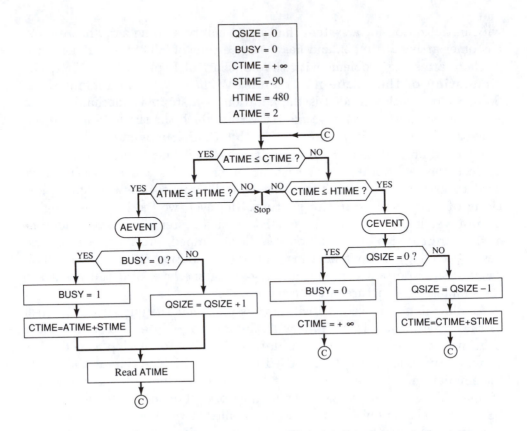

Figure 1.2–5. Single-server queue with initialization and stopping.

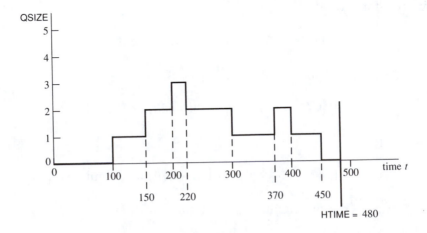

Figure 1.2–6. A plot of QSIZE as a function of time.

Since the QSIZE values, and the proportion of the 480 minutes it had each of those values, were

QSIZE	0	1	2	3	≥ 4
PROPORTION OF (0,480)	130/480	170/480	160/480	20/480	0

the average QSIZE over $(0, 480)$—say, AQUEUE—was

$$\text{AQUEUE} = 0 \cdot \frac{130}{480} + 1 \cdot \frac{170}{480} + 2 \cdot \frac{160}{480} + 3 \cdot \frac{20}{480} = 1.1458333.$$

Thus, if we were to keep track of each time the QSIZE changed and the times at which those changes occurred, from that information we could, at the end of the simulation, find AQUEUE. However, for an active simulation with thousands of arrivals over the length of the simulation, this would require a large amount of storage, **so we seek a simpler solution.**

If we rewrite AQUEUE, we note that it can be expressed as

$$
\begin{aligned}
\text{AQUEUE} = \{&(0)((100 - 0) + (480 - 450)) \\
+ &(1)((150 - 100) + (370 - 300) + (450 - 400)) \\
+ &(2)((200 - 150) + (300 - 220) + (400 - 370)) \\
+ &(3)(220 - 200)\}/480.
\end{aligned}
$$

The numerator is just the **area under the QSIZE curve** over the time period 0 to 480 minutes. Thus, if we can keep track of that area—say, QAREA—then, at the end of the simulation, we can simply set

$$\text{AQUEUE} = \text{QAREA/HTIME.}$$

We can accomplish this (keeping track of QAREA) by initializing QAREA to 0, and then updating this variable each time the QSIZE curve jumps. Let LTIME designate the last time we updated QAREA. LTIME is needed because, when we are updating QAREA, we need to know the previous time we updated it in order to find the area of the new block, which is

$$((\text{current time}) - (\text{last update time})) \cdot (\text{current queue size}).$$

At an arrival (in AEVENT), we perform

$$
\begin{aligned}
\text{QAREA} &= \text{QAREA} + \text{QSIZE} \times (\text{ATIME} - \text{LTIME}) \\
\text{LTIME} &= \text{ATIME}
\end{aligned}
$$

while, at a service completion, we perform

$$\text{QAREA} = \text{QAREA} + \text{QSIZE} \times (\text{CTIME} - \text{LTIME})$$
$$\text{LTIME} = \text{CTIME}.$$

Since the QSIZE will not always be zero at HTIME, we update QAREA before calculating AQUEUE. Figure 1.2–7 shows these modifications.

It is often desirable to obtain results on the simulated system over a **number of time periods.** For example, instead of simply the average queue size at the end of one day (of 480 minutes), we may wish to see the results at the end of each day through the end of a week (of 5 working days). A simple addition of a loop to the flowchart of Figure 1.2–7, shown in Figure 1.2–8, can accomplish this. We need to carefully label the output with the time at which it was obtained. It is necessary to update LTIME just before stopping (in what is now called the "sampling event" or SEVENT part of the program) in order to obtain correct periodic outputs. Choice of the **number of time periods of simulation needed to satisfy a goal** (such as to estimate the mean queue size at day's end within 5.0 with probability 0.95) is considered in Section 6.1, where precise methods appropriate to this choice of replication number are given.

1.2.3   Generation of Arrival Times and Other Attributes

Above, we needed arrival times (ATIMEs) of successive customers and their service times (STIMEs) in order to run the simulation. We solved these problems by assuming a known constant service time for each customer (such as STIME= 0.90 minute), and by assuming that some "black box" is available that yields successive ATIMEs of customers. In the case of arrival times, in most examples it is reasonable to have a "black box" that generates the **interarrival times** (the times between customer arrivals), say $A_1, A_2, A_3, \ldots$; then the successive ATIMEs are $A_1, A_1 + A_2, A_1 + A_2 + A_3, \ldots$. But how can one obtain the A_i's? In Problem 1.1, we obtain A_i's by taking

$$X_i = (MULT * X_{i-1}) \bmod M \quad (i = 1, 2, 3, \ldots) \tag{1.2.1}$$
$$A_i = X_i/M \quad\quad\quad\quad\quad (i = 1, 2, 3, \ldots), \tag{1.2.2}$$

where $X_0 = 32767$, $M = 32768$, and $MULT = 5**13$. When (1.2.2) is reasonable will be discussed below and in later chapters, as will alternatives to (1.2.2). Here "$a \bmod b$" denotes the remainder when a is divided by b (for example, $15 \bmod 8 = 7$, since $15/8 = 1 + 7/8$ for a remainder of 7). Those not familiar with this notation may find it helpful to think of $a \bmod b$ as

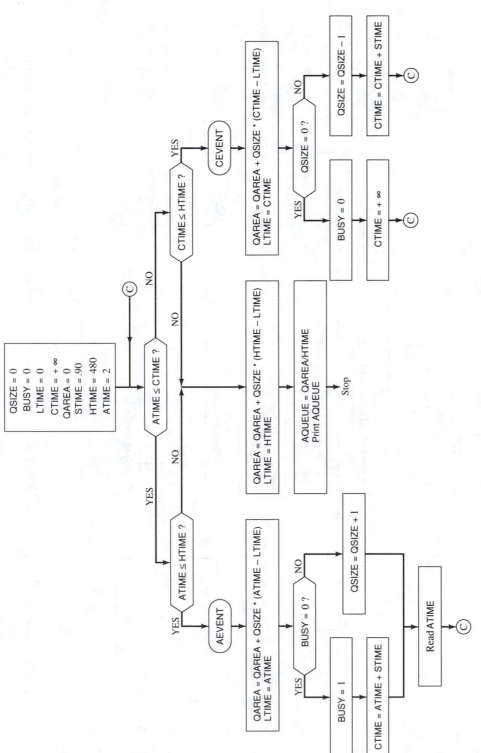

Figure 1.2–7. Single-Server Queue with initialization, observation, and stopping.

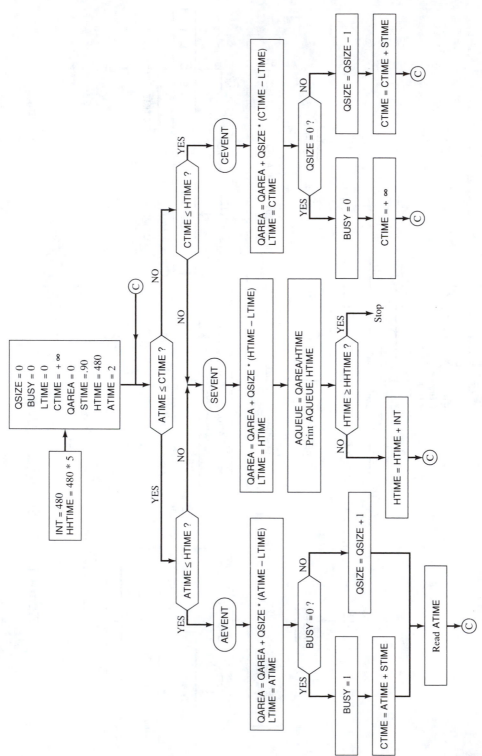

Figure 1.2–8.   Single-Server Queue with initialization, observation for multiple periods, and stopping.

the result of subtracting b from a as many times as possible without having the result become negative. (However, this is **not** a desirable computational method for calculating $a \bmod b$.)

The method described by (1.2.1) and (1.2.2) is called URN38; the more detailed study of URN38 provided in Chapter 3 reveals that it is not a good random number generator. It is used here merely for its ease of programming with the values of X_0, M, and $MULT$ noted below (1.2.2). Instructors may wish to instead use URN36 or URN37, both of which are shown to be good generators in Chapter 3, if the larger value of $MULT$ for these generators does not present difficulties with the available computational environments.

The calculation of (1.2.1) can be a problem if $MULT$ and X_{i-1} are large, since then we may **overflow** the word size on the computer we are working with. This is especially likely on microcomputers, and also on some mini-computers. A method that allows us to cope with large values of $MULT$ and X_{i-1} is based on the fact that

$$a * b * c \bmod M = ((a * b \bmod M) * c) \bmod M. \qquad (1.2.3)$$

To see that (1.2.3) is valid, note that $a * b$ can be expressed as $dM + r$ for some r between 0 and $M - 1$. Then the right-hand side of (1.2.3) equals

$$((dM + r \bmod M) * c) \bmod M = rc \bmod M.$$

On the other hand, the left-hand side of (1.2.3) equals

$$((dM + r) * c) \bmod M = (dcM + rc) \bmod M = rc \bmod M,$$

hence the equality of (1.2.3) has been established.

For example, this means that in computing $(32767)(5**13) \bmod 32768$, one may note that $5**13 = (15625)(15625)(5)$ and then do the computation as $5 * (15625 * (15625 * 32767 \bmod 32768) \bmod 32768) \bmod 32768$, avoiding overflow on many computers. Of course, some computers may have a word size so large that this trick is not needed; however, it is a method used widely on microcomputers (see Dudewicz, Karian, and Marshall (1985)).

Method (1.2.2) is often used to attempt to generate $A_1, A_2, A_3, \ldots$ which are **random numbers**; i.e., independent random variables that are each uniform on the interval of real numbers $(0, 1)$. We will see later, when we deal with random number generation, whether this is a good method. **If** such randomness were the true nature of $A_1, A_2, A_3, \ldots$, then the queueing system we are dealing with would be the one known in the queueing literature as the **U/D/1 queue**, a queue with uniform independent interarrival times (hence the "U"), deterministic service times (hence the "D"), and one server (hence the "1"). In fact, that queue is very difficult to deal with in terms of finite-time results, such as the average queue size after 8 hours of operation. Therefore, it does make sense to simulate, as we do in Problem 1.1, to find the characteristics of the queue.

For the U/D/1 queue with STIME = 0.90, we can immediately say some things about its characteristics. First, note that the **ratio of the average service time to the average interarrival time** is $0.90/0.50 = 1.8$. Thus, on the average, 1.8 jobs will arrive in the time it takes to serve 1. It is then clear that the queue will grow as time proceeds and that there should be a large queue at the end of an 8-hour day. One should expect something like $(8)(60)(2) = 960$ jobs to arrive in that time, while at most $(8)(60)/0.90 = 533$ can be served, resulting in a queue of size $960 - 533 = 427$.

The above **method of replacing random times by their means**, which we call **deterministic simulation**, is **at best** one that yields some qualitative insight on how the system may behave. It is easy to see that replacing all random variables by their means **will in most cases yield wrong answers**. (The exception is the case where, e.g., the variable being replaced by its mean has no, or very small, variability, which occurs in some manufacturing systems.)

Consider a queue with 1 server, service times all 0.90, and average interarrival times of 2.00. The deterministic simulation method would yield a system with arrivals at times $2, 4, 6, 8, \dots$. The arrival at time 2 would leave at time 2.9; the arrival at time 4 would leave at time 4.9; etc. Thus, there would never be any queue (the average queue size would be exactly zero). In fact, the deterministic simulation would yield zero queue size whenever the average interarrival time was greater than the average service time. **However, this would say that all queues** (except those few that are oversaturated, having average interarrival time smaller than the average service time, and that have queue sizes that become large as time progresses) **have queue size equal to zero at all times.** We know from practical experience, from queueing theory, and from simulations that **this is false**.

When dealing with large systems, the falsity of deterministic simulation will not be so easy to see. Consequently, the inappropriate use of deterministic models has been popularized, for example, under names which imply that such deterministic models can show the "dynamics" of a system. For this reason, we believe it is important for us to expose it as a fallacy: whether one deals with queues (where the fallacy is clear, as shown above), or with systems of differential equations whose random coefficients are replaced with their means (where the fallacy is veiled by complexity), the method is one unworthy of any work that is to be called "scientific."

1.3 The Single-Server Queue: Additional Goals

In our consideration of the single-server queue in Section 1.2, we ultimately were able to determine the average queue size after $1, 2, 3, 4$, and 5 days of operation. In this section, we will examine how we may keep the statistics needed to answer other, and more complex, questions.

While the average queue size is a simple measure of system performance, we may wish more detailed information, such as the **queue size distribution**; i.e., what proportion of the time the queue was of size $0, 1, 2, \ldots$. If we define a vector variable H(I + 1) as the length of time the queue has had size I during the simulation, then the proportion of time (up to HTIME) that the queue was of length I can be simply calculated as H(I + 1)/HTIME. For example, for the realization (i.e., the simulation results) shown in Figure 1.2–6, at the end of the simulation we would have

$$
\begin{aligned}
\text{H(1)} &= (100 - 0) + (480 - 450) = 130, \\
\text{H(2)} &= (150 - 100) + (370 - 300) + (450 - 400) = 170, \\
\text{H(3)} &= (200 - 150) + (300 - 220) + (400 - 370) = 160, \quad\quad (1.3.1)\\
\text{H(4)} &= 220 - 200 = 20, \\
\text{H(5)} &= \text{H(6)} = \cdots = 0,
\end{aligned}
$$

so if Q(I) denotes the proportion of time (up to HTIME) that the queue had size I, then we have

$$
\begin{aligned}
\text{Q(0)} &= \text{H(1)}/480 = 130/480 = 0.2708, \\
\text{Q(1)} &= \text{H(2)}/480 = 170/480 = 0.3542, \\
\text{Q(2)} &= \text{H(3)}/480 = 160/480 = 0.3333, \\
\text{Q(3)} &= \text{H(4)}/480 = 20/480 = 0.0417, \\
\text{Q(J)} &= \text{H(J + 1)}/480 = 0/480 = 0.0000 \text{ for J} = 4, 5, 6, \cdots.
\end{aligned}
$$

Note that H(1) + H(2) + H(3) + H(4) + H(5) + $\cdots$ = HTIME, and Q(0) + Q(1) + Q(2) + $\cdots$ = 1 in all simulations. The set of numbers Q(0), Q(1), Q(2),..., called the queue size distribution, is in fact a **discrete probability distribution** on the numbers $0, 1, 2, \ldots$, since the Qs are nonnegative numbers that sum to 1.

To keep track of the H(I)s during the simulation, we can proceed as follows. First, in the initialization block, set H(I) = 0 for all I, since at the start of the simulation we have each queue size for length of time zero. Then, in each of the events AEVENT, CEVENT, and SEVENT, respectively, increase H(QSIZE + 1) by ATIME − LTIME, CTIME − LTIME, and HTIME − LTIME. This will keep the H(I)s updated to the same time to which QAREA had been updated. This needs to be done before variables such as LTIME are updated (e.g., right after the update of the QAREA). The following are some of the finer points of this approach.

First, we let H(I + 1) denote the length of time the queue was of size I, rather than using H(I) for this length of time, since in some computer programming languages the subscript zero is not allowed for a vector variable. If you are using a language that does allow a zero subscript, then H(I) can be used to denote the length of time the queue was of length I.

Second, in many languages it is necessary to declare the length of the vector H(·) in advance, e.g. to note a dimension of 6. That would mean you would have available only H(1), H(2), H(3), H(4), H(5), H(6) in which to keep track of times, and hence would keep track of the length of time the queue had size $0, 1, 2, 3, 4, \geq 5$. From this information, one could compute the proportion of time the queue had length $0, 1, 2, 3, 4, \geq 5$ but would not be able to break down the proportion ≥ 5 into the proportions of time for the individual lengths $5, 6, 7, \ldots$ that comprise it. If one is using a language that has **dynamic allocation of storage**, then the vector H(·) need not be dimensioned in advance, and one can add components as needed. However, even in this case one usually would not go beyond some preset maximum dimension, in order not to record (at a large cost in storage space) sizes that might occur only infrequently.

Third, note that one can compute the AQUEUE as the sum

$$\sum_{I=0}^{\infty} H(I+1)I/HTIME. \tag{1.3.2}$$

However, this is exactly correct only if one has dynamic allocation of storage and has not set a maximum on the H(·) vector's dimension.

Finally, when printing out the proportions of time that the queue had size $0, 1, 2, \ldots$, one can (for the simple queueing system we are dealing with) stop printing after the first zero is encountered, since (with at most one customer arriving at any time) the queue can never be J + 1 or larger unless it has been J. This, of course, is not true for systems that have batch arrivals; for example, where a delivery truck may bring several jobs into the system simultaneously. In practice, one often **either** sets in advance the sizes one will print the proportions of time for, **or** prints until the proportions sum to at least 0.99.

Another item of information that might be desired for management consideration is the **proportion of time the server is idle**. If TIDLE is a variable that is to contain the length of time the server has been idle, we can initialize TIDLE to ATIME, since the server will be idle from the start of the simulation until the arrival of the first customer. TIDLE can then be updated in CEVENT (since an idle server can occur only at a service completion), where (when QSIZE = 0, the case where the server will be going idle) we update with

$$\text{TIDLE} = \text{TIDLE} + (\text{ATIME} - \text{CTIME}) \tag{1.3.3}$$

since we know that from this completion time (CTIME) until the next arrival (which will be at ATIME) the server will be idle. In SEVENT, the proportion of time the server has been idle is not simply TIDLE/HTIME. Note that TIDLE has been updated to ATIME. If the server is busy at HTIME, then there is no additional increment to make, and TIDLE/HTIME is correct. However, if the server is idle at HTIME, then the proportion of idle time **up to** HTIME is

$$(\text{TIDLE} - (\text{ATIME} - \text{HTIME}))/\text{HTIME} \tag{1.3.4}$$

since TIDLE had previously been updated beyond HTIME (and thus an adjustment is needed to make the proportion correct up to HTIME).

The final item of information that management might desire, which we will discuss in this section, is the **mean waiting time** of the customers. There are, of course, many additional measures of system performance, some of which will be discussed in the problems. A thorough understanding of the methods presented should allow the reader to formulate efficient methods for many other measures of system performance. Note that some managers have a tendency to ask for many measures, as all have some **information value**. However, those same measures also each have an **information cost** in terms of programming time, model complexity, and computer time. In choosing which measures to include in any simulation, one needs to carefully weigh the cost as well as the value in terms of the goals of the simulation experiment; generally this will result in the inclusion of most of the measures that might be on a first list of "measures of possible interest." We will define the "waiting time" of a customer to be the time from arrival at the queueing system until the start of service, as shown in Figure 1.3–1. Then one need only save the arrival time of a customer until the customer is to start service, and the difference in the time of starting service and the arrival time will equal the waiting time of that customer.

The following are some notes on the above method. **First**, this method will, unless one has dynamic allocation of storage, work only for systems that have an upper limit on the number of customers who can be waiting at any time, since one will need to dimension the vector in which the waiting times of the customers on the queue will be kept. The storage locations can, of course, be reused once a customer starts service. **Second**, for the first-come,

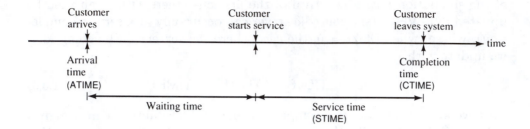

Figure 1.3–1. Definition of "waiting time" of a customer.

first-served queue discipline, one can, if the service times of the customers in the queue are all known, easily find the waiting time a new arrival to the system will suffer; namely,

$$\left(\begin{array}{l}\text{Sum of the service times of all} \\ \text{customers already in the queue}\end{array}\right) + (\text{CTIME} - \text{ATIME}). \qquad (1.3.5)$$

The newly arriving customer will need to wait for all those already in the queue to finish (hence the first term in (1.3.5), which is the sum of the service times of those customers), and also for the customer already being served to finish service (since that will occur at CTIME and the current time is the time of arrival of the new customer, ATIME, the time the customer already being served needs to finish service equals the second term, CTIME − ATIME). With this second method, one can keep track of the total waiting time (say, TW) incurred by all customers who have arrived at the system, and of how many customers incurred that total waiting time (say, NW). Then the average waiting time at any instant is the ratio TW/NW.

1.4   GPSS Model of the Single-Server Queue

GPSS was first released as a product by IBM in 1961. Since then it has been made available on many IBM mainframes (the 704, 7090, and 360/370 among others), and more recently on a variety of minicomputers such as the VAX-11 system. With the advent of powerful microcomputers, "dialects" of GPSS have been developed for microcomputers during the last few years. These include GPSS/PC^{TM}, an implementation of the GPSS language together with an interactive simulation environment distributed by Minuteman

Software; GPSS/H (an enhanced version of GPSS) and an interactive version of GPSS/H distributed by Wolverine Software; and GPSSR/PC, from Simulation Software Ltd.

GPSS uses dynamic entities called **transactions**, which move through the system, as well as **queues**, **facilities**, and **storages**, which model waiting lines, servers, and equipment pools. The GPSS simulation process consists of the creation of transactions, the allocation of static entities such as simulated service facilities, and the eventual removal of transactions from the simulation.

Since many transactions may be present in a system at the same time, during their "life span" transactions may compete for facility and/or storage entities. The GPSS processor manages all contentions through the use of linked lists called **chains**, which by default provide resources to transactions on a FIFO basis. The GPSS entities that alter the internal state of the simulated system are called **blocks**. Unless intentionally diverted, the natural path for transactions is through the sequence of blocks that embody the logical structure of the simulation. Blocks are inert in the sense that they do not cause any activities to occur by their presence; they are activated only when a transaction moves into them.

In almost all cases, without any explicit request from the modeler, GPSS provides output regarding the state of various components of the simulation at the end of the simulation run. In fact, if no output is desired, the modeler must explicitly suppress output.

The movement of transactions through the GPSS blocks of the various segments of the model is governed by an internally simulated clock that is used to control and schedule all internal activities within a GPSS simulation. Beginning with the creation of transactions by the GENERATE block, the clock is automatically advanced to the time when the next model event is to occur. The clock registers only integer values; hence, suitable increments of real time must be used to avoid fractional time units.

In the case of the simple single-server queueing model, for example, the point of view of the customer (transaction) consists of

1. Arrival

2. Waiting for service facility to become idle

3. Obtaining service

4. Departure.

As customers move through these steps, they may encounter delays at various stages. Following arrival, they may have to wait for the service facility to become available to receive service.

If we assume that the time lapse between successive customer arrivals is a uniformly distributed random variable between 9 and 15, and that the service time required by customers is a uniformly distributed random variable between 6 and 14, then Figure 1.4–1 gives a block diagram describing the single-server queueing model. A GPSS **block diagram** is a graphic display of the flow of transactions through the simulated system. A GPSS **program** consists of a sequence of control and executable statements. The executable statements have a one-to-one correspondence with the blocks of the block diagram. The GPSS program corresponding to the diagram of Figure 1.4–1 is given in Figure 1.4–2 and the program output is shown in Figure 1.4–3.

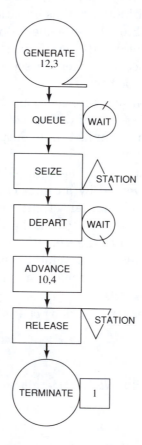

Figure 1.4–1. A GPSS block diagram for a single-server queue.

```
SIMULATE

GENERATE      12,3            Customers arrive
QUEUE         WAIT            Enter waiting line
SEIZE         STATION         Access facility
DEPART        WAIT            Leave waiting line
ADVANCE       10,4            Obtain service
RELEASE       STATION         Return facility
TERMINATE     1               Leave the system


START         50              Simulate for 50 customers
END
```

Figure 1.4–2. A GPSS model of the single-server queue.

RELATIVE CLOCK 627		ABSOLUTE CLOCK 627			
BLOCK	CURRENT	TOTAL	BLOCK	CURRENT	TOTAL
1	1	51			
2	0	50			
3	0	50			
4	0	50			
5	0	50			
6	0	50			
7	0	50			

QUEUE	MAXIMUM CONTENTS	AVERAGE CONTENTS	TOTAL ENTRIES	ZERO ENTRIES	PERCENT ZEROS
WAIT	1	0.102	50	29	58.000
		AVERAGE TIME/TRANS	$AVERAGE TIME/TRANS	TABLE NUMBER	CURRENT CONTENTS
WAIT		1.280	3.048		0

FACILITY	AVERAGE UTILIZATION	NUMBER ENTRIES	AVERAGE TIME/TRANS	SEIZING TRANS. NO.	PREEMPTING TRANS. NO.
STATION	0.809	50	10.140		

Figure 1.4–3. Output of the program given in Figure 1.4–2.

Notice that the program does not have any "output" statement, because GPSS provides some minimal output by default. A more detailed discussion of the program syntax, the function of each statement within the program, and the output associated with this program will be given in the next chapter. However, even a casual survey indicates that all the measures encountered in Sections 1.2 and 1.3 are included in the program output. The average queue length is AVERAGE CONTENTS = 0.102, the proportion of idle time is (1 − AVERAGE UTILIZATION) = 0.191, and the average waiting time is AVERAGE TIME/TRANS = 1.280.

A typical GPSS block diagram or, equivalently, a typical GPSS program, consists of one or more separate segments each of which describes the movement of transactions through portions of the simulated system. The GPSS processor executes the logic embedded within these separate segments in parallel by synchronizing the actions associated with each segment.

1.5   Single-Server Queue with Priority Classes: Data Structures

In our consideration of the single-server queue with first-in, first-out queue discipline in Section 1.2, we had little trouble accumulating the information needed to determine the average queue size observed after each of 1, 2, ..., 8 hours of simulated operation. When we studied how to accumulate the information needed to evaluate additional measures of system performance in Section 1.3, we found a need to maintain more, and more complex, information such as the arrival times (ATIMES) of customers until they start service. A method for storing, retrieving, and updating such information is called a **data structure**. While a simplistic data structure can be used for these problems, it will be at considerable cost in terms of the computation time and storage needed. For efficient simulation, and in the special simulation languages, more sophisticated methods are used. This section considers certain data structures associated with queueing systems that have priority classes. In Chapter 5, we will take a detailed look at the corresponding data structures used in GPSS.

1.5.1   Use of Ranked List Structures

Suppose we have a single-server queueing system where the queue may be at most N customers long. If a customer arrives when the queue is full, that customer is turned away and the next arrival is scheduled, with FIFO discipline. To find the mean waiting time, we will need (as discussed in Section 1.3) to keep track of the arrival times (ATIMEs) of customers on the queue from their arrival until they enter service. Suppose that for this purpose we use a vector A(·) that is dimensioned to N. With a **ranked list** data structure, we will keep the ATIME of the customer who is I-th in the queue in A(I). **If a customer arrives to the system**, it is easy to update this structure: we simply set

$$A(QSIZE + 1) = ATIME$$
$$QSIZE = QSIZE + 1.$$

If QSIZE = N before an arrival, we turn this arrival away and schedule the next; this is accomplished by a simple test of whether QSIZE < N. **When a customer is leaving the queue to enter service**, we need to

1. Use the arrival time, A(1), to compute the customer's waiting time, CTIME − A(1).

2. Discard the entry in A(1) and move the information in A(2) to A(1), . . ., A(QSIZE) to A(QSIZE − 1) so that, after this selection from the queue, the information on the customers left on the queue will still be in ranked order starting in the first location of the vector A(·).

3. Decrement QSIZE.

This needed updating of A(·), shown in Figure 1.5–1, is inefficient because at the time of selection from the queue, we need to move QSIZE − 1 numbers.

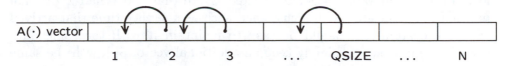

Figure 1.5–1. Updating the ranked list at selection time.

1.5.2   Use of Floating File Structures

The first entry on the queue, after the selection of the customer whose arrival time is in A(1), is in A(2); successive customers to be taken have their arrival times stored in A(3),. . .,A(N). **We can eliminate the need to move items in A(·) if we maintain a pointer variable, TAKE, that tells where to take the next time from.** TAKE should be initialized to 1 and should be incremented after a customer goes into service. This is called a **floating file** data structure, and the updating needed is illustrated in Figure 1.5–2. The name is given by analogy to the office system wherein files "float" on a metal rod, so that after one file is taken, the next is at hand rather than further towards the back of a large drawer.

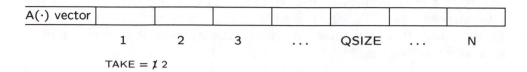

TAKE = $\not{1}$ 2

Figure 1.5–2. Updating the ranked list at selection time.

In the floating file system, it is simple **to select a customer for service.** If the QSIZE is at least 2, one sets TAKE = TAKE + 1. If, however, QSIZE = 1, the queue will be empty after the selection, in which case one may set TAKE = 0. The next customer to be added to the queue will then have its time stored in A(TAKE + 1) = A(1). Since the floating file "wraps around," TAKE + 1 may become greater than N; in this case, we change the N + 1 to 1. This is illustrated in Figure 1.5–3.

To add a customer to the queue, one checks that QSIZE is less than N (if not, the customer is turned away and the next arrival is scheduled); the new arrival time is placed into A(TAKE + QSIZE) (if this is greater than N, then the file has wrapped around, and we place it in location A(TAKE + QSIZE – N)). This system is simple to maintain, and is considerably more efficient than the ranked list—in essence it is a ranked list with the ranked list's location being movable rather than fixed. Note that if one had additional information associated with a customer, such as a service time that arrived with the customer, it would be easy to maintain it with the same data organization; one could simply keep an additional vector S(·) with the same dimension as the vector A(·), and whenever a customer had its arrival time stored in A(J), it would have its service time stored in S(J)—so there would be no additional calculation needed to find the correct index J to use.

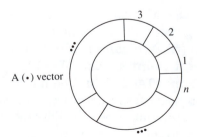

Figure 1.5–3. The wraparound nature of the floating file.

The ranked list/floating file system can be modified for use in other queue disciplines, as we will now show in two cases. First, a simple case is **the last-in, first-out or LIFO** queue discipline, where the last to arrive is the first to be served. This discipline is used in many storage systems where newly arriving stock is placed in front of stock already on the shelf; then, when orders arrive, stock is taken from the front of the shelf, so we have LIFO service. One important question for such a system is how often to rotate the stock to avoid spoilage, and simulation is often used to answer such questions. Here, placing information on an arrival into A(QSIZE + 1), and taking the next to be served from A(QSIZE) will maintain a ranked list that is the reverse of the ranking that FIFO used.

1.5.3   Linked-List Priority Queues

A more general case is that of the priority queue where each arriving customer or job has associated with it a **priority number** and the lowest number is to be selected first for service. If there are ties for priority, we will break them by taking first that one with the lowest priority number which arrived first. Such priority service systems are in wide use. For example, in an airline system, a first-class passenger may be served before others; in a restaurant, a customer with a reservation may be seated first; in an office, inquiries arriving by telephone may be taken before those from individuals who come in person. In processing work through systems, priority may be the due date of the job, the time required to finish the job, or some other calculated quantity based on job characteristics. One may simulate various priority-assignment rules to see which would yield the largest profit for the facility. Note that FIFO (take priority to be arrival time) and LIFO (take priority to be the negative of arrival time) queues are special cases of the general priority queue.

To use a ranked list in the general case where arriving customers or jobs have associated **with** them **a priority number**, an arrival time, and a service time, and the queue can be at most N items long, suppose we have vectors $P(\cdot)$, $A(\cdot)$, and $S(\cdot)$ in which this information will be stored respectively. Then the first-arriving job to the queue can have its information stored in P(1), A(1), and S(1), and index $J = 1$ will always be the one to be used when an item is taken off the queue. After **removal of an item from the queue,** we will need to move the jobs still on the queue to the left one index. This is inefficient and can be made unnecessary by using a floating file. **To add an item to the queue,** we need to search for where it fits in on priority and use arrival times to break priority ties. Then items are moved to make room for the new entry.

We saw that, with the FIFO queue discipline, the floating file concept made the ranked list an efficient data structure for the queue by making the moving of items at times of selection from the queue unnecessary. However, when applied to the general priority queue, we still have much movement needed when items are added to the queue. **Thus, we now seek another data structure for the general priority queue which will not have this drawback of need for movement of entries to make room for a new arrival to the queue.** The structure we will introduce for this purpose is called a **linked list** or an **associative list**. In this structure, we do not move items to make room for new entries, but instead add a vector that points to the next entry. This is like having in each file a note stating which file is the next on the list. The key to this data structure is a vector POINT($\cdot$) of dimension N with

$$\text{POINT(I)} = \begin{pmatrix} \text{Index of the job to be taken next after the} \\ \text{one whose attributes are stored in index I} \end{pmatrix} \quad (1.5.2)$$

with POINT(I) = 0 if index I pertains to the last item on the queue. **To remove an item from the queue,** we go to index TAKE, find the information P(TAKE), A(TAKE), and S(TAKE), and then set TAKE = POINT(TAKE).

1.5.4   Double-Pointing Structures

To add an item to the queue, first find an index not in use for the current queue, say index J, and place the priority, arrival time, and service time of the new arrival into P(J), A(J), and S(J). Next, search through the current queue to find where the new arrival fits in on priority and arrival time by comparing the new priority P(J) with the first on the queue, P(TAKE), then with the second on the queue, etc., until we find for the first time that P(J) is not $\geq$ the entry it is being compared with. Now change that entry's POINT($\cdot$) so that it has value J, and set POINT(J) to the value that the one being compared with had; care in coding is needed for the first item on the queue, and also if arrival times are not in increasing order, as e.g., in a job shop where arrival to a queue may be with a time that is not clock time at arrival to that queue, but rather time of first arrival to the total facility.

Above, we noted that we needed a search to find an index not in use when an arrival was to be added to the queue. A simple method of accomplishing this is called **double pointing**, in which we keep two lists in POINT($\cdot$): the list of the queue, and also a list of the empty storage locations. If we start with an empty queue, then we can have the number of the first free location as FREE = 1, and POINT(I) = I + 1 (I = 1, 2, ..., N – 1), POINT(N) =

0 (the zero denoting the last free location). To add to the queue, we store job characteristics in index FREE, then update to FREE = POINT(FREE). To remove from the queue, we add the index freed up, TAKE, to the list of free locations by POINT(TAKE) = FREE, FREE = TAKE. Of course, TAKE must also be updated, so a temporary variable must be used to keep track of POINT(TAKE), which will be the new value of TAKE.

To illustrate the double-pointing linked list, suppose that at the current time our queue has data structures as in Figure 1.5–4. There, TAKE = 5 since the lowest priority number in P(·) is a 12; the jobs stored in columns 5 and 8 have priority 12, and of those the one in column 5 arrived first. POINT(5) = 8 since the other priority 12 job is to be taken after the one in column 5. Next comes the job in column 2, having arrived earlier than the job in column 9, which has the same priority 13, so POINT(8) = 2 and POINT(2) = 9. After those comes the priority 14 job in column 7, so POINT(9) = 7 and POINT(7) = 3. The column 3 job is the last one, so POINT(3) = 0.

Also, FREE = 1 since column 1 is not in use. Then POINT(1) = 4, since column 4 is also free. Threading together the free columns, POINT(4) = 10, POINT(10) = 6, POINT(6) = N, and POINT(N) = 0 since column N is the last free column.

Suppose that what happens next is that a job with priority 14 **arrives** at time 1300, and its service time is 5. We store its values in column 1, so P(1) = 14, A(1) = 1300, and S(1) = 5, since FREE = 1. Next, FREE is updated to POINT(FREE) = 4. To find the value of POINT(1), we compare P(1) with P(TAKE), and we have P(1) = 14 while P(TAKE) = P(5) = 12, so the job in column 1 comes later than that in column 5. Continuing, we find that P(1) exceeds P(POINT(5)) = P(8) = 12, as well as P(POINT(8)) = P(2) = 13 and P(POINT(2)) = P(9) = 13. However, P(1) equals P(POINT(9)) = P(7) = 14; since A(1) = 1300 while A(7) = 1100, the job in column 1 will come after that in column 7. Going on, P(POINT(7)) = P(3) = 15, so the job in column 1 will come before that in column 3. Thus, we update to POINT(7) = 1 and POINT(1) = 3. The updated linked list is shown in Figure 1.5–5.

Suppose that next the server is completing service of a job, and needs to **take one from the queue**. The job in column TAKE = 5—which has priority P(5) = 12, arrival time A(5) = 200, and service time S(5) = 5—is taken. We then update to TAKE = POINT(TAKE) = POINT(5) = 8, which is the first of those left on the queue. We set POINT(5) = FREE, and FREE = 5 to make column 5 the first on the list of available storage locations. This process is illustrated in Figure 1.5–6.

A time line, showing the jobs which we have discussed in Figures 1.5–4, 1.5–5, and 1.5–6 (but not other jobs that may have come to, and departed

	1	2	3	4	5	6	7	8	9	10	⋯	N
P(·)		13	15		12		14	12	13			
A(·)		100	800		200		1100	900	400			
S(·)		2	1		4		7	1	5			
POINT(·)	4	9	0	10	8	N	3	2	7	6		0

TAKE = 5   FREE = 1

Figure 1.5–4. Value of a linked list at a particular time.

	1	2	3	4	5	6	7	8	9	10	⋯	N
P(·)	14	13	15		12		14	12	13			
A(·)	1300	100	800		200		1100	900	400			
S(·)	5	2	1		4		7	1	5			
POINT(·)	4̶ 3	9	0	10	8	N	3̶ 1	2	7	6		0

TAKE = 5   FREE = 1̶ 4

Figure 1.5–5. The linked list of Figure 1.5–4, updated after a priority 14 and service time 5 arrival at time 1300.

	1	2	3	4	5	6	7	8	9	10	⋯	N
P(·)	14	13	15		1̶2̶		14	12	13			
A(·)	1300	100	800		2̶0̶0̶		1100	900	400			
S(·)	5	2	1		4̶		7	1	5			
POINT(·)	4̶ 3	9	0	10	8̶	N	3̶ 1	2	7	6		0

TAKE = 5̶ 8   FREE = 1̶ 4̶ 5

Figure 1.5–6. The linked list of Figure 1.5–5, updated after a selection from the queue.

from, the system), is shown in Figure 1.5–7.

In programming the linked-list data structure for updating the queue, **it is helpful to also have a second pointer vector that tells us the previous customer on the queue** (just as POINT($\cdot$) tells us the next on the queue), i.e., a vector BPOINT($\cdot$) with

$$\text{BPOINT(I)} = \left(\begin{array}{l} \text{Index of the job to be taken just before the} \\ \text{one whose attributes are stored in index I} \end{array} \right) \quad (1.5.3)$$

and BPOINT(I) = 0 if index I pertains to the first entry on the queue. We may keep a variable END which points to the last entry on the queue. For the linked list of Figure 1.5–6, we illustrate the back pointer in Figure 1.5–8. Note that for any index J on the queue, BPOINT(POINT(J)) = J.

While to this point we have considered the **linked list for a single-server** queue, it is often the case that a system (e.g., of tellers in a bank, checkers in a supermarket, pumps in a gas station, etc.) will have **multiple servers**. If such a facility had 30 servers with a maximum queue size of 4000 in the total facility, keeping a separate linked list for each server would use a considerable amount of memory. The POINT($\cdot$) vector alone would need $4000 \times 30 = 120,000$ words of memory. This amounts to $4 \times 120,000 = 480,000$ bytes (almost 0.5 megabyte) of memory on systems that use 4 bytes to store pointers.

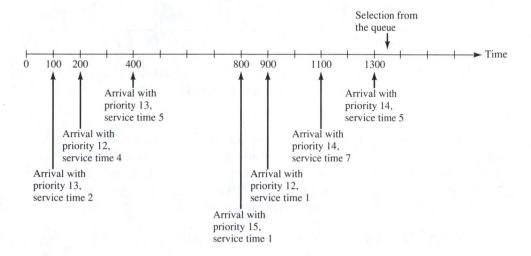

Figure 1.5–7. Time-line of jobs discussed in Figures 1.5–4 through 1.5–6.

	1	2	3	4	5	6	7	8	9	10	⋯	N
P(·)	14	13	15		12		14	12	13			
A(·)	1300	100	800		200		1100	900	400			
S(·)	5	2	1		4		7	1	5			
POINT(·)		9	0		8		3 1	2	7			
	4 3			10	4	N				6		0
BPOINT(·)	7	8	1				9	0	2			

TAKE = 8   END = 3   FREE = 5

Figure 1.5–8. The linked list of Figure 1.5–6, with the addition of a back pointer vector BPOINT(·).

	1	2	3	4	5	6	7	8	9	10	⋯	N
P(·)	E		A	F	I	B		G	C	D		H
A(·)												
S(·)												
POINT(·)	0		6	8	0	9		N	0	1		5

START(1) = 3   START(2) = 10   START(3) = 4

Figure 1.5–9. More than one active list in a linked list structure, for the 3-server queueing system of Figure 1.5–10.

Since often we will not need to allow for all 4,000 jobs to be on each of the queues, we often use **more than one active list in one linked list**. Thus, we have a data structure like those seen before, illustrated in Figure 1.5–9. If there are 3 queues, as in Figure 1.5–1, then in a vector START(·) we have the beginning customer on each queue:

$$\text{START}(J) = (\text{Index of the first job on queue for server } J) \qquad (1.5.4)$$

for J ranging over the number of machines (1, 2, 3 in our example of Figure 1.5–9). In this pooled system waiting space, the example of 30 servers with

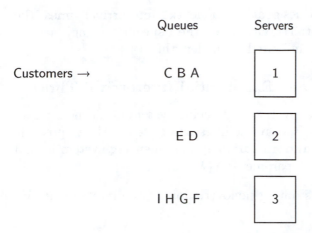

Figure 1.5–10. A 3-server queueing system, with 3, 2, 4 jobs waiting in queue at servers 1, 2, and 3.

up to 4000 jobs in the system now requires only 4000 words, or approximately 16K bytes of memory in our computer model of the system.

We could also have a pooled back-pointer structure for this system. Furthermore, while customers can move from queue to queue, **a customer cannot be on more than one queue at a time** with this space-saving structure. This important restriction will be found in virtually all simulation languages.

1.6   Theoretical Results about the Single-Server Queue

In Section 1.4 we saw how a single-server queueing system can be simulated in GPSS; in Sections 1.2, 1.3, and 1.5 we developed general algorithms for simulating single-server queues. Here, we consider some of the theoretical results associated with single-server queueing systems. These results can be used to validate a simulation model for a queueing system by comparing the results of a preliminary simulation with theoretical results (after which the validated simulation model may be used for cases where theoretical results are not available). We will confine this discussion to single-server queues with exponentially distributed interarrival and service times. Following the notation used in Section 1.2, we will use the notation M/M/1/∞ to refer to such queues. (The first M designates exponential interarrival times, the second M designates exponential service times, the 1 specifies a single server, and ∞ indicates capacity for unlimited queue length.)

1.6.1   Poisson Arrivals and Exponential Interarrival Times

Let X denote the number of arrivals at a queueing system in a time interval of length 1. The variable X has probabilities associated with its possible values, and is thus called a **random** variable. The queueing system is said to have **Poisson arrivals** with parameter λ ($\lambda > 0$) if

1. The numbers of arrivals during nonoverlapping time intervals are independent.

2. The probability of one arrival in a small time interval of length h is λh.

3. The probability of 2 or more arrivals on a sufficiently small time interval is zero.

In a queueing system with Poisson arrivals, the unit time interval can be divided into n subintervals, each of length $1/n$. If n is sufficiently large (we will eventually make sure n is sufficiently large by letting n go to infinity), then by property 3, the probability of 2 or more arrivals on any subinterval will be zero and we can restrict our attention to 0 or 1 arrival during each subinterval. If we interpret arrival on a given subinterval as a success and non-arrival as a failure, then X, already defined as the number of arrivals on the unit time interval, will have approximately a binomial distribution with parameters n and $p = \lambda/n$. Accordingly,

$$P[X = k] \simeq \binom{n}{k} p^k (1-p)^{n-k} = \binom{n}{k} \left(\frac{\lambda}{n}\right)^k \left(1 - \frac{\lambda}{n}\right)^{n-k}. \qquad (1.6.1)$$

To obtain the exact value of $P[X = k]$, we let $n \to \infty$ in (1.6.1).

$$P[X = k] = \lim_{n \to \infty} \binom{n}{k} \left(\frac{\lambda}{n}\right)^k \left(1 - \frac{\lambda}{n}\right)^{n-k}$$

$$= \lim_{n \to \infty} \frac{n(n-1)\dots(n-k+1)}{k!} \frac{\lambda^k}{n^k} \frac{(1 - \frac{\lambda}{n})^n}{(1 - \frac{\lambda}{n})^k}$$

$$= \frac{\lambda^k}{k!} \lim_{n \to \infty} \left(\frac{n}{n} \cdot \frac{n-1}{n} \dots \frac{n-k+1}{n}\right) \frac{(1 - \frac{\lambda}{n})^n}{(1 - \frac{\lambda}{n})^k}$$

$$= \frac{\lambda^k}{k!} \lim_{n \to \infty} \left(1 \left(1 - \frac{1}{n} \right) \left(1 - \frac{2}{n} \right) \cdots \left(1 - \frac{k-1}{n} \right) \right) \frac{(1 - \frac{\lambda}{n})^n}{(1 - \frac{\lambda}{n})^k}.$$

We now observe that the limits of $1 - 1/n, \ldots, 1 - (k-1)/n$, and $(1 - \lambda/n)^k$ are all 1. Thus,

$$P[X = k] = \frac{\lambda^k}{k!} \lim_{n \to \infty} \left(1 - \frac{\lambda}{n} \right)^n. \tag{1.6.2}$$

From elementary mathematics we know that the limit in (1.6.2) is $e^{-\lambda}$ and

$$P[X = k] = \frac{\lambda^k}{k!} e^{-\lambda} \quad \text{for } k = 0, 1, \ldots. \tag{1.6.3}$$

The formulation given in (1.6.3) can now be recognized as the probability density function of the Poisson random variable with parameter λ. Detailed discussion of the Poisson distribution and its properties can be found in Dudewicz and Mishra (1988). We merely note here that the mean and variance of the Poisson distribution are both equal to λ. Furthermore, if Y were the random variable representing the number of arrivals on a time interval of length a, then Y would be Poisson distributed with parameter (hence also mean and variance) $a\lambda$.

We now consider the relationship between X, the number of arrivals during a unit time interval and interarrival times (times between successive arrivals). If we let the random variable W be the time lapse between two successive arrivals, then for any $w > 0$,

$$P[W \le w] = 1 - P[W > w]$$
$$= 1 - P(0 \text{ arrivals on an interval of length } w).$$

Since, as noted in the previous paragraph, the number of arrivals on an interval of length w is Poisson distributed with parameter $w\lambda$,

$$P[W \le w] = 1 - \frac{(w\lambda^0)e^{-w\lambda}}{0!} = 1 - e^{-w\lambda}. \tag{1.6.4}$$

This gives the distribution function

$$F(w) = \begin{cases} 1 - e^{-w\lambda}, & w > 0 \\ 0, & \text{otherwise.} \end{cases} \tag{1.6.5}$$

A simple differentiation now gives us the probability density function (**p.d.f.**)

$$f(w) = \begin{cases} \lambda e^{-\lambda w}, & w > 0 \\ 0, & \text{otherwise.} \end{cases} \tag{1.6.6}$$

We recognize (1.6.6) as the p.d.f. of the exponential distribution with mean $E(W) = 1/\lambda$ and variance $\sigma^2 = 1/\lambda^2$. The following theorem summarizes the results of this section.

Theorem 1.6.7. *If arrivals at a queueing system are Poisson with parameter λ, then*

1. *The random variable X, the number of arrivals during a unit of time, is Poisson distributed with probability density function*

$$f(x) = \begin{cases} (e^{-\lambda}\lambda^x)/x!, & x = 0, 1, \ldots \\ 0, & otherwise. \end{cases}$$

2. *The random variable W, the interarrival time, has exponential distribution with probability density function*

$$f(x) = \begin{cases} \lambda e^{-\lambda x}, & x \geq 0 \\ 0, & otherwise. \end{cases}$$

1.6.2   Number of Customers in M/M/1/∞ Queueing Systems

Throughout this and succeeding sections, we will assume that arrivals are Poisson with parameter λ and service times are exponential with parameter μ. From these assumptions, we conclude that if Δt is a sufficiently small time increment, then on the interval from t to $t + \Delta t$ we need not consider the possibility of more than one arrival or more than one departure. This leaves four possibilities on the interval from t to $t + \Delta t$:

$$\begin{aligned} E_1 &: \quad \text{1 arrival and 0 departure} \\ E_2 &: \quad \text{0 arrival and 1 departure} \\ E_3 &: \quad \text{1 arrival and 1 departure} \\ E_4 &: \quad \text{0 arrival and 0 departure.} \end{aligned}$$

The probability of E_1 can be obtained by

$$P(E_1) = (\text{Probability of 1 arrival }) \times (\text{ Probability of 0 departure})$$
$$= (\lambda(\Delta t)e^{-\lambda\Delta t}) \times (e^{-\mu\Delta t}) = \lambda(\Delta t)e^{-\Delta t(\lambda+\mu)} . \tag{1.6.8}$$

In a similar manner,

$$P(E_2) = \mu(\Delta t)e^{-\Delta t(\lambda+\mu)} , \tag{1.6.9}$$
$$P(E_3) = \lambda\mu(\Delta t)^2 e^{-\Delta t(\lambda+\mu)} , \tag{1.6.10}$$

and,

$$P(E_4) = 1 - P(E_1) - P(E_2) - P(E_3)$$
$$= 1 - (\Delta t)e^{-\Delta t(\lambda+\mu)} (\lambda + \mu + (\Delta t)\lambda\mu). \qquad (1.6.11)$$

For simplicity, we let $A = e^{-\Delta t(\lambda+\mu)}$ and rewrite (1.6.8) through (1.6.11) as

$$P(E_1) = \lambda A(\Delta t)$$
$$P(E_2) = \mu A(\Delta t)$$
$$P(E_3) = \lambda\mu A(\Delta t)^2$$
$$P(E_4) = 1 - A(\Delta t)(\lambda + \mu + (\Delta t)\lambda\mu). \qquad (1.6.12)$$

If we let $P_n(t)$ designate the probability of n customers in the system (in the server and in the queue) at time t, then for $n > 0$ (the case of $n = 0$ is taken up in Problem 1.12),

$$P_n(t + \Delta t) = P_{n-1}(t)P(E_1) + P_{n+1}(t)P(E_2)$$
$$+ P_n(t)P(E_3) + P_n(t)P(E_4). \qquad (1.6.13)$$

Using (1.6.12),

$$P_n(t + \Delta t) = P_{n-1}(t)\lambda A(\Delta t) + P_{n+1}(t)\mu A(\Delta t) + P_n(t)\lambda\mu A(\Delta t)^2$$
$$+ P_n(t)(1 - \lambda A(\Delta t) - \mu A(\Delta t) - \lambda\mu A(\Delta t)^2). \qquad (1.6.14)$$

Moving $P_n(t)$ in equation (1.6.14) to the left side and dividing both sides by Δt, we obtain

$$\frac{P_n(t + \Delta t) - P_n(t)}{\Delta t} =$$
$$A\left(\lambda P_{n-1}(t) + \mu P_{n+1}(t) - \lambda P_n(t) - \mu P_n(t)\right). \qquad (1.6.15)$$

It is not at all clear that the number of customers in the system, or for that matter $P_n(t)$, will stabilize as the queueing operation continues over time. For example, if the arrival rate, λ, exceeds the rate at which customers receive service, μ, then the number of customers will increase without bound. However, if we assume eventual stability, then although the number of customers in the system will change with time, $P_n(t)$ will not change for t sufficiently large; say, $t > T$. This implies that

$$\frac{dP_n(t)}{dt} = 0 \quad \text{for } t > T \text{ and } n = 1, 2, \dots . \qquad (1.6.16)$$

Thus, under the assumption of stability, generally referred to as **steady-state**, from (1.6.15) and (1.6.16) we obtain

$$0 = \frac{dP_n(t)}{dt} = \lim_{\Delta t \to 0} \frac{P_n(t + \Delta t) - P_n(t)}{\Delta t}$$

$$= \lim_{\Delta t \to 0} A \left(\lambda P_{n-1}(t) + \mu P_{n+1}(t) - \lambda P_n(t) - \mu P_n(t) \right). \qquad (1.6.17)$$

Noting that $\lim_{\Delta t \to 0} A = 1$, we have

$$\lambda P_{n-1}(t) - (\lambda + \mu) P_n(t) + \mu P_{n+1} = 0 \quad \text{for } n = 1, 2, \ldots . \qquad (1.6.18)$$

Since steady-state is assumed, $P_k(t)$ does not change when $t > T$, and we can drop the t in (1.6.18) and conclude that in steady-state

$$\lambda P_{n-1} - (\lambda + \mu) P_n + \mu P_{n+1} = 0 \quad \text{for } n = 1, 2, \ldots . \qquad (1.6.19)$$

The simpler case for $n = 0$ (taken up in Problem 1.12) yields

$$\mu P_1 - \lambda P_0 = 0. \qquad (1.6.20)$$

The following theorem provides a simple formula for P_n.

Theorem 1.6.21. *If in an M/M/1/∞ queueing system $\lambda < \mu$ (i.e., the service rate is higher than the arrival rate), then in steady-state, $P_k = \rho^k P_0$ for $k = 0, 1, \ldots$ where $\rho = \lambda/\mu$.*

Proof. For $k = 0$, $P_k = \rho^k P_0$ is trivially true and for $k = 1$ it follows from (1.6.20). Using mathematical induction, we assume that the conclusion of the theorem is true for $k \le i$ or

$$P_k = \rho^k P_0 \quad \text{for } k = 0, 1, \ldots, i.$$

For $k = i + 1$ we have, by substitution of $n = i$ in (1.6.19),

$$\lambda P_{i-1} - (\lambda + \mu) P_i + \mu P_{i+1} = 0.$$

Solving for P_{i+1},

$$\begin{aligned}
P_{i+1} &= \left(\frac{\lambda + \mu}{\mu} \right) P_i - \left(\frac{\lambda}{\mu} \right) P_{i-1} \\
&= \left(\frac{\lambda}{\mu} + 1 \right) P_i - \left(\frac{\lambda}{\mu} \right) P_{i-1} \\
&= (\rho + 1)\rho^i P_0 - \rho \cdot \rho^{i-1} P_0 \\
&= \rho^{i+1} P_0 + \rho^i P_0 - \rho^i P_0 = \rho^{i+1} P_0,
\end{aligned}$$

concluding the induction step and the proof of the theorem.

1.6.3 Idle Time in M/M/1/∞ Queues

The idle time of a queueing system is the proportion of time that the server is idle. Since the server is idle if and only if there are 0 customers in the system, P_0 represents idle time of M/M/1/∞ queues that are in steady-state.

Theorem 1.6.22. *In steady-state, the idle time of an M/M/1/∞ queueing system is $P_0 = 1 - \rho$.*

Proof. From Theorem 1.6.21 and the fact that $\sum_{n=0}^{\infty} P_n = 1$, we have

$$1 = \sum_{n=0}^{\infty} \rho^n P_0 = P_0 \sum_{n=0}^{\infty} \rho^n.$$

By the steady-state assumption, $\rho < 1$. Therefore, the geometric series $\sum_{n=0}^{\infty} \rho^n$ converges to $1/(1 - \rho)$ and we obtain $P_0 = 1 - \rho$.

Note that using $P_0 = 1 - \rho$ in Theorem 1.6.21 gives us the following explicit formula for P_n:

$$P_n = \rho^n(1 - \rho), \quad n = 0, 1, \ldots. \tag{1.6.23}$$

1.6.4 Expected Number of Customers in M/M/1/∞ Queues

To derive an expression for L, the expected number of customers in the system, we will need the following lemma.

Lemma 1.6.24. *If $|a| < 1$, then $\sum_{i=0}^{\infty} ia^i = a/(1 - a)^2$.*

Proof. Consider

$$
\begin{aligned}
(1 - a) \sum_{i=0}^{N} ia^i &= \sum_{i=0}^{N} ia^i - \sum_{i=0}^{N} ia^{i+1} \\
&= (0 + a + 2a^2 + \cdots + Na^N) \\
&\quad - \left(0 + a^2 + 2a^3 + \cdots + (N-1)a^N + Na^{N+1}\right) \\
&= a + a^2 + \cdots + a^N - Na^{N+1} \\
&= \sum_{i=1}^{N} a^i - Na^{N+1} = a \sum_{i=0}^{N-1} a^i - Na^{N+1}.
\end{aligned}
$$

By summing the finite geometric series on the right, we conclude that

$$(1 - a) \sum_{i=0}^{N} ia^i = a \left(\frac{1 - a^N}{1 - a}\right) - Na^{N+1}. \tag{1.6.25}$$

To obtain the result of the lemma, we convert the finite sum in (1.6.25) to an infinite series by taking limits as $N \to \infty$ and noting that $\lim_{N \to \infty} a^N = 0$. Thus,

$$(1 - a) \sum_{i=0}^{\infty} i a^i = \frac{a}{1 - a}$$

and $\sum_{i=0}^{\infty} i a^i = a/(1 - a)^2$.

Theorem 1.6.26. *In an M/M/1/∞ queueing system that is in steady-state,*

$$L = \rho/(1 - \rho)$$

where L is the expected number of customers in the system.

Proof. By definition, $L = \sum_{n=0}^{\infty} n P_n$. Therefore, by (1.6.23),

$$L = \sum_{n=0}^{\infty} n \rho^n (1 - \rho) = (1 - \rho) \sum_{n=0}^{\infty} n \rho^n.$$

Lemma 1.6.24 now gives $L = \rho/(1 - \rho)$.

1.6.5   Expected Number of Customers in the Facility and in the Queue of an M/M/1/∞ System

At any time, the facility is either idle or contains a customer. The probability that it is idle is $P_0 = 1 - \rho$ (Theorem 1.6.22); hence, the probability that it is busy must be $1 - P_0 = \rho$. Therefore, L_f, the expected number in the facility, is given by

$$L_f = 0 \times (1 - \rho) + 1 \times \rho = \rho \tag{1.6.27}$$

and L_q, the expected number of customers in the queue, is the difference, $L - L_f$. Therefore,

$$L_q = L - L_f = \frac{\rho}{1 - \rho} - \rho = \frac{\rho^2}{1 - \rho}. \tag{1.6.28}$$

1.6.6   Expected Time Spent Waiting and in the System

The expected number of customers in the system, L, and the expected time spent by a customer in the system, w, have the relationship

$$L = \lambda w \tag{1.6.29}$$

which was given by Little (1961). Since (1.6.29) is essential to the derivation
of an expression for w, we give a heuristic argument in support of it.

In an empirical setting, if we plot the number of customers in the system
as a function of time, we would obtain a graph similar to the one given in
Figure 1.6–1. In the particular case described by Figure 1.6–1, T_0, the total
time during which there were 0 customers in the system is 3 (from 0 to 2
and from 11 to 12 units of time). The total time during which there was
1 customer in the system, T_1, is 13 (from 2 to 6, 8 to 11, 12 to 13, and 19
to 24 time units). Similarly, $T_2 = 7$ and $T_3 = 1$. In general, the empirical
average, $\hat{L}$, of the number of customers in the system will be

$$\hat{L} = \left(\frac{1}{T}\right) \sum_i iT_i \qquad (1.6.30)$$

where the sum is taken over the total time span, T, of empirical observation.
The area under the graph of Figure 1.6–1 can be represented by $\sum_i iT_i$.

A different interpretation of the area under the graph of Figure 1.6–1
can be obtained by considering $w_1, w_2, \ldots$, the times spent by successive
customers in the system. In this interpretation, the first customer arrives at
time 2 and leaves at time 8 so its total time, 6, is represented by the block
with $2 \le t \le 8$ and vertically between 0 and 1. The second customer arrives
at time 6 and leaves at time 11. Its time, 5, is represented by two blocks;
the first vertically between 1 and 2 and with $6 \le t \le 8$, the second vertically
between 0 and 1 and with $8 \le t \le 11$. Continuing this way, we can see that
the average time spent in the system is

$$\hat{w} = \left(\frac{1}{N}\right) \sum_i w_i \qquad (1.6.31)$$

where N is the total number of customers and $\sum_i w_i$ is the area under the
graph. Therefore, in (1.6.30) we can substitute $\sum_i w_i$ for $\sum_i iT_i$ to get

$$\hat{L} = \left(\frac{1}{T}\right) \sum_i w_i = \left(\frac{N}{T}\right)\left(\frac{1}{N}\right) \sum_i w_i = \hat{\lambda}\hat{w}.$$

Theorem 1.6.32. *In an $M/M/1/\infty$ queueing system that is in steady-
state, w and w_q, the expected times customers spend in the system and in
the queue, respectively, are*

$$w = 1/(\mu - \lambda)$$

and

$$w_q = \rho/(\mu - \lambda).$$

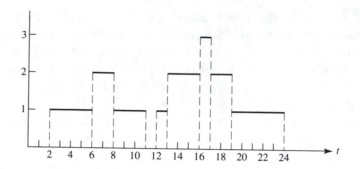

Figure 1.6–1. Number of customers in an M/M/1/∞ system
as a function of time, t.

Proof. With the assumption of (1.6.29),

$$w = L/\lambda$$

and by Theorem 1.6.26,

$$w = \left(\frac{1}{\lambda}\right)\left(\frac{\rho}{1-\rho}\right) = \frac{1}{\mu - \lambda}.$$

To obtain the result for w_q, we note that

$$w = w_q + \text{expected service time.}$$

Hence,

$$w_q = w - \frac{1}{\mu} = \frac{1}{\mu - \lambda} - \frac{1}{\mu} = \frac{\rho}{\mu - \lambda}.$$

1.6.7 Summary of Results and an Example

The analysis of Sections 1.6.1 through 1.6.6 was confined to the M/M/1/∞
queueing systems. More general results and extensions of what was presented
there can be found in a variety of sources: Banks and Carson (1984), Cooper
(1981), Ross (1985), and Taylor and Karlin (1984). For ease of reference, we
summarize the results that were derived in Sections 1.6.1 through 1.6.6 in
Table 1.6–2.

Example. A store operates one checkout counter. The time to check out a
customer is exponentially distributed with mean time 3 minutes. Customers
arrive at the counter according to a Poisson distribution with a mean rate
of 15 per hour. Find:

Table 1.6–2. Summary of M/M/1/∞ queueing results.

NOTATION	INTERPRETATION	VALUE
λ	Average arrival rate	
μ	Average service rate	
ρ	Ratio of arrival rate to service rate	λ/μ
P_0	Idle time	$1 - \rho$
P_k	Probability of k customers in system	$\rho^k(1-\rho)$
L	Average number in system	$\rho/(1-\rho)$
L_f	Average number in facility	ρ
L_q	Average number in queue	$\rho^2/(1-\rho)$
w	Average time in system	$1/(\mu-\lambda)$
w_q	Average time in queue	$\rho/(\mu-\lambda)$

1. The mean time a customer spends waiting to be checked out.

2. The mean time a customer spends at the checkout counter (waiting time plus checkout time).

3. The mean number of customers at the checkout counter.

4. The average service rate that would reduce the customer waiting time to half its current value (we assume that we have no control over arrivals).

From the specifications of the problem, we know that $\lambda = 15$ arrivals per hour. Since other time units in the problem are in minutes, we convert λ to 1/4 arrivals per minute. A mean service time of 3 minutes specifies $1/\mu = 3$ minutes or $\mu = 1/3$. Therefore, $\rho = \lambda/\mu = 3/4$. Part 1 simply asks for w_q and

$$w_q = \frac{\rho}{\mu - \lambda} = \frac{3/4}{(1/3 - 1/4)} = 9 \text{ minutes.}$$

Part 2 asks for w, which we obtain by adding 3 (average service time) to w_q or by

$$w = \frac{1}{\mu - \lambda} = \frac{1}{1/3 - 1/4} = 12 \text{ minutes.}$$

Part 3 seeks L, which can be calculated by

$$L = \frac{\rho}{1 - \rho} = \frac{3/4}{1/4} = 3.$$

In Part 4, we make $w_q = 9/2$ by keeping λ fixed at $1/4$ but altering μ.

$$9/2 = w_q = \rho/(\mu - \lambda).$$

Substituting λ/μ for ρ, cross-multiplying, and simplifying, we get

$$36\mu^2 - 9\mu - 2 = 0$$

which yields

$$\mu = (3 + \sqrt{41})/24 \approx 0.39.$$

1.7   Use of Efficient Sorting Techniques in Simulation

The need for efficient sorting of a list arises in at least two ways in simulations. First, **in many simulations a list (often a very long list) needs to be sorted in many replications** of a simulation run. Second, in complex simulations, often **the events that need to occur, and the times at which they need to occur, will be kept in a list, which needs to be kept sorted.** Since, as we will see below, **the difference in computer time required for various methods can be on the order of hours, the choice of sorting method can be critical** for our ability to obtain the computer time needed to run our simulation.

One of the obvious ways to sort a list is by exchanges, in what is called a **bubble sort**. Suppose our list consists of numbers $R_1, R_2, \ldots, R_N$. This sort first interchanges R_1 and R_2 if they are out of order; then similarly for R_2 and R_3; R_3 and R_4; etc. One pass through is sure to move the largest number into position R_N. Repetitions get the appropriate items into positions $R_{N-1}, R_{N-2}, \ldots$. It has been shown (see Knuth (1973)) that the average running time of this algorithm is

$$5.75N^2 + O(N \ln N). \tag{1.7.1}$$

Here, $O(x)$ denotes a term which is said to be "of the order of big oh of x as $x \to \infty$," which means that

$$\lim_{x \to \infty} \frac{O(x)}{x} = c \tag{1.7.2}$$

for some constant c. (The order is said to be $o(x)$, "little oh of x," if $c = 0$.) Some authors have called the bubble sort, which derives its name from the fact that it "bubbles up" the large items, a "trick of the trade" (e.g., McQuay (1973), p. 187); we will see that it is more of a "trick" (on us).

The basic problem with the bubble sort is that, if a sorting algorithm moves items only one position at a time, its average running time (against a random order of $R_1, R_2, \ldots, R_N$) will be at best proportional to N^2 for the sorting of N items. Suppose that $a_1, a_2, \ldots, a_N$ is a random permutation of the integers $1, 2, \ldots, N$. Then the average amount these items are "out of position" is

$$E\left(|a_1 - 1| + |a_2 - 2| + \cdots + |a_N - N|\right)$$

$$= \sum_{j=1}^{N} E|a_j - j| = \sum_{j=1}^{N} \frac{1}{N}(|1 - j| + |2 - j| + \ldots + |N - j|)$$

$$= \sum_{j=1}^{N} \frac{1}{N}((j - 1) + (j - 2) + \cdots + (j - j) + ((j + 1) - j) +$$

$$\cdots + ((N - 1) - j) + (N - j))$$

$$= \sum_{j=1}^{N} \frac{1}{N}(jj - (1 + 2 + \ldots + j) - (N - j)j + ((j + 1) + \ldots + N))$$

$$= \sum_{j=1}^{N} \frac{1}{N}(2j^2 - 2(1 + 2 + \ldots + j) - Nj + (1 + 2 + \ldots + N)) \qquad (1.7.3)$$

$$= \sum_{j=1}^{N} \frac{1}{N}(j^2 - (N + 1)j + N(N + 1)/2)$$

$$= \frac{(N + 1)(2N + 1)}{6} - \frac{(N + 1)^2}{2} + \frac{N(N + 1)}{2}$$

$$= \frac{(N + 1)(N - 1)}{3} = \frac{N^2 - 1}{3} = \frac{N^2}{3} - \frac{1}{3}.$$

The following facts were used in the above derivation:

1. If a_j is chosen at random from $1, 2, \ldots, N$, then it has value i with probability $1/N$ for $i = 1, 2, \ldots, N$.

2. $1 + 2 + \ldots + n = n(n + 1)/2$ for any positive integer n.

3. $1^2 + 2^2 + \ldots + n^2 = n(n + 1)(2n + 1)/6$ for any positive integer n.

One may intuitively think of the above derivation as saying that the average item must travel about $N/3$ positions in the sorting process to reach its proper position in the sorted list, since

$$\frac{1}{N}\left(\frac{N^2}{3} - \frac{1}{3}\right) = \frac{N}{3} - \frac{1}{3N} \approx \frac{N}{3}.$$

Many alternatives have been proposed, such as insertion sort, quick sort, merge sort, heap sort, and Shell sort. These algorithms are all explained, with full computer code, in Dudewicz and Karian (1985), pp. 279–293, along with comparisons and reports of running times. Since the Shell sort is the simplest to code and yet is a very efficient (though not the most efficient) algorithm, we will restrict ourselves here to that algorithm and indicate in Table 1.7–1 the relative running times of certain sorting algorithms. The times (in milliseconds) given in Table 1.7–1 (source of data: Dudewicz and Karian (1985), p. 293) are the times required for sorting lists of length 1000 that are initially in random order, ascending order, and descending order. In all cases, the algorithms were coded in Pascal and executed on a VAX-11/780 under the VMS operating system.

Table 1.7–1. Times (in milliseconds) needed by six sorting methods.

Sorting Method	Random Order	Ascending Order	Descending Order
Bubble sort	14,745	4,165	25,145
Insertion sort	2,500	80	5,105
Quick sort	255	3,275	3,325
Shell sort	1,000	120	360
Heap sort	460	470	430
Merge sort	300	270	260

As many know from the experience of sorting student lists, it is easy to sort very small lists and hard to sort very large lists. For this reason, many algorithms sort large lists by breaking them into small lists, then reassembling the small lists into a large list. The FORTRAN code for the **Shell sort**, given in Figure 1.7–2, employs this strategy. Suppose there are 32 items to sort; we set $M = 32$ and immediately divide it by 2, obtaining $M = 16$. This divides the items to be sorted into 16 groups of 2 items each, and comparisons (and, if needed, interchanges) are made of R_1 with R_{17}, of R_2 with $R_{18}, \ldots,$ and of R_{16} with R_{32}. This completely sorts the 16 groups,

```
      SUBROUTINE SORT(R,N)

C***************************************************************************
C PURPOSE
C    TO SORT A GIVEN ARRAY.
C
C USAGE
C    "CALL SORT(R,N)"
C
C DESCRIPTION OF PARAMETERS
C    R.. IS THE NAME OF THE ONE-DIMENSIONAL ARRAY TO BE SORTED.
C    N.. IS THE DIMENSION OF THE ARRAY TO BE SORTED.
C
C REMARKS
C    THIS ALGORITHM IS MORE EFFICIENT THAN EITHER THE "BUBBLE SORT"
C    OR THE "INTERCHANGE SORT" METHODS. WHILE NOT OPTIMAL, IT
C    PROVIDES A SIMPLE AND RELATIVELY EFFICIENT SORTING TECHNIQUE
C    WHICH IS EASILY PROGRAMMED.
C
C SUBROUTINES AND FUNCTION SUBPROGRAMS REQUIRED
C    NONE
C
C METHOD
C    THE METHOD USED IS CALLED THE SHELL SORT AND WAS INTRODUCED BY
C    D. L. SHELL IN REFERENCE 1. FOR MORE RECENT DEVELOPMENTS SEE
C    ITEM2. AND ITS REFERENCES.
C    1.    SHELL, D. L.: "A HIGH SPEED SORTING PROCEDURE, "COMMUNICATIONS OF
C          THE ASSOCIATION FOR COMPUTING MACHINERY, VOL. 2, NO. 7(JULY,
C          1959), PP. 30-32.
C    2.    GHOSHDASTIDAR, D. AND ROY, M. K.: "A STUDY ON THE EVALUATION OF
C          SHELL'S SORTING TECHNIQUE," THE COMPUTER JOURNAL, VOL. 18 (1975),
C          PP. 234-235.
C***************************************************************************
C
      SUBROUTINE SORT(R,N)
      DIMENSION R(1)
      M=N
1     M=M/2
      IF(M.EQ.0) GOTO 5
      J=1
2     I=J
4     IF(R(I).LE.R(I+M)) GOTO 3
      SAV=R(I)
      R(I)=R(I+M)
      R(I+M)=SAV
      I=I-M
      IF(I.GE.1) GOTO 4
3     J=J+1
      IF(J.GT.N-M) GOTO 1
      GOTO 2
5     RETURN
      END
```

Figure 1.7–2. FORTRAN code for the Shell sort.

each of which contains only 2 items. Next M is changed by division by 2, to $M = 8$. We now sort the 8 groups of 4 items until each group is completely sorted. Then we move to 4 groups of 8 items, then 2 groups of 16 items, and finally 1 group of 32 items. Working through an example will convince most readers that the algorithm will indeed sort any list; hence, we recommend that exercise to readers at this point.

At the outset we noted that bubble sort was a "trick" while Shell sort was a "treat." If we examine Table 1.7–1, we see that bubble sort may take 25 seconds to sort a list where Shell sort takes 1 second. (This is with $N = 1000$ items to sort.) If this sort is needed 400 times in a simulation, the bubble sort will spend over 166 minutes (2 hours and 46 minutes), while Shell sort will take less than 7 minutes for the same tasks. It is clear that bubble sort is unsuitable, especially since $N = 1000$ is not nearly the largest we may expect (e.g., with simulations of nuclear attack strategies, with an interest in the remaining airfield capability of the United States, one deals with sorting lists of over 30,000 airfields, so $N = 30,000$; similar sizes are encountered in genetic simulations).

As a final remark, we note that the interaction between data structures and sorting methods in simulation event lists may be seen in such articles as those by Ulrich (1978), Franta and Maly (1977), and Vuillemin (1978).

Problems for Chapter 1

1.1 (Section 1.2) For the single-server queue with first-come, first-served discipline, write a simulation program (in any general-purpose scientific programming language) and run it with the following specifications:

- Start with no queue and the server idle.

- Assume all customers have service time STIME= 0.85 minute.

- Assume the goal is to determine what the average queue size will be after each of 1, 2, 3, 4, 5, 6, 7, 8 hours of simulated operation and then terminate the simulation.

- ATIME for customers is to be determined as follows:
    Let $X_0 = 32767$, $MULT = 5{**}13 = (15625)(15625)(5)$,
    $M = 2{**}15 = 32768$.

$X_1 = (MULT * X_0) \bmod M,\ X_2 = (MULT * X_1) \bmod M,$
$\quad X_3 = (MULT * X_2) \bmod M, \ldots$
$A_1 = X_1/32768,\ A_2 = X_2/32768,\ A_3 = X_3/32768, \ldots$
Take the ATIMEs of the first, second, third, etc., customers
to be $A_1, A_1 + A_2, A_1 + A_2 + A_3, \ldots$

a. Calculate X_1, X_2, X_3 and also A_1, A_2, A_3 by hand.

b. Write a computer program that will calculate $A_1, A_2, A_3, \ldots$ Run your
program and print out the first 100 A_is; i.e., $A_1, A_2, A_3, \ldots, A_{100}$.

c. Run your simulation program.

Note: For this problem, the items to be submitted include the following: for
Part a, a page showing clearly the calculations and final answers for the 6
numbers noted; for Part b, the computer program and its output; for Part c,
the flowchart for your program, the input and call for execution, the program,
and the output. (**The computer program should contain sufficient
comments that it can be read**; a rule of thumb is one line of comments
for each line of code. **The output should be well documented**—you
should **not** need to write by hand on the final output if the output has been
well documented.)

1.2 (Section 1.2)

a. Adjust the algorithm described in Figure 1.2–7 to accommodate changing service times.

b. Generate values of ATIME and STIME through the algorithms (implemented as subprograms) ATIMEGEN and STIMEGEN as given below

```
ATIMEGEN
        input ATIMESEED, ATIME
        MULT ← 5**13
        M ← 2**15
        ATIMESEED ← (MULT*ATIMESEED) MOD M
        TEMP ← ATIMESEED/M
        ATIME ← ATIME + [9 + 7 * TEMP]
        output ATIME
    end of algorithm ATIMEGEN

STIMEGEN
        input STIMESEED
        MULT ← 5**13
```

```
            M ← 2**15
            STIMESEED ← (MULT*STIMESEED) MOD M
            TEMP ← STIMESEED/M
            STIME ← [12 + 9 * TEMP]
            output STIME
      end of algorithm STIMEGEN
```

  c. Write a program in a general-purpose programming language that uses the specifications given in parts a and b to implement the algorithm of Figure 1.2–7. Do this with HTIME = 200.

  d. Turn in your program listing and program output consisting of values of ATIME, STIME, CTIME, BUSY, QSIZE, QAREA, LTIME prior to the invocation of either of the arrival or service completion routines.

1.3 (Section 1.3) This problem continues Problem 1.1 by adding additional output. For the single-server queue with first-come, first-served discipline, write a simulation program (in any general-purpose scientific programming language) and run it with the following specifications:

- Start with no queue and the server idle.

- Assume all customers have service time STIME= 0.85 minute.

- ATIME for customers is to be determined as in Problem 1.1.

- The goal is to determine each of the following performance measures after 1, 2, 3, 4, 5, 6, 7, and 8 hours of simulated operation and then terminate the simulation.

  1. Average queue size.
  2. Distribution of queue size.
  3. Proportion of idle time.
  4. Average number of customers in the system.
  5. Average waiting time.
  6. Distribution of waiting time. Note that waiting time is a continuous variable (unlike queue size) and we can specify its distribution by determining the proportions in the intervals $[0], (0, 1], (1, 2], (2, 3], \ldots, (59, 60], (60, +\infty)$, where units are minutes.
  7. Number of customers turned away.

- The waiting room in the service facility is such that only 10 customers can wait at any time; if a customer would arrive when the queueing area is full (queue size is 10), turn him or her away and schedule the next arrival.

Note: For this problem, the items to be submitted include the following: the flowchart for your program, the input and call for execution, the program, and the output. **The program should be well documented** (containing sufficient comments so that it can be read—a rule of thumb is **one line of comments for each line of code**), and **the output should also be well documented (you should not need to write by hand on the final output if the output has been well documented).**

1.4 (Section 1.5) In any scientific programming language, write a program that will maintain a list with a linked-list data structure. Detailed specifications are

- Items on the list have a priority number and an arrival time.

- When an item is to be withdrawn from the list, the item with the lowest arrival time among those items with the lowest priority number should be withdrawn.

- Construct your program so that it reads a variable ACT and (based on the value of ACT) then decides what list-keeping function needs to be performed next:

    1. If ACT = 0, withdraw an item from the list according to the withdrawal rule.
    2. If ACT = 1, read PRTY and ATIME, which will be the priority and arrival time of the next item, and place that item on the list.
    3. If ACT = 2, list the items on the list, in the order that they would be withdrawn if all were being withdrawn.
    4. If ACT = 3, stop your program.

- The list is to be allowed to be of up to 400 items.

Note: For this problem, the items to be submitted include the following: the flowchart for your program, the input and call for execution, the program, and the output. The program and the output should be well documented (for details on what this means, see Problem 1.3). For the run you hand in, use the following sequence of data for ACT (and, if ACT = 1, PRTY and ATIME):

				(continued)		
ACT	PRTY	ATIME		ACT	PRTY	ATIME
2				0		
2				0		
1	2	4		0		
1	3	1		0		
1	3	4		0		
1	4	11		2		
1	5	8		1	0	3
2				1	0	4
1	4	13		1	0	62
2				1	0	77
1	4	12		2		
1	−99	999		0		
1	99	−999		2		
2				1	0	−77
0				1	0	−4
2				2		
0				0		
0				2		
0				3		
2						

1.5 (Section 1.5) This problem continues Problem 1.3 by adding priority numbers. For the single-server queue as in Problem 1.3, now suppose that, on arrival, each customer has a priority number and service time (as well as an arrival time). Write a simulation program (in any general-purpose scientific programming language) and run it with the following specifications:

- Start with no queue and the server idle.

- Arrival times, priorities, and service times are to be read from a data file as needed.

- The goal is to determine the same seven performance measures as in Problem 1.3 after each of 100(100)2000 minutes of simulated operation and then to terminate the simulation.

- The waiting room can hold up to 10 customers (as in Problem 1.3).

- Use the linked-list data structure of Problem 1.4 to maintain the queue.

Note: For this problem, the items to be submitted are as in Problem 1.4. For the run you hand in, use the following data for the successive (arrival time, priority, service time) vectors:

20,0,10   40,0,30 60,0,20   80,0,30   120,0,30   140,0,5
160,0,35   180,0,1 190,0,1   301,0,20   302,0,1   303,0,1
304,0,1   305,0,1 306,0,1   307,0,1   308,0,1   309,0,1
310,0,1   311,0,1 312,0,1   380,0,10   410,0,30   411,−5,1
412,−6,1   2001,0,1

1.6 (Section 1.6) Arrivals at a single-channel, single-server queueing system are Poisson with an average of 8 per hour. The times required for service are exponentially distributed with an average of 6 minutes. After the system reaches steady-state,

   a. What is the probability that the service facility is idle?

   b. What is the average waiting time?

   c. What average number of arrivals (per hour) would halve the waiting time, if other features of the system are left intact?

1.7 (Section 1.6) How much would the mean service time have to be decreased in Problem 1.6 to cut the mean waiting time in half?

1.8 (Section 1.6) The time between successive arrivals at a car wash is exponentially distributed with an average time between arrivals of 0.1 hour. Cars are washed one at a time on a first-come, first-served basis. The time taken to wash a car is also exponentially distributed with mean 0.08 hour. Find:

   a. The mean time a car spends waiting to be washed.

   b. The mean total time in the system.

   c. The probability that a car has zero waiting time.

1.9 (Section 1.6) An M/M/1/∞ queueing system, with average arrivals of 6 per hour, has an average utilization (proportion of time the facility is busy) of 0.9. When the system is in steady-state,

   a. What is the mean service time?

   b. What is the probability that one or more customers will be waiting?

1.10 (Section 1.6) Patients arrive for treatment at a small clinic in a Poisson fashion with a mean rate of 4 per hour. Patients are treated one at a time on a first-come first-served basis. The time spent with a patient is exponentially distributed with mean time 12 minutes. A waiting room is to be provided with seating capacity such that the probability that a patient will have to stand upon arrival is 0.10. What should the seating capacity be?

1.11 (Section 1.6) An appliance repair shop guarantees that it can repair certain failures in appliances within an hour. The guarantee states that customers who wait more than an hour get the repair free. The average charge per repair is $5 and the average cost of the repair to the shop is $3, yielding an average profit of $2 per repair. Appliances are brought to the shop for repair in a Poisson fashion with mean rate of 8 per hour. Service time per repairman is exponentially distributed with a mean service time of 5 minutes. The shop employs one repairman. Can the shop operate profitably under this guarantee? (Ignore overhead and taxes in your answer.)

1.12 (Section 1.6) First, explain why equation (1.6.13) does not make sense if $n = 0$ and then derive (1.6.19) by

  a. Showing that $P_0(t + \Delta t) = P_0(t)P(E_1) + P_1(t)P(E_2)$ where E_1 is the event of 0 arrivals from t to $t + \Delta t$ and E_2 is the event of no arrivals and 1 departure.

  b. Justify $P(E_1) = 1 - P(1 \text{ arrival}) = 1 - \lambda(\Delta t)e^{-\lambda \Delta t}$.

  c. From a and b, obtain an expression for $dP_0(t)/dt$ and show that in steady-state this gives equation (1.6.19).

1.13 (Section 1.7) This problem compares the bubble sort with the Shell sort. The running time of a sorting algorithm is strongly influenced (nearly determined) by the number C of comparisons it makes and the number E of exchanges it makes. These quantities C and E depend on: the sorting algorithm to be used; the coding of that algorithm; the number N of items to be sorted; and the order of the items to be sorted. **Write a program** (in any general-purpose scientific programming language) that takes as input an array R of N numbers, sorts them with the bubble sort algorithm, and outputs C_N and E_N. **Also write a program** that performs the same functions using the Shell sort algorithm. (Note that if you choose to use FORTRAN, this can easily be done by modifying the program of Figure 1.4–2.) **Run** both of these programs with the following specifications:

  a. List items are in decreasing order, and each of $N = 10, 100, 1000$.

  b. List items are in increasing order, and each of $N = 10, 100, 1000$.

c. List items are in random order (see hint below), and each of $N = 10, 100, 1000$.

For this problem, the items to be submitted include the following: the flowchart for your program(s), the input and call for execution, the program, and the output. All programs and output should be well documented (for details of what this means, see Problem 1.3). Also, table the 18 quantities determined in the program runs, and interpret the table. If interchanges take 10 times as long as comparisons, we might wish to compare on $10E + C$; do this.

Hint: In Section 1.2 (also see Problem 1.1), we gave a method for determining numbers $A_1, A_2, A_3, \ldots$ that are between 0 and 1. A method for generating a **random order** for N numbers is as follows: generate a vector Y of N numbers by letting $Y(1) = A_1, Y(2) = A_2, \ldots, Y(N) = A_N$. Now sort the vector Y into increasing order, and when interchanging $Y(i)$ and $Y(j)$, also interchange $Z(i)$ and $Z(j)$. Then at the end of the sorting, Y is in increasing order, and Z contains the numbers 1 through N in an order such that $Y(Z(1))$ is the smallest number in the original Y vector, $Y(Z(2))$ is the next smallest, etc., and $Y(Z(N))$ is the largest. **The resulting vector Z is a random order of the integers 1 through N.**

1.14 (Section 1.7) This problem investigates efficient data structures for the priority queue studied in Problem 1.5, and comparing the linked-list plus bubble sort, with the linked-list plus the Shell sort. In Problem 1.5, a list must be kept of the queue for the server. As discussed in Section 1.3, a linked-list is an efficient and widely used method for achieving this. However, to insert a new arrival onto the queue, one searched through the current queue to find where to insert the new arrival; this amounts to a bubble sort (adding one new item to a list known to be in order). **Develop** a use of the Shell sort ideas (which move items more than one place at a time) to perform this addition to the queue. **Run** Problem 1.5 with linked-list with bubble-sort insertion, and with linked-list with Shell-sort insertion, keeping track of C, the number of comparisons needed, for each method. **Compare** the methods.

Note: For this problem, submit your flowchart for Problem 1.5, and on it show the modifications needed for the Shell sort ideas to be used. Also submit input, call for execution, and output (well documented, as discussed in Problem 1.3).

Chapter 2

Introduction to GPSS

This chapter discusses the structure of GPSS models and some of the basic features of GPSS. The following sections describe in greater detail the statements that were used in the single-server queueing program presented in Section 1.4, as well as the output associated with that program. Some of the common GPSS statements that we have not yet encountered are also included in this chapter. After mastering this chapter, you will be able to develop, code, execute, and interpret the simulations of simple problems. Considerations of the internal workings of the GPSS processor, which lead to a deeper understanding of GPSS, are postponed to Chapter 5.

The computing environment in which GPSS runs, and the user's interaction with that environment, differ among GPSS implementations that are currently available. The limited (educational) version of GPSS/PC, along with the major GPSS models encountered in the book, are on the floppy disk attached to this book. In particular, the three versions of the example of Section 2.10 are on this disk so that you may experiment with your own variations of these programs. An introduction to using the limited version of GPSS/PC is given in Appendix A.

2.1 GPSS Program Structure

GPSS programs consist of a sequence of control and executable statements. The executable statements, called **blocks**, describe the logic of the flow of transactions through the simulation. The mere presence of executable statements within a GPSS program, however, does not cause any action to occur. Only when a transaction enters a given block will any changes in the model take place. Blocks may contain up to four fields: **label**, **operator**, **operand**, and **comment**. In earlier GPSS versions, fields were constrained to specific column locations within statements (and were implemented on punched cards). More recent free-form implementations of GPSS require only that fields be separated by one or more blanks. Comments cannot be embedded within the text of any field except for the comment field.

The leftmost field is the label field which is used only if it is needed for program branching. The label is a user-defined name or **identifier** that consists of alphabetic and/or numeric characters, with an alphabetic leading character. The maximum number of characters that can be employed for a user-defined identifier is limited by the particular implementation of GPSS; most GPSS processors allow the use of at least 7 characters.

The operator field, located immediately to the right of the label field, specifies the action(s) that should be taken when a transaction enters or attempts to enter a particular block. The operand field, located to the right of the operator field, makes the action of the operator more specific. The number of operands, as well as the type of each operand, will vary with each block. Consider, for example, the block

ADVANCE   10,4

from the illustration in Section 1.4 (Figure 1.4–2). No label or comment is associated with this block; its operator, ADVANCE, stipulates that the transaction that enters this block will be held there a specified amount of time. The specification 10,4, given in the operand field, indicates that the transaction will be held in the ADVANCE block for a length of time that is randomly generated from the uniform distribution on the interval from $10-4$ to $10+4$. This means that the delay time will be one of the integers $6, 7, \ldots, 14$, each with a probability of $1/9$.

In most situations, it is possible to default (not provide any specification for) an operand; the defaulted operand will be assigned a value, called its default value, by the GPSS processor. Operands are separated by commas and are referred to as the A, B, C, etc., operands, depending on their position within the sequence of operands. An operand may be defaulted by simply being left out of the sequence. Care must be taken to make the position of the operand being defaulted explicit. For example,

GENERATE   5,,,17

specifies the values 5 and 17 for the A and D operands, respectively, and defaults the B, C, E, F, and G operands. The default values that will be assigned to operands B, C, E, F, and G will be considered in Section 2.3. In this chapter and several that follow, we will consider in detail many of the GPSS blocks. A list of GPSS V blocks is provided in Appendix B.

In contrast to blocks, **control statements** set up initial conditions and define the environment in which the simulation will execute. For example, control statements specify seeds for the random number generators and define functions and variables. The SIMULATE, START, and END statements of the program shown in Figure 1.4–2 are control statements. The next section describes these statements.

In addition to the comment that can be appended to a block or a control statement (as the rightmost field within the block or control statement), it is possible to include entire lines of documentation in a GPSS program. Such lines must start with an asterisk.

2.2 The SIMULATE, START, and END Statements

It may be desirable, for debugging purposes, to compile a GPSS program and check it for syntax errors without producing executable code. In order to actually produce executable code, and then execute it, a program must contain the SIMULATE control statement. This statement, which consists of the word SIMULATE in the operator field, has no operands. It must precede all other program statements that generate code so that the GPSS processor will be notified that code generation should not be suppressed. Typically, when SIMULATE is used, it is the first statement in a program.

To run a program, in addition to SIMULATE, a START control statement must be placed following all the blocks and control statements. START indicates the point at which the GPSS processor will start executing the code that has been generated. It is possible to redefine and/or reinitialize various model parameters following program execution and to reexecute a program. This, of course, would require additional START statements. Unlike the SIMULATE statement, START can have up to four operands:

Operand A: This operand initializes the **termination counter** prior to program execution. The termination counter is an allocated memory location that is initialized and modified by the GPSS processor. As long as the termination counter is positive and there are transactions in the simulated system, GPSS programs continue to execute. One way to stop a simulation is to initialize the termination counter to some positive integer and, during execution of the program, decrement the counter so that it eventually becomes zero or negative. When this happens, execution stops. The ways in which the termination counter can be decremented will be considered in the next section. The default value of operand A is 0.

Operand B: All GPSS programs automatically produce some output. If the B operand contains NP (literally), then all output will be suppressed.

Operands C and D: These operands deal with special types of output, which will be discussed in Section 5.2.

The END control statement, the last statement in the program, returns control to the host operating system. Thus, a GPSS program and its execution can be controlled by the following arrangement of statements with items 2 and 3 repeated any number of times if repeated executions are desired.

1. A SIMULATE statement.

2. A sequence of control statements for definitions and initializations.

3. A START statement.

4. An END statement.

2.3   The GENERATE and TERMINATE Blocks

The GENERATE block introduces transactions into the simulated system and the TERMINATE block removes the transactions that enter it from the system. The GENERATE block has 7 operands:

Operand A: Operand A is the average time lapse before the arrival of the next transaction, generally referred to as the average interarrival time. The default value of operand A is 0, which implies the introduction of an "unlimited" number of transactions through this GENERATE block when operand A is defaulted.

Operand B: This is the half-spread of the interarrival time. If the A and B operands of a GENERATE block are 12 and 3, respectively, as was the case in the program given in Figure 1.4–2, the half-spread of the time lapse between successive transaction arrivals will be 3, which will make the time lapse between successive arrivals a random integer between $12 - 3$ and $12 + 3$. The default value of operand B is also 0; when operand B is defaulted, as in

GENERATE   12

transactions will enter the system exactly 12 units of time apart.

Operand C: This operand specifies the arrival time of the first transaction. GPSS schedules the first transaction through the GENERATE block at the time given by operand C. If defaulted, operand C is ignored and the first transaction arrives at a time stipulated by operands A and B.

Operand D: Operand D gives the total number of transactions to be created by the GENERATE block. After the stipulated number of transactions have been generated, at the times prescribed by operands A, B, and C, no additional transactions will enter the system through this GENER-ATE block. The effective default value for this operand is infinity in the sense that transactions will continue to be introduced into the system indefinitely, according to the dictates of the A, B, and C operands.

Operand E: Operand E specifies the priority of transactions created through the GENERATE block. It is possible, through the use of this operand, to assign a priority level (an integer between 0 and 127) to transactions being created by a GENERATE block. Priority values can subsequently be used in placing a transaction in an advantageous position within a list such as a queue. The default value of operand E is 0.

Operand F: Each transaction has a number of integer values, or parameters, uniquely associated with it. The number of parameters associated with a transaction is given by the F operand of the GENERATE block that creates it. Suppose that a modeler wishes to find out the time taken by transactions to move through a certain portion of a simulation. This can be done by attaching the entry time into that portion to each transaction as a parameter; the current time, when each transaction leaves this portion of the simulation, could then be compared to the stored parameter value to determine the elapsed time over the particular section of the simulation. Typically, the number of parameters associated with a transaction is limited to 100. The default value of operand F is 12.

Operand G: Some versions of GPSS distinguish between halfword (16-bit) parameters designated by H and fullword (32-bit) parameters designated by F. Due to bit-size limitations, a halfword parameter can assume values in the range $-32,768$ to $32,767$ and fullword parameters are restricted to $-2,147,483,648$ to $2,147,483,647$. In most versions of GPSS, the default parameter type is halfword. Some of the more recent versions (e.g., GPSS/PC) have adopted only one type of parameter.

Examples of the GENERATE block follow. The block

GENERATE   7

will create a transaction every 7 units of time, starting at time 7. The block

GENERATE   7,2

will produce transactions with interarrival times sampled randomly from the uniform distribution on the interval from 5 to 9. The block

GENERATE   20,3,,15

will introduce 15 transactions with interarrival times randomly chosen from the distribution indicated by the A and B operands. The block

GENERATE   20,3,,15,2,25

will create the same pattern of transactions as in the previous example. However, these transactions will have priority levels of 2, and 25 fullword parameters will be associated with each transaction.

The purpose of the TERMINATE block is to remove the transactions that enter it from the simulation. Thus, TERMINATE performs the opposite function as GENERATE. The TERMINATE block has a single A operand.

Operand A: When transactions enter a TERMINATE block, the termination counter is decremented by the amount specified by the A operand of the TERMINATE block. Recall that the termination counter is initialized by the A operand of the START statement. If, as a result of this decrementation, the value of the termination counter becomes 0 or negative, the simulation stops. The default value of this parameter is 0.

The program structure given in Figure 2.3–1 shows that a GPSS model may have several TERMINATE blocks, each with a different A operand. There is, however, only one termination counter, and all TERMINATE blocks cause this single counter to be decremented. The simulation shuts off when the decrementations from all the TERMINATE blocks make the value of the termination counter 0 or negative. As illustrated in Figure 2.3–1, the overall structure of a GPSS model can be viewed as a sequence of segments (k segments in this case) bracketed by GENERATE–TERMINATE blocks.

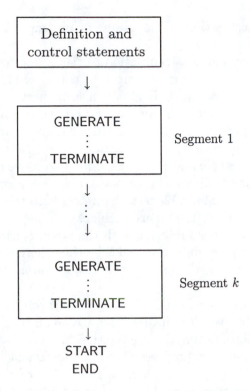

Figure 2.3–1. A k-segment GPSS program.

2.4 Facilities and the SEIZE and RELEASE Blocks

Entities within GPSS models that represent providers of services are called facilities. A transaction "captures" a facility in order to receive service for some length of time. A teller in a bank, a gas station attendant, a telephone operator, a repairman, a technician in a mammography center, or the CPU (central processing unit) of a computing system perform services; hence, they would probably be modeled as facilities in appropriate simulations. Since it is possible to have more than one facility in a simulation, user-defined identifiers are employed to give unique names to facilities.

Facilities can serve at most one transaction at a time. Therefore, it is quite possible that when a transaction wishes to use a facility, that facility may already be in use by another transaction. In such situations, the transaction that is trying to access the facility must wait until it becomes available, branch to another portion of the model, or preempt the current user of the facility.

The SEIZE block allows a transaction to "seize" or "capture" a facility for service. When a transaction attempts to enter a SEIZE block, the facility may or may not be in use. If it is not in use, the transaction is permitted to move sequentially through the blocks, and the facility is immediately allocated to the transaction, with its status changed from idle to busy. If the facility is in use, however, the transaction is kept in its current block until the facility becomes idle and the transaction can move on. While a transaction is waiting for a facility to become available, additional transactions may attempt to enter the SEIZE block to gain access to the facility. Such transactions are also kept in their current blocks. When the facility eventually becomes available, the transaction with the highest priority that has waited the longest is advanced into the SEIZE block. Therefore, if all transactions have the same priority, the facility is allocated on a first-come, first-served basis.

A transaction that has seized a facility spends some time at the facility to receive service (this is done through the ADVANCE block, which is discussed in the next section). After the transaction has been served, it vacates the facility and continues its progress through subsequent blocks. This is done by the RELEASE block, which changes the status of the facility from busy to idle. Entry into a RELEASE block is always allowed, provided the facility to be released was one that had previously been seized by the transaction entering the RELEASE block. If a transaction attempts to RELEASE a facility that is not allocated to itself, a run-time error will occur. The SEIZE and RELEASE blocks each have a single operand.

Operand A: The A operand of a SEIZE or RELEASE block gives the user-defined name of the facility to be seized or released. Since a model may have more than one facility, this operand cannot be defaulted; an error will occur if it is not specified.

2.5 The ADVANCE Block

When a transaction enters this block, it is kept there for a time period that is specified by the A and B operands of the block. Following this explicit delay, the transaction attempts to move to the block sequentially following the ADVANCE block.

Operand A: This is the average time delay before the transaction attempts to move. The default value of operand A is 0.

Operand B: This operand gives the half-spread of the delay time. The default value of this operand is 0. The two operands act in exactly the same way as the A and B operands of the GENERATE block to determine the necessary time lapse before the transaction can move forward. Thus,

ADVANCE   10

will hold transactions for exactly 10 units of time and

ADVANCE   23,4

will hold transactions for a randomly determined period of time chosen from the uniform distribution on the interval between $23 - 4$ and $23 + 4$.

2.6 Queues and the QUEUE and DEPART Blocks

When a facility is busy and additional transactions try to gain access to it, the GPSS processor constructs a queue, or waiting line, of transactions that are waiting to use the facility. In addition to such **implicit** queues, the analyst can **explicitly** use queues for a variety of reasons, including scheduling the use of facilities. If queues are used explicitly through the QUEUE and DEPART blocks, statistics regarding waiting times, number of transactions in the queue, etc., will be kept and printed out at the end of the simulation run. The QUEUE and DEPART blocks have the same two operands.

Operand A: This operand is the queue name; it cannot be defaulted, because without explicit mention of the queue name it will not be clear which queue a transaction is to join or depart from. It is possible that at a given time a transaction may belong to several queues.

Operand B: The B operand specifies the amount by which the content of the queue will be altered (incremented in the case of QUEUE and decremented in the case of DEPART). The default value of the B operand is 1; in most cases, the B operand may be defaulted to this value. It would be unusual to want to increment or decrement the number of entries in a queue by a value other than 1 when a single transaction joins or leaves it.

A common error among GPSS beginners is to associate queue membership with the location of the transaction in the QUEUE block. After entering a QUEUE block, a transaction becomes a member of the queue specified by the A operand of the QUEUE block. If possible, the transaction moves on through the successive blocks within the model without relinquishing its membership in the queue. Only when it encounters a DEPART block with the same A operand as that of the QUEUE block is the membership of the transaction in the queue terminated.

2.7   GPSS Program Output

The execution of any GPSS program will produce some minimal output unless output is suppressed by the NP condition of operand B in the START statement. This minimal output will contain information on the blocks within the program, as well as statistics captured on various model entities such as facilities and queues. If more information than that made available through the automatic minimal output is desired, special data gathering and output features of GPSS can be used. Only the default output associated with blocks, facilities, and queues will be considered in this section.

For illustration, consider the GPSS program of Section 1.4, reproduced here for convenience as Figure 2.7–1. The first portion of GPSS output typically consists of a reformatted listing of the source file with a number identifying each block of the program, as in Figure 2.7–2. This is followed by the values of the relative and absolute simulation clocks at the termination of the simulation run (the distinction between relative and absolute clocks will be discussed in Section 5.4). Next comes a printout of the current and total content of each block of the program (Figure 2.7–3). The current count is the number of transactions in a block when execution stops; the total count is the total number of transactions that have moved into a given block during the simulation run.

The output associated with the WAIT queue (Figure 2.7–4) gives the following information:

Maximum Contents: The maximum number of entries in the queue at any time during the simulation.

```
          SIMULATE

          GENERATE    12,3          Customers arrive
          QUEUE       WAIT          Enter waiting line
          SEIZE       STATION       Access facility
          DEPART      WAIT          Leave waiting line
          ADVANCE     10,4          Obtain service
          RELEASE     STATION       Return facility
          TERMINATE   1             Leave the system

          START       50            Simulate for 50 customers
          END
```

Figure 2.7–1. A GPSS program for a simple queueing model.

```
1
2                   SIMULATE
3
4      1    GENERATE    12,3          Customers arrive
5      2    QUEUE       WAIT          Enter waiting line
6      3    SEIZE       STATION       Access facility
7      4    DEPART      WAIT          Leave waiting line
8      5    ADVANCE     10,4          Obtain service
9      6    RELEASE     STATION       Return facility
10     7    TERMINATE   1             Leave the system
11
12          START       50            Simulate for 50 customers
```

Figure 2.7–2. Listing of the program in Figure 2.7–1.

RELATIVE CLOCK 627   ABSOLUTE CLOCK 627

BLOCK	CURRENT	TOTAL
1	1	51
2	0	50
3	0	50
4	0	50
5	0	50
6	0	50
7	0	50

Figure 2.7–3. Clock and block count of the program in Figure 2.7–1.

QUEUE	MAXIMUM CONTENTS	AVERAGE CONTENTS	TOTAL ENTRIES	ZERO ENTRIES	PERCENT ZEROS
WAIT	1	0.102	50	29	58.000

QUEUE		AVERAGE TIME/TRANS	$AVERAGE TIME/TRANS	TABLE NUMBER	CURRENT CONTENTS
WAIT		1.280	3.048		0

Figure 2.7–4. Queue output of the program in Figure 2.7–1.

Average Contents: The average number of entries in the queue for the duration of the simulation. This is a time-weighted average (described in Section 1.2.2). Thus, if the queue contained 0 transactions for 2 units of time and 1 transaction for 5 units of time during a simulation run of 7 units' time duration, the average would be not 0.5, but

$$\frac{0 \times 2 + 1 \times 5}{7} = 0.714.$$

Total Entries: Total number of transactions that enter the queue during the simulation.

Zero Entries: The number of transactions that enter the queue and immediately leave it because they reach the appropriate DEPART block in zero time. These are the transactions that do not wait in the queue.

Percent Zeros: The percentage of the total entries in the queue that are zero entries.

Average Time/Transaction: The average of the times that transactions spend in the queue. Zero entry transactions are included in this figure.

$Average Time/Transaction: The average of the times that transactions spend in the queue. Zero entry transactions are excluded from this figure.

Table Number: The table number in which the wait times associated with this queue are being kept. (The use of tables will be discussed in Section 7.8.)

Current Contents: The number of transactions in the queue at the end of the simulation.

The output associated with the STATION facility (Figure 2.7–5) gives the following information:

Average Utilization: Proportion of the time that the facility was in use.

Number Entries: Total number of transactions that used the facility.

Average Time/Transaction: The average length of use by transactions that seize the facility.

Seizing Transaction Number: The identification number of the transaction that is using the facility when the simulation stops; the facility may be idle at that time (this is why this column is blank in Figure 2.7–5).

Preempting Transaction Number: Preemption will be considered in Section 7.9.

FACILITY	AVERAGE UTILIZATION	NUMBER ENTRIES	AVERAGE TIME/TRANS	SEIZING TRANS. NO.	PREEMPTING TRANS. NO.
STATION	0.809	50	10.140		
13	END				

Figure 2.7–5. Facility output of the program in Figure 2.7–1.

2.8 Storages and the ENTER and LEAVE Blocks

To simulate the use of parallel servers, GPSS provides a storage entity that represents a predetermined number of identical servers. A transaction can capture a server as long as there is one available; furthermore, a transaction can capture several servers simultaneously if the needed number of servers are available.

To establish the name, and a fixed number of identical servers to be referenced by that name, the STORAGE control statement is used. The label portion of the statement contains a user-defined identifier that will be used as the reference name. The operator portion of the statement is the word STORAGE. The A operand specifies the number of servers that are to be made available. Thus,

TELLER   STORAGE   5

establishes 5 tellers that are indistinguishable from one another to a transaction attempting to capture one or more tellers.

Servers are allocated to, and deallocated from, transactions through the ENTER and LEAVE blocks, respectively. If the request exceeds the number of items available at the time of the request, the transaction remains in its current block and attempts to move into the ENTER block when more units become available. As with the request for entry into a SEIZE block, transactions eventually move through ENTER blocks on a first-come, first-served basis. If so desired, a queue could be explicitly constructed to gather information on waiting times associated with access to an ENTER block.

The LEAVE block, which is the counterpart of the ENTER block, returns to the storage the number of units specified by its B operand. A run-time error will result if a transaction attempts to return more storage items than it had allocated to it. The ENTER and LEAVE blocks each have the same two operands:

Operand A: The A operand gives the name of the storage. Since a given transaction may need items from different storages at the same time, this operand cannot be defaulted.

Operand B: The B operand specifies the number of items that are to be taken from, or returned to, the storage. The default value of the B operand in the ENTER and LEAVE blocks is 1.

As an illustration of the ENTER and LEAVE blocks, suppose that the interarrival times of patients at a medical facility are uniformly distributed between 15 and 29 minutes. When patients arrive, they require the attention of one X-ray technician (for general preparation) for 5 ± 2 minutes (assume uniform distribution). The facility has a single X-ray machine, which requires two technicians for its operation. If 15 ± 5 minutes (again assume uniform distribution) are required to take X-rays of patients, then the program given in Figure 2.8–1 describes the work pattern of the medical facility with interarrival times of 22 ± 7 minutes and three X-ray technicians.

Note that patients are modeled as transactions; the X-ray machine, XR-MACH, as a facility; and the pool of technicians, XRTECH, as a storage. This program does not use explicit queues in connection with the allocation of XRMACH or XRTECH; therefore, the output given in Figure 2.8–2 does not contain any information regarding the waiting times of patients.

The interpretation of the output associated with the storage XRTECH of Figure 2.8–2 follows.

Capacity: The number of items in the storage that are available for this simulation run.

1		SIMULATE		
2	XRTECH	STORAGE	3	Establish 3 technicians
3				
4	1	GENERATE	22,7	Patients arrive,
5	2	ENTER	XRTECH,1	get a technician,
6	3	ADVANCE	5,2	prepare for X-ray,
7	4	ENTER	XRTECH,1	get a second technician,
8	5	SEIZE	XRMACH	get the X-ray machine,
9	6	ADVANCE	15,5	get service,
10	7	RELEASE	XRMACH	
11	8	LEAVE	XRTECH,2	
12	9	TERMINATE	1	and leave.
13				
14		START	40	

Figure 2.8–1. Simulation of an X-ray facility.

RELATIVE CLOCK 882		ABSOLUTE CLOCK 882
BLOCK	CURRENT	TOTAL
1	1	41
2	0	40
3	0	40
4	0	40
5	0	40
6	0	40
7	0	40
8	0	40
9	0	40

FACILITY	AVERAGE UTILIZATION	NUMBER ENTRIES	AVERAGE TIME/TRANS	SEIZING TRANS. NO.	PREEMPTING TRANS. NO.
XRMACH	0.700	40	15.425		

STORAGE	CAPACITY	AVERAGE CONTENTS	TOTAL ENTRIES	AVERAGE TIME/TRANS	AVERAGE UTILIZ.
XRTECH	3	1.643	80	18.112	0.548

			CURRENT CONTENTS	MAXIMUM CONTENTS	
XRTECH			0	3	

15	END

Figure 2.8–2. Output of the X-ray facility simulation in Figure 2.8–1.

Average Contents: The time-weighted average of the number of items that were in use during the simulation.

Total Entries: The accumulated number of times that the items in the storage were used.

Average Time/Transaction: The average length of time that the storage units were used.

Average Utilization: The proportion of time that the storage items were in use.

Current Contents: The number of storage items in use at the end of the simulation run.

Maximum Contents: The maximum number of storage items that were in use at any time during the simulation.

2.9   The TRANSFER Block

The purpose of the TRANSFER block is to interrupt the sequential movement of transactions by directing them to a different block. There are three GPSS transfer modes:

1. Unconditional mode: Movement to a specified location is attempted, regardless of internal model conditions.

2. Statistical mode: A transaction is randomly directed to one of two locations, with a fixed probability as to which location the transaction be directed to.

3. Conditional mode: A transaction is directed to a preferred block; if entry into this block is denied, it is then directed to an alternate block.

In all cases, a transaction is never denied entry into a TRANSFER block, and once in the block the transaction immediately attempts to move to the destination specified by the TRANSFER block. The transaction remains in the TRANSFER block if entry into its destination block is denied. The A operand determines the mode of the transfer block.

Operand A: If this operand is

a. Defaulted, the transfer mode is unconditional and the transaction attempts to move to the block specified by the B operand.

b. A decimal fraction, p, between 0 and 1, the transfer mode is statistical and the transaction attempts to move to the blocks designated by the B and C operands, with probabilities $1-p$ and p, respectively. An internal random number generator is used to determine whether the destination is as specified by the B or C operand. Random number generators will be considered in Chapter 3, and the details of the GPSS random number generators will be discussed in Section 5.3.

c. Literally BOTH; then the transfer mode is conditional and the transaction attempts to move into the block given by the B operand. If it is not successful, it attempts to move to the block specified by the C operand. If both attempts fail, the transaction remains in the TRANSFER block.

Operands B and C: These are user-defined identifiers that, if not defaulted, must appear as labels in some block within the program. The B operand may not be defaulted in the event of an unconditional transfer, and its default in other cases is equivalent to specifying the block sequentially following the TRANSFER block. The C operand is not used when the transfer mode is unconditional, and it cannot be defaulted in the statistical and conditional modes.

The relationship between the operands of a TRANSFER block and the mode that is used by that block is summarized in Table 2.9–1.

Table 2.9–1. Summary of the TRANSFER block.

Operand A	Mode	Operand B	Operand C
null	Unconditional	Block, no default	Not used
p	Statistical	Block, probability $1-p$, default: next block	Block, probability p, no default
BOTH	Conditional	Preferred block, default: next block	Alternate block, no default

The program section given in Figure 2.9–2 illustrates the use of an unconditional transfer mode where transactions immediately move from the TRANSFER to the ADVANCE block, without entering any of the blocks in between.

$$\vdots$$

```
                TRANSFER        ,SHOP
                  ⋮
   SHOP          ADVANCE         252,12
                  ⋮
```

Figure 2.9–2. The unconditional transfer mode.

The program section in Figure 2.9–3 shows the use of the statistical transfer mode where transactions are randomly directed to the SEIZE block 25% of the time and to the ADVANCE block 75% of the time. Note that unless a transaction that has taken the REPAIR option encounters another TRANSFER statement, it will "fall through" to the SHOP portion of the model. The program section given in Figure 2.9–4 illustrates the use of the TRANSFER block in the conditional mode. In this example, the transaction entering the TRANSFER block prefers CLRK1. However, if CLRK1 is not available, the transaction will settle for CLRK2. Note that in this situation the unconditional transfer

```
                TRANSFER    ,DONE
```

is used to ensure that exactly one of CLRK1 or CLRK2 is captured.

```
                  ⋮
                TRANSFER        .25,SHOP,REPAIR
                  ⋮
   REPAIR        SEIZE           TOOLBOX
                  ⋮
   SHOP          ADVANCE         252,12
                  ⋮
```

Figure 2.9–3. The statistical transfer mode.

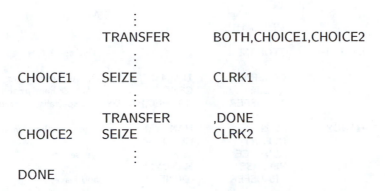

Figure 2.9-4. The conditional transfer mode.

2.10   Example: An Appliance Repair Shop

In addition to support personnel, a small appliance repair shop is staffed by a handyman who can perform a variety of tasks and by two highly trained specialists who take care of difficult repairs. Seventy percent of the customers who visit this shop can be served by the handyman, and one of the two specialists is needed in the remaining 30% of more complex repairs. Past shop records show that

1. The time (in minutes) between successive customer arrivals is uniformly distributed on the interval from 6 to 16.

2. The service times of those customers who do not require a specialist is uniformly distributed between 8 and 18 minutes.

3. Those who need a specialist spend 4 ± 2 minutes (again, uniformly distributed) with the handyman and use the specialist for 62 ± 20 minutes (this time is also uniformly distributed).

4. Approximately 45 customers are served in a given day.

To determine how busy the repairmen are and how long customers wait for the handyman and the specialist, the GPSS model shown in Figure 2.10-1 is used to simulate the arrival of customers, their waiting times, and the service patterns in this shop.

```
1                       SIMULATE
2
3        SPCLST         STORAGE       2              Two specialists.
4
5     1                 GENERATE      11,5,,45       Produce customers.
6     2                 QUEUE         GENQ           Wait for handyman.
7     3                 TRANSFER      .70,SPEC,HANDY Is a specialist needed?
8
9     4   HANDY         SEIZE         HANDYM         If not, get handyman,
10    5                 DEPART        GENQ
11    6                 ADVANCE       13,5           get service,
12    7                 RELEASE       HANDYM
13    8                 TRANSFER      ,DONE          and leave.
14
15    9   SPEC          SEIZE         HANDYM         If yes,
16    10                DEPART        GENQ
17    11                ADVANCE       4,2            get service from handyman,
18    12                RELEASE       HANDYM
19    13                QUEUE         SPECQ          wait for specialist,
20    14                ENTER         SPCLST,1       get a specialist,
21    15                DEPART        SPECQ
22    16                ADVANCE       62,20          obtain service,
23    17                LEAVE         SPCLST,1       and leave.
24
25    18   DONE         TERMINATE
26
27                      START         1
```

No new event in system. All CEC transactions blocked. FEC empty.
Transaction 0. Block number 18. Clock time 594. PC=%X00012EE2.
Field A = 0, Field B = 0, Field C = 0.

Figure 2.10–1. A program for simulating the repair shop.

The handyman is modeled as a facility, whereas the specialists are modeled as a storage, with the STORAGE declaration establishing the presence of the two specialists. The model uses two queues, GENQ (for those waiting to see the handyman) and SPECQ (for those waiting for service from one of the specialists), to gather the waiting-time information associated with the handyman and the specialists, respectively.

The START statement of the program given in Figure 2.10–1 initializes the termination counter to 1; however, the counter is never decremented, since the TERMINATE block defaults its A operand to 0. The simulation stops executing because all transactions (customers in this case) produced by the GENERATE block, with its operand $D = 45$, have moved through the system and there is nothing more to be done. This condition is flagged as a run-time error at the end of the simulation, and an appropriate message is printed, following the program listing, as shown in Figure 2.10–1.

A review of the output of this program, given in Figure 2.10–2, shows that

1. It took 594 minutes (9.9 hours) to serve all 45 customers.

2. Of the 45 customers, 33, or 73%, did not require special repairs.

3. The average simulated waiting times for the handyman and the specialists are 13.756 and 13.167 minutes, respectively.

4. The handyman was busy 83.7% of the time and the two specialists 65.7% of the time.

RELATIVE CLOCK 594			ABSOLUTE CLOCK 594		
BLOCK	CURRENT	TOTAL	BLOCK	CURRENT	TOTAL
1	0	45	11	0	12
2	0	45	12	0	12
3	0	45	13	0	12
4	0	33	14	0	12
5	0	33	15	0	12
6	0	33	16	0	12
7	0	33	17	0	12
8	0	33	18	0	45
9	0	12			
10	0	12			

QUEUE	MAXIMUM CONTENTS	AVERAGE CONTENTS	TOTAL ENTRIES	ZERO ENTRIES	PERCENT ZEROS
GENQ	3	1.042	45	7	15.556
SPECQ	2	0.266	12	6	50.000
		AVERAGE TIME/TRANS	$AVERAGE TIME/TRANS	TABLE NUMBER	CURRENT CONTENTS
GENQ		13.756	16.289		0
SPECQ		13.167	26.333		0

FACILITY	AVERAGE UTILIZATION	NUMBER ENTRIES	AVERAGE TIME/TRANS	SEIZING TRANS. NO.	PREEMPTING TRANS. NO.
HANDYM	0.837	45	11.044		

STORAGE	CAPACITY	AVERAGE CONTENTS	TOTAL ENTRIES	AVERAGE TIME/TRANS	AVERAGE UTILIZ.
SPCLST	2	1.313	12	65.000	0.657
			CURRENT CONTENTS	MAXIMUM CONTENTS	
SPCLST			0	2	
31	END				

Figure 2.10–2. The output of the repair shop simulation.

We need to be aware that there is an initial lag time (time at the start of the simulation when all queues are empty and facilities and storages are idle). This is a reasonably faithful representation of a repair shop at the start of a workday. There is, however, a serious modeling problem, which may distort the average wait times and the utilization times observed above. This problem concerns the latter part of the workday following the arrival of the 45th customer. While no new arrivals were allowed, the system was winding down by processing all the transactions that were in it. This phenomenon stretches the time span over which averages are computed and could distort such averages. One way to deal with this problem would be to default the D operand of the GENERATE block, initialize the termination counter to 45 with the START statement, and decrement the counter by 1 with the TERMINATE block.

As a variation of the original problem, suppose that the data concerning the two types of customers had been kept separately and had produced the following results:

1. The interarrival times of customers who did not need a specialist were uniformly distributed between 11 and 21 minutes.

2. The interarrival times of customers who did need a specialist were uniformly distributed between 28 and 42 minutes.

As indicated in Figure 2.10–3, the movement of each type of customer can now be modeled in a separate segment with its own GENERATE and TERMINATE blocks. Remember that when a simulation is run, all of its segments execute in parallel. Since the arrival patterns were altered, we should not expect the results given in Figure 2.10–4 to be the same as, or even close to, the corresponding results from the previous simulation. In fact, significant differences in waiting time and utilization figures can be observed between the two runs.

In the last variation on this problem, given in Figure 2.10–5, we shut off the simulation after 9 hours of operation rather than after the 45th customer. This can be done by introducing a third segment whose purpose is to generate a transaction at time 540 minutes and decrement the termination counter to zero. For this method to function properly, the customer transactions should no longer decrement the termination counter. Since the arrival and service patterns of this and the preceding simulation are identical and only the termination time has changed, the output of the program of Figure 2.10–5 closely parallels the output given in Figure 2.10–4.

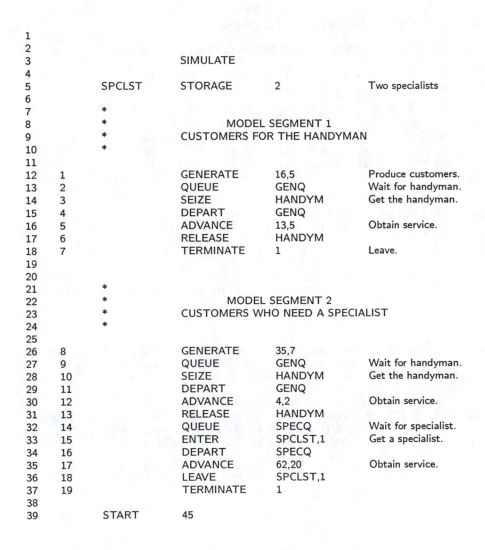

```
1
2
3                              SIMULATE
4
5         SPCLST      STORAGE         2            Two specialists
6
7            *
8            *                 MODEL SEGMENT 1
9            *       CUSTOMERS FOR THE HANDYMAN
10           *
11
12    1                GENERATE        16,5         Produce customers.
13    2                QUEUE           GENQ         Wait for handyman.
14    3                SEIZE           HANDYM       Get the handyman.
15    4                DEPART          GENQ
16    5                ADVANCE         13,5         Obtain service.
17    6                RELEASE         HANDYM
18    7                TERMINATE       1            Leave.
19
20
21           *
22           *                 MODEL SEGMENT 2
23           *       CUSTOMERS WHO NEED A SPECIALIST
24           *
25
26    8                GENERATE        35,7
27    9                QUEUE           GENQ         Wait for handyman.
28    10               SEIZE           HANDYM       Get the handyman.
29    11               DEPART          GENQ
30    12               ADVANCE         4,2          Obtain service.
31    13               RELEASE         HANDYM
32    14               QUEUE           SPECQ        Wait for specialist.
33    15               ENTER           SPCLST,1     Get a specialist.
34    16               DEPART          SPECQ
35    17               ADVANCE         62,20        Obtain service.
36    18               LEAVE           SPCLST,1
37    19               TERMINATE       1
38
39         START       45
```

Figure 2.10–3. A two-segment model of the repair shop.

RELATIVE CLOCK 546   ABSOLUTE CLOCK 546

BLOCK	CURRENT	TOTAL	BLOCK	CURRENT	TOTAL
1	1	34	11	0	15
2	0	33	12	0	15
3	0	33	13	0	15
4	0	33	14	1	15
5	0	33	15	0	14
6	0	33	16	0	14
7	0	33	17	2	14
8	1	16	18	0	12
9	0	15	19	0	12
10	0	15			

QUEUE	MAXIMUM CONTENTS	AVERAGE CONTENTS	TOTAL ENTRIES	ZERO ENTRIES	PERCENT ZEROS
GENQ	2	0.498	48	12	25.000
SPECQ	1	0.044	15	10	66.667

QUEUE		AVERAGE TIME/TRANS	$AVERAGE TIME/TRANS	TABLE NUMBER	CURRENT CONTENTS
GENQ		5.667	7.556		0
SPECQ		1.600	4.800		1

FACILITY	AVERAGE UTILIZATION	NUMBER ENTRIES	AVERAGE TIME/TRANS	SEIZING TRANS. NO.	PREEMPTING TRANS. NO.
HANDYM	0.914	48	10.396		

STORAGE	CAPACITY	AVERAGE CONTENTS	TOTAL ENTRIES	AVERAGE TIME/TRANS	AVERAGE UTILIZ.
SPCLST	2	1.502	14	58.571	0.751

STORAGE			CURRENT CONTENTS	MAXIMUM CONTENTS	
SPCLST			2	2	

40   END

Figure 2.10–4. The output of the two-segment repair shop model.

```
 1
 2
 3                         SIMULATE
 4
 5        SPCLST          STORAGE         2                Two specialists
 6
 7            *
 8            *                    MODEL SEGMENT 1
 9            *           CUSTOMERS FOR THE HANDYMAN
10            *
11
12    1                   GENERATE        16,5             Produce customers.
13    2                   QUEUE           GENQ             Wait for handyman.
14    3                   SEIZE           HANDYM           Get the handyman.
15    4                   DEPART          GENQ
16    5                   ADVANCE         13,5             Obtain service.
17    6                   RELEASE         HANDYM
18    7                   TERMINATE
19
20
21            *
22            *                    MODEL SEGMENT 2
23            *           CUSTOMERS WHO NEED A SPECIALIST
24            *
25
26    8                   GENERATE        35,7
27    9                   QUEUE           GENQ             Wait for handyman.
28   10                   SEIZE           HANDYM           Get the handyman.
29   11                   DEPART          GENQ
30   12                   ADVANCE         4,2              Obtain service.
31   13                   RELEASE         HANDYM
32   14                   QUEUE           SPECQ            Wait for specialist.
33   15                   ENTER           SPCLST,1         Get a specialist.
34   16                   DEPART          SPECQ
35   17                   ADVANCE         62,20            Obtain service.
36   18                   LEAVE           SPCLST,1
37   19                   TERMINATE
38
39            *
40            *                    MODEL SEGMENT 3
41            *           TIMER AT THE END OF 9 HOURS
42            *
43
44   20                   GENERATE        540
45   21                   TERMINATE       1
46
47                        START           1
```

Figure 2.10–5. A three-segment model of the repair shop.

Problems for Chapter 2

2.1 (Section 2.3) Give a GENERATE block that will cause arrivals to occur

  a. Every 5 units of time.

  b. Every 5 units of time and limit the number of arrivals to 20.

  c. With interarrival times of 6, 7, 8, 9, 10, each with a probability of 1/5 and a priority of 2.

  d. As in c and in addition each arrival should have 17 fullword parameters.

2.2 (Section 2.3) Give a GENERATE, TERMINATE, START combination that will cause a GPSS simulation to stop when

  1. The 100th transaction is terminated.

  2. Exactly 100 transactions have been introduced and completely processed in the simulation.

How do a and b differ? What are some differences you are likely to observe in the output associated with the two methods of program termination?

2.3 (Section 2.3) At what simulated time will the program given below stop executing?

```
SIMULATE
GENERATE     5
TERMINATE    2
GENERATE     15
TERMINATE    1
START        9
```

2.4 (Section 2.6) Consider the following two-segment GPSS program:

```
SIMULATE                    GENERATE     25
GENERATE     3              TERMINATE    1
QUEUE        Q              START        1
SEIZE        F
DEPART       Q
ADVANCE      6
RELEASE      F
TERMINATE
```

a. How many transactions will have been through the system when the simulation ends?

b. How many transactions will be in the queue at the end of the simulation?

2.5 (Section 2.6) At a one-barber barbershop, the interarrival times (in minutes) between customers are uniformly distributed on the range of integer values 20 ± 5, and the service times are similarly distributed on 23 ± 10. Write a GPSS program that simulates the barbershop's operation over a 9-hour day.

2.6 (Section 2.7) What will the TOTAL BLOCK count of the QUEUE block be at the end of the simulation in Problem 2.4?

2.7 (Section 2.7) In the standard GPSS output associated with queues, the value of AVERAGE TIME/TRANS will never be greater than the value of $AVERAGE TIME/TRANS. Why?

2.8 (Section 2.7) Would the program given in Figure 2.7–1 remain syntactically correct if the QUEUE and DEPART blocks were removed? If so, in what manner would the output of the program change?

2.9 (Section 2.7) Do Problem 1.2 in GPSS. However, instead of using the routines ATIMEGEN and STIMEGEN, let GPSS determine appropriate values through GENERATE and ADVANCE blocks. As in Problem 1.2, run the simulation for 200 units of time, and then run it again for 2000 time units. Compare your results with the corresponding results from Problem 1.2.

2.10 (Section 2.8) What is the difference between a facility and a storage? Can all facilities in a GPSS program be changed to storages without affecting the logic of the program? Can facilities be substituted for storages in a similar manner?

2.11 (Section 2.8) What is the relationship between CAPACITY, AVERAGE CONTENTS, and AVERAGE UTILIZ in the output associated with storages?

2.12 (Section 2.8) What would happen if the GENERATE block of the program in Figure 2.8–1 were changed to GENERATE 1,1? If you cannot answer by looking at the program, run the simulation with this modification and see what happens, then explain the results.

2.13 (Section 2.8) The phenomenon that you should heve observed in Problem 2.12 is called deadlock. Deadlock occurs when transactions hold all available resources (XRTECH, in the case of the program in Figure 2.8–1) and need additional resources before they are able to return what they hold. This produces a situation where transactions end up waiting for each other indefinitely. Give several examples where deadlock can occur.

2.14 (Section 2.8) Referring to the issue of deadlock in Problem 2.12, suggest a change in the logic of the program in Figure 2.8–1 that would prevent the possibility of deadlock. What are the consequences of your solution?

2.15 (Section 2.9) Suppose that the 85% of the customers who arrive at the barbershop described in Problem 2.5 need only a haircut and the remaining 15% require both a haircut and a shave. Assume that haircuts take 20 ± 7 minutes (uniformly distributed) and shaves take 20 ± 5 minutes (again, uniformly distributed). Write a GPSS program that simulates the barbershop's operation under these circumstances for a 9-hour day.

2.16 (Section 2.9) Suppose that the interarrival patterns of the two types of customers described in Problem 2.15 are known to be 18 ± 5 minutes for customers who need only a haircut and 102 ± 30 minutes for those who need both a haircut and a shave. Write a GPSS program that simulates this situation with the service times given in Problem 2.15 for a 9-hour day.

2.17 (Section 2.9) A tax-consulting office is staffed by a single consultant and support personnel. Eighty percent of the clients who visit the office can be served by the attending consultant, and a specialist needs to be called in for the remaining 20% of the clients. Assume:

  a. The time between successive client arrivals is in the range 22 ± 12 minutes.

  b. Clients who do not need a specialist require 22 ± 15 minutes of the attending consultant's time; those who require a specialist take 5 ± 3 minutes of the attending consultant's time, wait 30 ± 20 minutes for the specialist to arrive, then spend 45 ± 20 minutes with the specialist.

Simulate the operation of this office for an 8-hour day to determine:

  a. How busy is the consultant (i.e., what portion of the time is the consultant with a client)?

  b. How long do clients have to wait to see the consultant?

2.18 (Section 2.9) The publishers of *Sleaze* magazine decide to increase their subscriptions through prime-time TV advertising that encourages viewers to call a toll-free number right away to obtain a year's subscription at a substantial discount. From previous experience, the advertiser expects the interarrival time between calls to be a random number in the range 20 ± 10 seconds and that the calls will last 2 ± 1 minutes.

Of those who call and get a busy signal, 80% give up after their initial attempt and the remaining 20% redial in 10 minutes, at the next commercial break. As was the case with the initial call, 20% continue to redial at 10-minute intervals.

Through a 2-hour simulation, determine the number of telephone lines that the switchboard at the toll-free number should have in order to guarantee that at least 90% of all callers get through.

Chapter 3

Random Number Generation and Testing

We have already seen that, to run in the computer, a simulation needs a source of random variables. For example, random variables were used to provide interarrival times in the algorithms of Section 1.2 and in the GPSS programs of Chapter 2. A sequence of random variables $U_1, U_2, U_3, \ldots$ that are statistically independent and each of which has a uniform probability density function on the interval of real numbers $(0, 1)$, is called a sequence of **random numbers**. As a consequence of various disastrous results and their widespread dissemination (e.g., see the article by Browne (1993) in the *New York Times*) the need for the best random number generators, in an age when supercomputers may use 10,000,000 random numbers per second, has become more widely recognized. Indeed, "... the production of random numbers is a multibillion dollar industry around the world involving considerable research and sophisticated high speed computers" (Rao (1989), p. 4). In this chapter, we deal with the question of how one can provide random numbers for a simulation.

The respective topics dealt with in this chapter are: history of random number generation (Section 3.1); measures of quality of random numbers (Section 3.2); statistical quality measures for random numbers (Section 3.3); theoretical tests (Section 3.4); specific generators, including microcomputer random number generators and generators with very large periods, and their test results (Section 3.5); notes on the use of the TESTRAND package of random number generator tests (Section 3.6).

Parts of this chapter require a modest background in integral calculus. This chapter can be used as a reference on random number generators. Section 3.5.2 contains specific recommendations on the choice of a random number generator (see the paragraph just before "Some remarks on shuffling") and Section 3.6 gives explicit procedures for testing generators. The results in Sections 3.5 and 3.6 are written to be readily accessible without previously digesting the earlier sections of this chapter.

While it is, in general, not the case that the random variables needed for a simulation are independent and uniform on $(0, 1)$—for example, this is typically not true of interarrival times to a queueing system—such random numbers are the virtually universal base from which one works to obtain random variables with other distributions. For example, suppose that U is a uniform random variable on $(0, 1)$. Then by taking $X = -2\ln(U)$, one obtains a new random variable which has distribution function

$$F_X(x) = P(X \le x) = P(-2\ln(U) \le x) = P(2\ln(U) \ge -x)$$

$$= P(\ln(U) \ge -x/2) = P(e^{\ln(U)} \ge e^{-x/2})$$

$$= P(U \ge e^{-x/2}) = \int_{e^{-x/2}}^{\infty} f_U(u)\,du$$

where $f_U(u)$ is the probability density function of a random variable that is uniform on $(0,1)$. Since $f_U(u) = 1$ on $[0, 1]$ and 0 elsewhere, we find that

$$F_X(x) = \begin{cases} \displaystyle\int_{e^{-x/2}}^{1} 1\,du + \int_{1}^{\infty} 0\,du = 1 - e^{-x/2}, & \text{if } x \ge 0 \\[2ex] \displaystyle\int_{e^{-x/2}}^{\infty} 0\,du, & \text{if } x < 0. \end{cases}$$

Therefore, the probability density function of X is

$$f_X(x) = \frac{d}{dx}F_X(x) = 0.5e^{-x/2}$$

if $x \ge 0$ and zero otherwise. However, this is the well-known exponential distribution with mean

$$E(X) = \int_{-\infty}^{\infty} x f_X(x)\,dx = \int_{0}^{\infty} 0.5x e^{-x/2}\,dx\,.$$

The transformation $y = x/2$ yields

$$E(X) = 2\int_{0}^{\infty} y e^{-y}\,dy$$

which, through integration by parts, evaluates to

$$E(X) = 2(-ye^{-y}\big|_{0}^{\infty} - \int_{0}^{\infty}(-e^{-y})\,dy)$$

$$= 2(0 + \int_{0}^{\infty} e^{-y}\,dy) = 2(0 + 1) = 2.$$

Thus, from a source of random numbers $U_1, U_2, U_3, \ldots$ we can obtain a sequence of independent exponential random variables $X_1, X_2, X_3, \ldots$ by taking $X_i = -2 \ln(U_i)$. If we use these as the interarrival times in a queue with deterministic service time, we will be studying the M/D/1 queue.

In this chapter, we deal with the generation and testing of random numbers; in the next chapter we deal with transformations from the uniform distribution to other distributions.

3.1 History and Variety of Random Number Generation Methods

One of the earlier uses of simulation and Monte Carlo techniques was that of W. S. Gosset in 1908. Writing under the pseudonym "Student" (as in "Student's t-distribution"), Gosset used simulation to study his t-distribution, for which the theoretical analysis was then incomplete. At that time, the random numbers needed to run studies were obtained by **observational methods**. For example, one might take telephone numbers (ignoring the exchange) to form a collection of random integers on the range $0, 1, 2, 3, 4, 5, 6,$ $7, 8, 9$. Thus, if the first telephone number is 486-2127, one discards the "486" and the first four digits are 2127. Proceeding through successive telephone numbers, one obtains a collection of digits such as

$$2127210128891172\ldots. \qquad (3.1.1)$$

If one wishes to form random numbers on the interval $(0, 1)$ to eight decimal places, one then takes successive groups of 8 of the digits and places a decimal point to their left. For example, from the sequence in (3.1.1) we would obtain the random numbers

$$U_1 = .21272101, \quad U_2 = .28891172. \qquad (3.1.2)$$

Since there are only 16 digits in (3.1.1), we can obtain only two 8-digit numbers. The first table of form (3.1.1) was given by L. H. C. Tippett in 1927, and contained $41,600$ digits (enough to make 5200 8-digit numbers). Since 5200 random numbers is not enough for many studies, later, in 1939, a larger table of $100,000$ digits capable of providing $12,500$ random numbers was developed by Kendall and Babington-Smith. The tabling efforts culminated in 1955 with the publication by the Rand Corporation of a table of $1,000,000$

random digits. For detailed references, see Dudewicz and Karian (1985), pp. 4–5.

Now with $1,000,000$ digits, one can make only $1,000,000/8 = 125,000$ random 8-digit decimal numbers and, as we will see at the start of Section 3.2, this is insufficient for even a simple simulation. Even $1,000,000$ digits take many pages to display: at 30 lines per page and 50 digits per line, they take over 666 pages. Thus, whatever source one might choose for obtaining observational random digits (telephone numbers, license plate numbers, an electronic roulette wheel as did the Rand Corporation, or some other method), one is faced with certain problems:

1. The generation is very time consuming. If one could produce numbers at a rate of 1 per second, producing $1,000,000$ digits would take over 277 hours.

2. If one decides to produce numbers once and for all in a table, a large volume will be required to store these numbers.

3. If somehow numbers are stored (perhaps on magnetic tape, disk, or some other medium for computer use), one is faced with a relatively slow reading process that ties up the CPU while numbers are being read into memory.

One could conceive of these problems being solved with today's large, fast computers. However, that has not happened yet (though for several years some scientists have claimed it is possible), and it is unlikely to occur. Just as one no longer stores tables of constants, such as logarithms, in the computer for reading as needed (rather, one computes them as needed), so similarly we do not expect to see a resurrection of the tabling efforts that ended over three decades ago.

The literature of tables contains hints on how to make the numbers "go further" by reuse, or what we might today call "**shuffling**." For example, after using the rows of the table, one may next use the columns. This points up **another problem of tabling: by its nature, it focuses concern on quantity of numbers, on memory requirement, and on speed.** It consequently **deemphasizes the issue of quality: do the numbers obtained pass tests of randomness?**

A second proposal for a source of random numbers was an **internal physical source** in the computer, such as noise in an electronic circuit. A serious problem with this proposal, and the reason for its demise, is that it **makes debugging very difficult** since a different input sequence is obtained every time the simulation program is run. Even if one has a debugged program, often one wishes, in different simulation runs, to study different alternatives. In such a setting there is sampling efficiency in using the same input sequence in the different alternatives, so again one faces the problem of a need for reproducibility. If one tries to overcome this problem by keeping a list of the numbers used, one is back to tabling numbers, the faults of which have already been discussed.

The failure of the two methods above led to a search for seeking an **internal numeric source** of random numbers. Some proposals, and the reasons they were discarded, are

1. **The use of successive digits of certain transcendental numbers such as** π **or** e. Here the generation process is long. If one decides to generate and table the numbers, one is back to the problems associated with observational methods.

2. **The Mid-Square Method.** Here one squares a $2n$-digit number and takes the middle $2n$ digits of the $4n$ digit product as the next number. How long such a sequence can proceed before repeating depends on the starting number in complex ways (and can be quite short); hence, this method did not develop far. For example, if one has $2n = 2$ and starts with $x_0 = 10$, then the square is 0100, whose middle $2n$ digits are 10, so the process degenerates immediately.

3. **Additive Congruences**. Here one specifies a starting set of numbers $x_0, x_1, \ldots, x_{k-1}$ and then for $n = k, k+1, \ldots$ uses

$$x_n = (x_{n-1} + x_{n-k}) \bmod m$$

for some integer m. While this method can produce many numbers, it requires storing numbers in memory and accessing them. This method was initially rejected for its use of memory and lack of speed when these were critical factors in making simulation feasible. Furthermore, if k is small, the numbers tend to be nonrandom. This is particularly true with what we will later describe as the "runs test," since the numbers tend to increase for a while and then decrease. The case $k = 2$ is called the **Fibonacci series method**.

4. **Linear Recurrences.** Here one again specifies $x_0, x_1, \ldots, x_{k-1}$ and then for $n = k, k+1, \ldots$ uses

$$x_n = a_1 x_{n-1} + a_2 x_{n-2} + \ldots + a_k x_{n-k} + c \bmod m$$

for some $a_1, a_2, \ldots, a_k, c, m$. This method includes the **feedback shift generators** that were rejected for lack of good randomness properties of their numbers (see Dudewicz and Ralley (1981)). The general class has not been studied in detail, although it includes as a special case the most-used generators: those with $k = 1$.

The most frequently used random number generators are the mixed-congruential generators, which set x_0, a, c, m as integers and then generate via

$$x_{i+1} = ax_i + c \mod m \qquad (3.1.3)$$

for $i = 1, 2, \ldots$. Here a and m are positive integers and c is a nonnegative integer. Historically, this generator arose from the method that takes

$$x_n = x^n \mod m \qquad (3.1.4)$$

for some integers x and m. For large n this would cause numerical problems; however, it is easy to show that (3.1.4) is equivalent to

$$x_n = x \cdot x_{n-1} \mod m, \qquad (3.1.5)$$

which is easy to generate. The proof uses ideas similar to those used at the end of Section 1.2. Since $a * b \mod m = (a * (b \mod m)) \mod m$,

$$\begin{aligned}
x_n &= x^n \mod m \\
&= x * x^{n-1} \mod m \\
&= (x * (x^{n-1} \mod m)) \mod m \\
&= x * x_{n-1} \mod m.
\end{aligned}$$

The generators with $c = 0$ are called **multiplicative congruential** (or **power residue**) **generators**, and the numbers used in the simulation are $U_i = x_i/m$. These were once called pseudo-random numbers, but in more recent years the prefix "pseudo" has been dropped. Although they are called "random numbers," that does not mean they are; for carefully chosen x_0, a, c, m they do pass tests of randomness but as we will see in Section 3.3, for many other choices they fail miserably. Most computers today use binary storage, so it is easiest to implement (3.1.5) with $m = 2^e$ for some integer e; often e is taken to be the word size of the computer. As we will see in Theorem 3.2.7, if we choose a multiplier $a = \pm 3 \mod 8$, i.e., if $a = 8t \pm 3$ for some $t = 1, 2, 3, \ldots$, and x_0 odd, then $x_n = x * x_{n-1} \mod 2^e$ will (if e is at least 3) produce 2^{e-2} terms before repeating.

As an example, consider the generator $x_n = 3x_{n-1} \mod 32$. This has $m = 2^5$ (i.e., $e = 5$) and 3 is equal to $\pm 3 \pmod 8$, hence the period is $2^{5-2} = 8$. If we take $x_0 = 1$, we find that the numbers generated, and their binary representations, are:

$$
\begin{array}{ll}
x_0 = 1 & 0\ 0\ 0\ 0\ 1 \\
x_1 = 3 & 0\ 0\ 0\ 1\ 1 \\
x_2 = 9 & 0\ 1\ 0\ 0\ 1 \\
x_3 = 27 & 1\ 1\ 0\ 1\ 1 \\
x_4 = 81 \bmod 32 = 17 & 1\ 0\ 0\ 0\ 1 \\
x_5 = 51 \bmod 32 = 19 & 1\ 0\ 0\ 1\ 1 \\
x_6 = 57 \bmod 32 = 25 & 1\ 1\ 0\ 0\ 1 \\
x_7 = 75 \bmod 32 = 11 & 0\ 1\ 0\ 1\ 1.
\end{array}
$$

Continuing, we obtain $x_8 = 33 \bmod 32 = 1$, and the sequence repeats with a period of 8, as claimed. Recall that a 4-digit number in binary form is represented as a sum of some of the terms 2^4, 2^3, 2^2, 2^1, 2^0; for example,

$$
\begin{aligned}
11 &= 0 * 16 + 1 * 8 + 0 * 4 + 1 * 2 + 1 * 1 \\
&= 0 * 2^4 + 1 * 2^3 + 0 * 2^2 + 1 * 2^1 + 1 * 2^0 \\
&= 0 \quad\quad 1 \quad\quad 0 \quad\quad 1 \quad\quad 1.
\end{aligned}
$$

The bits in such a generator have a pattern:

1. The rightmost bit is always 1.

2. The second bit from the right alternates between 0 and 1.

3. The third bit from the right is always 0. In general, the second and third bits from the right either do not change, or they alternate.

4. The other bits form patterns that then repeat. The leftmost bit forms a pattern with 2^{e-2} ($= 8$ in our example) terms, which then repeats (in our case, 00011110 in the first 8 terms, and then the next 8 have 00011110 again, and so on). The next bits have patterns formed in 2^{e-2-j} terms (for the j-th from the leftmost bit), which then repeat. In our example, the next bit over from the leftmost has $j = 1$, and we find the pattern 0011 in the first $2^{e-2-1} = 2^{5-2-1} = 2^2 = 4$ terms. That pattern is repeated in the next four terms, and so on.

Due to these patterns, such numbers should not be decomposed to form random bits (if such bits are needed). Rather, if one needs a random variable B_i such that $P(B_i = 0) = P(B_i = 1) = .5$, then this can be attained by setting $B_i = 0$ if $U_i \le 0.5$, and $B_i = 1$ otherwise. It is easy to see that if U_i is random on the interval $(0, 1)$, then B_i as just defined will be 0 or 1 with probability 0.5 each; i.e., it will be a random bit.

If x_0 is not chosen as an odd integer (as 1 above), then the period is cut in half. For example, if above we chose $x_0 = 2$ we would find

$$x_0 = 2, \quad x_1 = 6, \quad x_2 = 18, \quad x_3 = 54 \bmod 32 = 22,$$

after which $x_4 = 66 \bmod 32 = 2$, so a period of 4 terms (half the previous 8) is obtained before repeating. Other non-odd seeds x_0 may yield even shorter sequences (for example, if $x_0 = 24$, then $x_1 = 8$, and then the sequence repeats; if $x_0 = 16$, then $x_1 = 16$, for a period of only 1).

As the above discussion has noted, **quality** of a random number generator has traditionally meant first **speed** and low **memory requirements** (since without these, in the days of slower computers, simulation would not have been possible), second a long **period** (since without this a simulation will produce the same output repeatedly), and finally **good randomness properties**. Much of the literature contains hints, results, and folklore— sometimes valid, sometimes not—oriented to providing speed first, and quality next. For example, one is often told to choose the multiplier a in (3.1.3) so that its binary representation contains few 1s, in order to obtain a faster multiplication; however, in modern computers multiplication is hard wired and not software emulated, so $a * x_i$ takes the same time regardless of what two numbers a and x_i are. Hence, this traditional guideline is not a valid one today. As another example, one is often told to take a close to $\sqrt{m}$; if $m = 2^{32}$, this would mean near $2^{16} = 65536$, such as $a = 65539$. However, that is exactly what a generator called RANDU does, and it has been shown to be unsuitable (see URN07, URN08, and URN09 in Section 3.5).

Our recommendations on random number generators are:

1. Choose a generator with speed and memory needs you can tolerate, and with period at least 1 billion.

2. Make sure the generator can pass the TESTRAND tests (Dudewicz and Ralley (1981)) of Section 3.5.2.

3. Never trust a generator that comes with your computer (most are put in almost as an afterthought, and we almost never see a good one), and do not construct your own. If you don't know where a generator came from, replace it with one that has passed the TESTRAND tests.

4. Several generators have passed the TESTRAND tests, but many have failed. One of the simplest that passed sets x_0 as any odd integer, and

$$x_{i+1} = 663,608,941 * x_i \bmod 2^{32},$$

$$U_{i+1} = x_{i+1}/2^{32}.$$

This generator, called URN13, has period $1,073,741,824$ (as is shown following Theorem 3.2.8). URN13 satisfies points 1, 2, and 3 above.

TESTRAND is a package of random number generators—called URN01, URN02, ..., URN35—and tests of random number generators. The complete TESTRAND code is available in Dudewicz and Ralley (1981). The original version of TESTRAND, designed for IBM 360/370 and compatible computers, was available from The Ohio State University, and a later version, for VAX-11 computers was available from Denison University (see Dudewicz and Karian (1988)). Today, efforts are underway in the U.S.A., Belgium, Argentina, and Finland to produce TESTRAND in various forms, including one that can be used on Windows-based PC's. Since your instructor may have arranged for TESTRAND to be available to you for course work, we will note below which of its routines may be used for certain calculations. (If it has not been made available to you, simply skip over those notes.) TESTRAND will be discussed in greater detail, with sample runs, in Section 3.6. In the next two sections we will see details of how one finds the period of a generator (Section 3.2), then some of the statistical tests that go into testing a random number generator (Section 3.3).

3.2 Traditional Quality Measures of Random Number Generators

As noted briefly above, until fairly recent times "quality" of random number generators was evaluated by such measures as:

1. Speed of generation.

2. Memory requirements.

3. Quantity of numbers available.

While **speed of generation** is still at times an important consideration (for example, in real-time studies using microcomputers), in most cases today's computers are so fast that (barring extremely time-consuming methods) this consideration is not the prime concern. Similarly, **memory requirements of the generation method** will even more rarely still be of any great concern to the simulationist who is using internal numeric methods to generate random numbers. However, the **quantity of numbers available** will be a concern.

In the single-server queue studied in Chapters 1 and 2, there exist many examples, such as package delivery systems, where an arrival rate of one customer every 30 seconds would not be unusual. This means that in a simulation of one day's operation one expects to need $(8)(60)(2) = 960$ jobs. Suppose that each job needs a generated arrival time, service time, and priority number, and that each of these requires only one random number for its generation. We will see in the next chapter that often more random numbers are required to generate one random quantity; however, in this simplified case we will need about $(960)(3) = 2880$ random numbers for a simulation of one day. For simulating a year's operation for a 5-day-a-week system, we then need about $(2880)(52)(5) = 748{,}800$ random numbers. Since one realization of the simulation provides us with no measure of variability of the output, often we will make a number of runs and average the performance measures observed in them. If 100 runs is used, we need a total of about $(748{,}800)(100) = 74{,}880{,}000$ random numbers. This was for a simple system; complex systems can easily require 10 times as many, or about **1 billion random numbers**.

In statistics, studies involving large sets of random numbers include those of Konishi (1991), where $1{,}000{,}000$ bivariate normal random observations (hence $2{,}000{,}000$ random numbers) were used, and the work of Seo and Siotani (1992) where 100 replications of $10{,}000$ samples of $p = 8$-variate normals with $m = 15$ variates per sample were used, requiring

$$100 \times 10{,}000 \times 8 \times 15 = 120{,}000{,}000$$

random numbers.

In the rest of this section we will examine how many random numbers a mixed congruential generator can provide; other measures of quality of numbers will be examined in the next section. A mixed congruential generator starts with an initial integer x_0, and a sequence of integers is defined by

$$x_{i+1} = ax_i + c \pmod{m}, \qquad i = 0, 1, 2, \ldots \tag{3.2.1}$$

where a and m are fixed positive numbers, c is a nonnegative integer, and x_{i+1} is the residue modulo m (and hence is between 0 and $m - 1$). Random numbers are generated by taking $U_i = x_i/m$ (which provides values between 0.0 and 1.0).

Since there are only m distinct residues modulo m, the terms of the sequence (3.2.1) must eventually repeat; i.e., there must exist integers r and s (with $0 \leq r < s < m$) for which $x_r = x_s$; hence, $x_{r+k} = x_{s+k}$ for all positive k. If s is the smallest positive integer for which $x_r = x_s$, then (3.2.1) consists of an initial segment $x_0, x_1, \ldots, x_{r-1}$, followed by a second segment $x_r, x_{r+1}, \ldots, x_s = x_r$ which is then repeated indefinitely. **The integer $d = s - r$ is called the period of the sequence.** The sequence is said to have **maximal period** if $d = m$, i.e., if each of the m distinct nonzero residues modulo m appears. In this section we will examine the period d as a function of a, c, m, x_0, drawing on results obtained in Dudewicz and Ralley (1981, Section 1.6).

First, let us look at some simple examples. If $x_0 = 1$, $a = 9$, $c = 0$, $m = 10$, then we find the sequence $x_1 = 9 \bmod 10 = 9$, $x_2 = 81 \bmod 10 = 1$, $x_3 = 9 \bmod 10 = 9$. Or, 1, 9, 1, 9, ... Here $r = 0$, $s = 2$, and the period is $d = s - r = 2 - 0 = 2$.

As a second example, if $x_0 = 1$, $a = 5$, $c = 0$, $m = 10$, then we find the sequence $x_1 = 5 \bmod 10 = 5$, $x_2 = 25 \bmod 10 = 5, \ldots$. Or, 1, 5, 5, ... Here $r = 1$, $s = 2$, and the period is $d = s - r = 2 - 1 = 1$. This generator has an initial segment "1", followed by the second segment "5" repeated.

As a third example, if $x_0 = 1$, $a = 2$, $c = 0$, $m = 10$, we find $x_1 = 2 \bmod 10 = 2$, $x_2 = 4 \bmod 10 = 4$, $x_3 = 8 \bmod 10 = 8$, $x_4 = 16 \bmod 10 = 6$, $x_5 = 12 \bmod 10 = 2$, $x_6 = 4 \bmod 10, \ldots$. Or, 1, 2, 4, 8, 6, 2. Here $r = 1$, $s = 5$, and the period is $d = s - r = 5 - 1 = 4$. We have an initial segment "1" followed by the segment "2, 4, 8, 6" followed by itself ad infinitum. Table 3.2–1 shows the generators with $c = 0$ and $m = 10$ for all possible x_0 and a. (The initial segment is not underscored, but the segment that repeats is underscored.) Of course, choosing $a = m = 10$ makes the sequence degenerate into 0 at the second step (x_0 is followed by $x_1 = 0 = x_2 = \ldots$). Choosing any a of 11 or more is the same as choosing that number modulo 10 (e.g., with $a = 33$ and $x_0 = 3$ we find 3,9,7,1,). Thus, Table 3.2–1 gives us a complete picture of the generators with $c = 0$ and $m = 10$. The largest period available with this generator (for any a, x_0) is $d = 4$, though poor choices of a and x_0 may reduce this period considerably. **We will therefore next examine results from number theory that can help us choose a generator for which d is large.**

Theorem 3.2.2. *If a and m are relatively prime, then there is a positive integer d for which $x_0 = x_d$.*

Note that "a and m are relatively prime" means that the only positive integer that divides evenly into both is 1. This is sometimes stated as "$\gcd(a, m) = 1$", i.e., the greatest common divisor of a and m is 1.

In our previous example of the generator with $m = 10$, the a's that are relatively prime to m are 1,3,7,9. We see from Table 3.2–1 that for all these cases x_0 was repeated by the generator. x_0 was also repeated by some of the

Table 3.2–1. Numbers resulting from $x_{i+1} = ax_i + c \pmod{m}$ with $c = 0$, $m = 10$.

x_0	1	2	3	4	5	6	7	8	9
1	1,	1,2,4,8,6,	1,3,9,7,	1,4,6,	1,5,	1,6,	1,7,9,3,	1,8,4,2,6,	1,9,
2	2,	2,4,8,6,	2,6,8,4,	2,8,	2,0,	2,	2,4,8,6,	2,6,8,4,	2,8,
3	3,	3,6,2,4,8,	3,9,7,1,	3,2,8,	3,5,	3,8,	3,1,7,9,	3,4,2,6,8,	3,7,
4	4,	4,8,6,2,	4,2,6,8,	4,6,	4,0,	1,4,	4,8,6,2,	4,2,6,8,	4,6,
5	5,	5,0,	5,	5,0,	5,	5,0,	5,	5,0,	5,
6	6,	6,2,4,8,	6,8,4,2,	6,4,	6,0,	6,	6,2,4,8,	6,8,4,2,	6,4,
7	7,	7,4,8,6,2,	7,1,3,9,	7,8,2,	7,5,	7,2,	7,9,3,1,	7,6,8,4,2,	7,3,
8	8,	8,6,2,4,	8,4,2,6,	8,2,	8,0,	1,8,	8,6,2,4,	8,4,2,6,	8,2,
9	9,	9,8,6,2,4,	9,7,1,3,	9,6,4,	9,5,	9,4,	9,3,1,7,	9,2,6,8,4,	9,1,

other cases, such as for $a = 8$ when $x_0 = 2$. But for no other "a" was x_0 repeated for all possible choices of x_0; e.g., for $a = 8$ we find that when we take $x_0 = 3$, x_0 is not repeated.

Theorem 3.2.3. *If $\gcd(a, m) = 1$, then the period of sequence (3.2.1) is the smallest positive integer d for which*

$$\sigma_d((a-1)x_0 + c) = 0 \pmod{m}, \tag{3.2.4}$$

where

$$\sigma_d = 1 + a + a^2 + \ldots + a^{d-1}. \tag{3.2.5}$$

For example, with the choice $a = 3$ and $m = 10$, we need to find the smallest positive d for which $\sigma_d(2x_0 + 0) = 0 \pmod{1}0$ in our previous example. With $x_0 = 1$ this is the smallest d for which $(1 + 3 + 3^2 + \ldots + 3^{d-1})(2) = 0 \pmod{1}0$. For $d = 1, 2, 3, 4$ the successive left-hand-side terms are 4, 8, 26, 80, so we see that $d = 4$. This is as we saw in Table 3.2–1.

If we choose $a = 3$ and $m = 10$ with $x_0 = 5$, we find the smallest d for which $(1 + 3 + 3^2 + \ldots + 3^{d-1})(10) = 0 \pmod{1}0$ is $d = 1$. This also agrees with Table 3.2–1.

Note that the result of Theorem 3.2.2 is not true for all cases where $\gcd(a, m) \neq 1$; for example, it holds when $a = 6$ and $x_0 = 2$, but fails if $x_0 = 1$. Thus, for its guarantee we must look only at generators for which $\gcd(a, m) = 1$.

We will not go into additional results that simplify the computation of the period d in Theorem 3.2.3, since those results involve substantial background in number theory; interested readers are referred to Section 1.6 of Dudewicz and Railey (1981). Those results are the basis of **computer routines that can be used to compute** d, such as routine TSTM2 of the TESTRAND package. The following result, which is of general interest, states when sequence (3.2.1) will have maximal period.

Theorem 3.2.6. *If* $\gcd(a, m) = 1$, *then the period of sequence (3.2.1) is* $d = m$ *if and only if: for each prime number* p *that divides* m *we have*

(i) $a = 1$ (mod p) *for each odd prime* p.

(ii) *If 4 divides* m, *then* $a = 1$ (mod 4).

(iii) $\gcd(c, m) = 1$.

As an example of the use of Theorem 3.2.6, consider the generator with $c = 0$, $m = 10$, and $a = 6$. Then 2 and 5 are the only primes that divide $m = 10$. For (i), the only odd prime is $p = 5$, and $6 = 1$ (mod 5) is true. For (ii), 4 does not divide 10, so there is nothing to check. For (iii), $\gcd(0, 10) = 10$. However the theorem does not apply since $\gcd(a, m) = \gcd(6, 10) = 2 \neq 1$. In fact, from our study of this generator earlier in this section we know its period is 1 (not 10). The two theorems that follow are of general interest.

Theorem 3.2.7. *If sequence (3.2.1) has* $c = 0$, $m = 2^e$ *for* $e \geq 3$, *and* $a = \pm 3$ (mod 8), *then its period is* $d = 2^{e-2}$ *if* x_0 *is an odd integer*.

Theorem 3.2.8. *Sequence (3.2.1) with* $c = 0$ *achieves the maximum possible period if* $\gcd(x_0, m) = 1$ *and, the smallest integer* L *for which* $a^L = 1$ (mod m) *is the largest for the value* a *(no other choice of* a *will yield a larger* L).

After extensive testing of generators (with methods to be covered in successive sections), Dudewicz and Ralley (1981, p. 134) recommended one use either of the generators called URN03 and URN13. URN13 is of type (3.2.1) with $c = 0$, $a = 663,608,941$, and $m = 2^{32}$. In terms of Theorem 3.2.7 we have $c = 0$, $m = 2^e$ with $e = 32$ (which is at least 3), $a - (82,951,118)(8) = -3$ so that $a = -3$ (mod 8), hence its period is $d = 2^{30} = 1,073,741,824$.

Many generators use an m of the form $2^e - 1$ for some prime number e. Numbers of this form, called **Mersenne numbers**, are named after the French monk and mathematician (1588–1648), who had been a classmate of Descartes. The first 27 Mersenne numbers that are prime are those for

$$e = 2, 3, 5, 7, 13, 17, 19, 31, 61, 89, 107, 127, 521, 607, 1279, 2203, 2281,$$
$$3217, 4253, 4423, 9689, 9941, 11213, 19937, 21701, 23209, 44497.$$

The first 8 of these were claimed to be prime by Father Martin Mersenne in 1644. Thus, $2^{31} - 1$, often used as m in random number generators, is a prime number. This fact was established by Euler in 1772. The larger numbers in this list are of much more recent date. For example, the three Mersenne numbers $2^e - 1$ for $e = 521$, 607, 1279 were established as primes at the National Bureau of Standards (now the National Institute of Standards and Technology) in 1952. There are currently 32 known Mersenne primes, the largest of which has $e = 756,839$, and was shown to be prime in 1992 by researchers at AEA Technology's Harwell Laboratory in Britain. This number is, in fact, the largest known prime and has 227,832 digits.

As a final note and a warning, observe that generators with large (and maximal) periods are easy to find. Consider sequence (3.2.1) with $a = 1$, $c = 1$, and any m, and take $x_0 = 0$. Then the successive terms of the sequence are $1, 2, 3, 4, \ldots, m - 1$, after which one again finds $0, 1, 2, 3, \ldots, m - 1$. Thus, its period of m is maximal. Unfortunately, the numbers $U_i = x_i/m$ are hardly random: they march from $0/m$, $1/m$, $2/m$, $\ldots$ through $(m - 1)/m$ and then start over. While we may not run out of numbers (e.g., if we take m very large, such as $m = 2^{500}$), the numbers are so nonrandom as to be useless. Thus the warning: **a large period is necessary, but not sufficient.** We need good numbers, as well as many. This leads us to the next section, where we will examine how to test the statistical quality of random numbers.

We point out that when $c = 0$, the period d for sequence (3.2.1) must be at most $m - 1$. This is so because we already know that the period is at most m, and now the value 0 must be avoided lest the sequence degenerate into $0, 0, \ldots$, which has a period of 1. Thus, "maximum possible period" in Theorem 3.2.8 means the largest one possible for the given m when c is set to zero.

3.3   Statistical Quality Measures of Random Number Generators

In Section 3.1, we noted the **considerations of speed, memory requirements**, and **quantity of numbers available** in evaluating random number generators. The last of these was (for mixed congruential random number generators) considered in more detail in Section 3.2, where we also saw that fast, low-memory, high-quantity generators exist that are totally unsuitable for use (last paragraph of Section 3.2).

In this section, we study statistical tests of whether a sequence of numbers $U_1, U_2, \ldots$ can be regarded as a sequence of independent uniform random variables on (0,1). Since there are many ways in which a sequence can deviate from randomness, there are many tests of randomness, each oriented toward detecting different types of departures from uniformity and independence. In many of these tests, one has k **prespecified categories**, and knows that under the hypothesis of randomness the **probability of an occurrence of category i is π_i $(i = 1, 2, \ldots, k)$.** Suppose we observe enough random numbers to allow us to see where n items fall and compute the frequencies, $o_1, o_2, \ldots, o_k$, in the k categories (where $o_1 + o_2 + \ldots + o_k = n$). Under randomness, we expect $n\pi_i$ occurrences in category i $(i = 1, 2, \ldots, k)$, since o_i has probability π_i of occurring and thus the number of occurrences in n trials is a binomial random variable with expected value $n\pi_i$; **denote this expected number by $e_i = n\pi_i$. A measure of the discrepancy between $o_1, o_2, \ldots, o_k$ and $e_1, e_2, \ldots, e_k$ can be defined as**

$$D = \frac{(o_1 - e_1)^2}{e_1} + \cdots + \frac{(o_k - e_k)^2}{e_k} = \sum_{i=1}^{k} \frac{(o_k - e_i)^2}{e_i}. \tag{3.3.1}$$

Under true randomness, o_i has expected value e_i, so $o_i - e_i$ has expected value zero, and we expect D to be small. Hence **we reject the hypothesis of randomness if D is "large"**. For level of significance 0.05, i.e., to have probability only 0.05 of rejecting randomness when in fact randomness is true, we need to set c so that

$$P_{H_0}(D > c) = .05, \tag{3.3.2}$$

where the subscript H_0 denotes that the probability is computed when the hypothesis of randomness is true. The exact distribution of D, even when H_0 is true, is very complicated; however, **if n is large enough that all e_i are at least 5, then D has approximately the chi-square distribution with $k - 1$ degrees of freedom.** It is preferable, for accuracy, that we have all e_i at least 10.

Values of c for which (3.3.2) holds, for various degrees of freedom $k - 1$, (i.e., where c is the value that a chi-square random variable with $k-1$ degrees of freedom exceeds with probability .05) are given in Table 3.3–1. For more extensive tables, see, e.g., Dudewicz and Mishra (1988), pp. 779–780, or Appendices E and F of this book.

Table 3.3–1. Values of c that solve (3.3.2).

$df = k - 1$	c	$df = k - 1$	c	$df = k - 1$	c
1	3.84146	16	26.2962	40	55.7585
2	5.99146	17	27.5871	50	67.5048
3	7.81473	18	28.8693	60	79.0819
4	9.48773	19	30.1435	70	90.5312
5	11.0705	20	31.4104	80	101.879
6	12.5916	21	32.6706	90	113.145
7	14.0671	22	33.9244	100	124.342
8	15.5073	23	35.1725		
9	16.9190	24	36.4150		
10	18.3070	25	37.6525		
11	19.6751	26	38.8851		
12	21.0261	27	40.1133		
13	22.3620	28	41.3371		
14	23.6848	29	42.5570		
15	24.9958	30	43.7730		

3.3.1 Uniformity of Distribution Test

This test is sometimes referred to as the "chi-square test". We will try to avoid this designation, because it confuses the test with the approximate distribution of the test statistic D. This test **splits the interval (0,1) into k disjoint and exhaustive subsets.** Under randomness, π_i, the probabilities of these subsets, are their lengths, and the test is done with a sample of n random numbers.

A common choice of the subsets is to take equal-length intervals, such as the 100 intervals $[0, .01], (.01, .02], (.02, .03], \ldots, (.98, .99], (.99, 1.00]$. Then

$$e_1 = e_2 = \ldots = e_{100} = n/100,$$

since each subinterval has probability 1/100 of containing a random number U_j under the hypothesis of randomness.

3.3.2 Coupon Collector's Test

Given a sequence of numbers $U_1, U_2, \ldots$, consider converting the numbers to integers $1, 2, \ldots, M$ by

> Replacing U_j by 1 if $0.0 \le U_j < 1/M$
> Replacing U_j by 2 if $1/M \le U_j < 2/M$
>
> $\vdots$
>
> Replacing U_j by M if $(M-1)/M \le U_j \le 1.0$.

Then the new numbers should be random integers from 1 to M if the U_js are truly random numbers. Take the new numbers successively until one has **all** of $1, 2, \ldots, M$ represented. Let Q denote the quantity of new numbers we need to look at to find a complete set $1, 2, \ldots, M$. Then the possible values of Q are $M, M+1, M+2, \ldots$, and under randomness the probabilities $P(Q = M)$, $P(Q = M+1)$, $P(Q = M+2), \ldots$ are known. If we calculate such Q values repeatedly, then test (3.3.1) may be used to test the randomness of the basic numbers $U_1, U_2, \ldots$. Some common choices of the categories for the test are as follows.

Case 1: $M = 5$ with categories $Q = 5$, $Q = 6, \ldots$, $Q = 19$, $Q \ge 20$.

Case 2: $M = 5$ applied to not $U_1, U_2, \ldots$ but to $FR(100U_1)$, $FR(100U_2)$, $\ldots$ where $FR(x)$ is the fractional part of x, e.g., $FR(36.83) = .83$. In this case the categories used are $Q = 5, Q = 6, \ldots, Q = 19, Q \ge 20$.

Case 3: $M = 10$ applied to $U_1, U_2, \ldots$ with categories $10 \le Q \le 19$, $20 \le Q \le 23$, $24 \le Q \le 27$, $28 \le Q \le 32$, $33 \le Q \le 39$, $Q \ge 40$.

To carry out the test, the category probabilities (i.e., the probabilities of each of the categories) under randomness are needed. One can show that **for Case 1 and Case 2 these category probabilities are**

$$
\begin{aligned}
P(Q = 5) &= .03840000 & P(Q = 6) &= .07680000 \\
P(Q = 7) &= .09984000 & P(Q = 8) &= .10752000 \\
P(Q = 9) &= .10450944 & P(Q = 10) &= .09547776 \\
P(Q = 11) &= .08381645 & P(Q = 12) &= .07163904 \\
P(Q = 13) &= .06011299 & P(Q = 14) &= .04979157 \\
P(Q = 15) &= .04086200 & P(Q = 16) &= .03331007 \\
P(Q = 17) &= .02702163 & P(Q = 18) &= .02184196 \\
P(Q = 19) &= .01760857 & P(Q \ge 20) &= .07144851.
\end{aligned}
$$

For Case 3, the category probabilities are

$$P(10 \leq Q \leq 19) \quad = \quad .17321155$$
$$P(20 \leq Q \leq 23) \quad = \quad .17492380$$
$$P(24 \leq Q \leq 27) \quad = \quad .17150818$$
$$P(28 \leq Q \leq 32) \quad = \quad .17134210$$
$$P(33 \leq Q \leq 39) \quad = \quad .15216056$$
$$P(Q \geq 40) \qquad\quad = \quad .15685380.$$

3.3.3   Gap Test

Let α and β be numbers between 0.0 and 1.0 with $\alpha < \beta$. Given a sequence of numbers $U_1, U_2, \ldots$, consider examining each number to see if it is between the numbers α and β (inclusive) or not. Replace each number that is between α and β by a 1, each number that is not by a 0. This converts the numbers into a sequence of 0s and 1s. Let $p = \beta - \alpha$. If the numbers are random, then the probability that j 0s will occur after a 1, before the next 1, is

$$p_j = p(1-p)^j, \quad j = 0, 1, 2, \ldots. \tag{3.3.3}$$

After observing how many times each category occurs, test (3.3.1) may be used to test randomness in accord with the probabilities (3.3.3).

This test, introduced by Kendall and Babington-Smith, was for sequences of the digits 0, 1, 2, 3, 4, 5, 6, 7, 8, 9; this is equivalent to choosing $\alpha = 0.0$, $\beta = 0.1$ here. Some common choices of α and β for this test are:

Case 1: $\alpha = 0.0$, $\beta = 0.5$ (called runs below the mean),

Case 2: $\alpha = 0.5$, $\beta = 1.0$ (called runs above the mean),

Case 3: $\alpha = .333$, $\beta = .667$ (the middle-third test).

3.3.4   Permutation Test

Given a sequence of numbers $U_1, U_2, \ldots$, consider taking successive sets of T of the numbers. Each such set has $T!$ possible orderings when we classify the T numbers as largest, second-largest,..., smallest. Under randomness, the probabilities of each of the $T!$ orderings is $1/T!$. With $T!$ categories, we observe how many of the T-tuples formed from successive sets of T of the numbers fall into each of the $T!$ ordering categories. Then test (3.3.1) may be performed to test randomness. The **cases** $T = 3$, $T = 4$, and $T = 5$ are commonly used.

3.3.5   Poker Test

Given a sequence of numbers $U_1, U_2, \ldots$, consider converting the numbers to integers 1,2,3,4,5,6,7,8,9,10 by

Replacing U_j by 1 if $0.0 \leq U_j < 0.1$
Replacing U_j by 2 if $0.1 \leq U_j < 0.2$
Replacing U_j by 3 if $0.2 \leq U_j < 0.3$
Replacing U_j by 4 if $0.3 \leq U_j < 0.4$
Replacing U_j by 5 if $0.4 \leq U_j < 0.5$
Replacing U_j by 6 if $0.5 \leq U_j < 0.6$
Replacing U_j by 7 if $0.6 \leq U_j < 0.7$
Replacing U_j by 8 if $0.7 \leq U_j < 0.8$
Replacing U_j by 9 if $0.8 \leq U_j < 0.9$
Replacing U_j by 10 if $0.9 \leq U_j \leq 1.0$.

Then the new numbers should be random integers from 1 to 10 if the U_js are random numbers, as in the Coupon Collector's Test with $M = 10$.

Case 1: Take successive sets of five integers. For each set, determine which of the following outcomes occurs:

- The same integer is repeated five times (we have chosen to denote this outcome by AAAAA, where A can be any one of the integers 1, 2, 3, 4, 5, 6, 7, 8, 9, 10).

- One integer is repeated four times and another appears once (denote this by AAAAB).

- One integer is repeated three times and another integer is repeated twice (denote this by AAABB).

- One integer is repeated three times and two other integers appear one time each (denote this by AAABC).

- One integer is repeated two times, another integer appears twice, and another integer once (denote this by AABBC).

- One integer is repeated two times and three other integers appear one time each (denote this by AABCD).

- Five distinct integers appear (denote this by ABCDE).

Exactly one of these possibilities will occur for each 5 integers examined. When true randomness prevails, the proportion of times each possibility occurs when we examine many groups of 5 integers will be given by the probabilities

$$P(AAAAA) = .0001$$
$$P(AAAAB) = .0045$$
$$P(AAABB) = .0090$$
$$P(AAABC) = .0720$$
$$P(AABBC) = .1080$$
$$P(AABCD) = .5040$$
$$P(ABCDE) = .3024.$$

Using the counts observed in these 7 categories (or partitions), the discrepancy test of (3.3.1) may be used. Often categories AAAAA and AAAAB are combined, so that we have **hands of 5** (sets of 5 integers which represent poker hands) **with 6 partitions**; the reason for combining these is that we desire the expected number of occurrences of each category to be at least 5. If AAAAA were kept separate, this would require that $(0.0001)(n/5) \geq 5$, or that we test $n \geq 250,000$ random numbers (which is larger than the 10,000 we wish to use). To make $(0.0046)(n/5) \geq 5$, we need $n \geq 5435$.

Case 2: The test above is also often used with the categories corresponding to "how many different integers are in the 5 selected." For this case, the probabilities can be obtained from those above as

$$P(1 \text{ Different}) = .0001$$
$$P(2 \text{ Different}) = .0135$$
$$P(3 \text{ Different}) = .1800$$
$$P(4 \text{ Different}) = .5040$$
$$P(5 \text{ Different}) = .3024.$$

In this setting, the first and second categories are often combined to make the expected cell counts all be at least 5.

Case 3: Because there is no need for us to restrict ourselves to hands of 5, often the test is done with hands of 4. In this case the relevant probabilities are

$$P(AAAA) = .001$$
$$P(AAAB) = .036$$
$$P(AABB) = .027$$
$$P(AABC) = .432$$
$$P(ABCD) = .504.$$

The first and second categories are usually combined.

3.3.6   Runs-Up Test

Given a sequence of numbers $U_1, U_2, \ldots,$ consider examining the lengths of "runs up" to see if they are of lengths $R = 1$, $R = 2$, $R = 3$, $R = 4$, $R = 5$, or $R \geq 6$. For example, if our sequence $U_1, U_2, \ldots$ is

$$\underbrace{.236, .603,}_{R=2} \underbrace{.188, .600, .692,}_{R=3} \underbrace{.410, .837,}_{R=2} \underbrace{.002, .033, .512,}_{R=3} \underbrace{.504, \ldots}_{\cdots}$$

then we see a run up (an increasing sequence of numbers) of length 2, followed by a run of 3, then a run of 2, then a run of 3, etc. One can thus see how many counts are in each of the 6 categories.

Case 1: Since the runs are not independent but are correlated random variables, one **cannot** simply compute statistic (3.3.1) and have a limiting chi-square distribution. To see the lack of independence between successive runs intuitively, note that a long run is likely to be followed by a short run: a long run may end with a large last number such as .898, and then the number that starts the next run will be in the range $(.000, .897)$. Whereas a short run will likely end with a small last number such as .308, in which case the next run will start with a number in the range $(.000, .307)$—which obviously "guarantees" a longer run, all else being equal.

The "correct" test statistic, derived by Jacob Wolfowitz in 1944, is often misused even today. To obtain the statistic one starts with the calculation of the expected counts in the 6 cells as

$$b_1 = n/6 + 2/3$$
$$b_2 = 5n/24 + 1/24$$
$$b_3 = 11n/120 - 7/60$$
$$b_4 = 19n/720 - 47/720$$
$$b_5 = 29n/5040 - 19/840$$
$$b_6 = n/840 - 29/5040,$$

where n is the number of random numbers $U_1, U_2, \ldots, U_n$ to be examined. Note that the b_is sum to $.5n + .498$, i.e., with n random numbers one expects to have about $n/2$ runs up. Next, compute the 6-by-6 symmetric matrix, C, with entries $C(i, j) = C(j, i)$ for $1 \leq i, j \leq 6$.

$$C(1, 1) = (23n + 83)/180$$
$$C(2, 1) = (-7n - 58)/360$$

$C(3, 1) = -5n/336 - 11/210$

$C(4, 1) = -433n/60480 - 41/12096$

$C(5, 1) = -13n/5670 + 91/25920$

$C(6, 1) = (-121n + 410)/181440$

$C(2, 2) = 2843n/20160 - 305/4032$

$C(3, 2) = (989n + 319)/20160$

$C(4, 2) = -7159n/362880 + 2557/72576$

$C(5, 2) = -10019n/1814400 + 10177/604800$

$C(6, 2) = -1303n/907200 + 413/64800$

$C(3, 3) = (54563n - 58747)/907200$

$C(4, 3) = -21311n/1814400 + 19703/604800$

$C(5, 3) = (-62369n + 239471)/19958400$

$C(6, 3) = (-7783n + 39517)/9979200$

$C(4, 4) = 886657n/39916800 - 220837/4435200$

$C(5, 4) = (-257699n + 1196401)/239500800$

$C(6, 4) = (-62611n + 360989)/239500800$

$C(5, 5) = 29874811n/5448643200 - 139126639/7264857600$

$C(6, 5) = -1407179n/21794572800 + 4577641/10897286400$

$C(6, 6) = 2134697n/1816214400 - 122953057/21794572800.$

Then, compute the matrix $A = (a_{ij}) = C^{-1}$. Now our test statistic (called V, to distinguish it from the previous D) which, for large n, has approximately a chi-square distribution with 6 (not 5) degrees of freedom, is

$$V = \sum_{i=1}^{6} \sum_{j=1}^{6} (R_i - b_i)(R_j - b_j) a_{ij}$$

where R_i is the number of runs of length i ($i = 1, 2, 3, 4, 5$) and R_6 is the number of runs of length 6 or more.

Case 2: This test is a commonly used variation of Case 1. Here the test is applied to $FR(10U_1)$, $FR(10U_2), \ldots, FR(10U_n)$, rather than to $U_1, \ldots, U_n$, where $FR(x)$ is the fractional part of x, e.g., $FR(6.873) = .873$.

Case 3: Another commonly used variation occurs when the test is applied to $FR(100U_1)$, $FR(100U_2), \ldots, FR(100U_n)$.

3.3.7 Serial Pairs Test

Given a sequence of numbers $U_1, U_2, \ldots$, let M be an integer ($M \geq 2$) and replace U_j by $1 + INT(M * U_j)$, where $INT(M * U_j)$ is the integer part of the product $M * U_j$ (but if $M * U_j = M$, we take $INT(M * U_j) = M - 1$). Then the new numbers should be random integers from 1 to M if the U_js are random. Now take successive pairs of integers from the converted sequence, and determine in which of the M^2 categories

$$
\begin{array}{cccccc}
(1,1), & (1,2), & (1,3), & \ldots, & (1, M-1), & (1, M) \\
(2,1), & (2,2), & (2,3), & \ldots, & (2, M-1), & (2, M) \\
\vdots & & & & & \\
(M,1), & (M,2), & (M,3), & \ldots, & (M, M-1), & (M, M)
\end{array}
$$

each lies. Under true randomness, the categories are equally probable with probability $1/M^2$ each, and test (3.3.1) can be used. Common cases used are **Case 1:** $M = 3$; **Case 2:** $M = 10$; and **Case 3:** $M = 20$.

3.3.8 Chi-Square on Chi-Squares Test (CSCS Test)

So far in this section we have discussed seven tests:

- Uniformity of Distribution Test

- Coupon Collector's Test

- Gap Test

- Permutation Test

- Poker Test

- Runs-Up Test

- Serial Pairs Test.

Each of these tests produces a statistic that has, under randomness of the sequence $U_1, U_2, \ldots$ being tested, approximately a chi-square distribution with a number of degrees of freedom previously discussed for each test. The **"shotgun of tests"** approach to testing a random number generator applies each of these tests to perhaps 1000 (or 10,000, or some other quantity) of the numbers $U_1, U_2, \ldots$. If all tests are passed at level of significance 0.05, the generator is accepted for use; if at least one test fails, the generator is rejected. **This test involves only a relatively small part of the random sequence, so this approach should not be taken.** In contrast, the TESTRAND tests (see Section 3.5.2 below) examine 10,000,000 random numbers in testing a generator.

A variant of the shotgun-of-tests approach is the **"subsequence testing"** approach. Here a few of the tests are selected, then applied to successive batches of (e.g.) 10,000 random numbers. This might be done for 100 (e.g.) batches. If the test is failed at level .05, then a batch is rejected; otherwise, it is accepted. If (e.g.) 93 of the 100 batches "pass" the test, then a list of the starting number x_0 is kept for each, and the code is constructed so that the generator used will move among these "acceptable" batches. Unless one tests 100,000 or more batches, this method will not have a long enough period (1 billion or more). Also, its overhead is larger, due to both its storage needs and its checking for batch end, at which time a new batch is invoked. **For these reasons alone, we strongly recommend against using this approach.**

Another reason for rejecting the shotgun-of-tests and subsequence-testing approaches is that, if $U_1, U_2, \ldots$ is truly random, then one should find a test statistic value in the upper 5% tail in about 5% of the cases in which the test is used. Thus, if one tests 100 batches of numbers at level 0.05, one expects, under true randomness, that 5%, or 5 of them, will lead to test statistics in the upper 5% tail. **If one rejects those 5 and keeps only the 95 that "passed," one will be "taming" the random sequence: the result will be a sequence without the large variations typical of true randomness.** This may well lead to simulations that never predict the wild swings that occur (though infrequently) in the real world and that are important aspects of most simulations. Taming a random number sequence by cutting off its tail is similar to taming a bull by cutting off some other part of his anatomy: the result is no longer a fertile generator.

The chi-square on chi-squares (or CSCS) test was introduced to deal with just such problems. For example, if one applies the permutation test with $T = 4$, there are $4! = 24$ categories and we have an approximate chi-square distribution with 23 degrees of freedom for the test statistic. If the test is applied to successive batches of 10,000 random numbers, then the resulting test statistics $D_1, D_2, \ldots$ should (under randomness of $U_1, U_2, \ldots$) follow the chi-square distribution with 23 degrees of freedom. This may be tested with the Uniformity of Distribution Test with (e.g.) 100 cells, whose boundaries are the 1%, 2%, ..., 99% points of the chi-square distribution with 23 degrees of freedom. If one takes 1000 batches, then $D_1, D_2, \ldots, D_{1000}$ are generated from 10,000,000 random numbers and **the test examines whether the**

**various possible ranges of values of D occur in the "right" propor-
tions.** A generator with many D_is beyond the 99% point will be rejected;
it is too wild. A generator with too few beyond the 99% point will also be
rejected; it is too tame. Similarly for the 1% point: too few or too many
counts here will lead to rejection. Note that the naive approach of looking
for small D values in each batch would never reject if all Ds were 0 (which
would indicate a very bad generator); this defect would be detected readily
by the CSCS test.

The chi-square percentile points needed to carry out the above test are
given on pp. 24–27 of Dudewicz and Ralley (1981) and in Appendix F of
this book, for 99 degrees of freedom, which corresponds to 100 cells. For
other degrees of freedom, the percentile points can be calculated with the
routine POINTS(V) for V degrees of freedom that they supply (see their pp.
446–447). Note that on pp. 24–27 the columns labeled "Cornish-Fisher
Approximation" are to be used. The test is usually carried out using the
calling sequence of TESTRAND, which is described in Section 3.6.

3.3.9 Entropy-Uniformity Test

The **differential entropy** (or **entropy**) of a random variable X with density
function f is defined as

$$H(X) = H(f) = -\int_{-\infty}^{\infty} f(x) \log f(x) \, dx \qquad (3.3.4)$$

and has properties that if $X \in [0, 1]$ with probability 1 then

$$H(f) \leq 0, \qquad (3.3.5)$$

and among all densities f concentrated on $[0, 1]$ the uniform density f_0 max-
imizes $H(f)$ to

$$H(f_0) = 0. \qquad (3.3.6)$$

The following two definitions, given by Dudewicz and van der Meulen (1981),
make use of attributes (3.3.5) and (3.3.6) of $H(f)$.

Definition 3.3.7. Two densities f_1 and f_2 are **entropy-distinguishable**
if $H(f_1) \neq H(f_2)$.

Definition 3.3.8. A density f^* is **entropy-unique** (or **e-unique**) in a
class C of densities if $f^* \in C$ and there does not exist any $f \in C$ with
$f \neq f^*$ and $H(f) = H(f^*)$.

By property (3.3.6) we know that in the class C of densities on $(0,1)$, f_0 is e-unique.

For $X_1, X_2, \ldots, X_n$ a random sample from an absolutely continuous distribution F with density f, let $X_{(1)}, X_{(2)}, \ldots, X_{(n)}$ be the order statistics of the sample. An estimator of $H(f)$ (Vasicek (1976)) is

$$H_{m,n} = n^{-1} \sum_{i=1}^{n} \log \left\{ \frac{n}{2m} \left[X_{(i+m)} - X_{(i-m)} \right] \right\} \qquad (3.3.9)$$

where $1 \leq m < n/2$, $X_{(j)} = X_{(1)}$ for $j < 1$ and $X_{(j)} = X_{(n)}$ for $j > n$.

Dudewicz and van der Meulen (1981) proposed the following test for uniformity based on the e-uniqueness of f_0.

Let $X_1, X_2, \ldots, X_n$ be a random sample from an absolutely continuous distribution F with density f concentrated on $[0,1]$ and let f_0 denote the $U(0,1)$ density. The level α test rejects $H_0 : f = f_0$ in favor of $H_A : f \neq f_0$ if and only if

$$H_{m,n} \leq H_\alpha^*(m,n) \qquad (3.3.10)$$

where $H_\alpha^*(m,n)$ is the 100α percentile point of the distribution of $H_{m,n}$ under f_0.

By the e-uniqueness of f_0 among all densities concentrated on $[0,1]$ and by the consistency of the estimator (Vasicek (1976)), it follows that the above test is consistent against all alternatives f on $[0,1]$. Further, in Dudewicz and van der Meulen (1981) it is shown that if f is concentrated on $[0,1]$ then, with probability 1, $H_{m,n} \leq 0$.

In Dudewicz and van der Meulen (1981), this test is studied in detail using both analytical and Monte Carlo techniques. The latter are necessary since the form of the distribution of $H_{m,n}$ appears to be analytically intractable. Dudewicz, van der Meulen, SriRam and Teoh (1995) also studied this test and applied it, in the above (one-sample) form as well as in a more sensitive form in combination with a chi-square test (similar to the one described below in Section 3.3.1), to 9 random number generators (from among those given in Section 3.5). The salient conclusions about these generators are noted in Section 3.5 below. An important general conclusion was that a cycling generator (e.g., one that cycles on .0000, .0001, .0002, ..., .9999), can have a sample entropy close to 0 even though it is not producing random numbers (i.e., independent uniform variables on $(0,1)$). For example, with $n = 10,000$ and $m = 1$ one finds, from (3.3.9),

$$H_{m,n} = \frac{1}{n} \cdot n \log \left\{ \frac{n}{2} \cdot 2 \cdot 10^{-4} \right\} = 0.$$

Thus, such a test is appropriate only for generators that pass other extensive tests such as the TESTRAND tests.

3.4 Theoretical Tests of Random Number Generators

We can, in some cases, evaluate a generator theoretically, rather than by looking at the numbers it produces. **In this section, we look at two tests of this type that are available for mixed congruential generators of type (3.2.1):**

$$x_{i+1} = ax_i + c \pmod{m}, \qquad i = 0, 1, 2, \ldots. \tag{3.4.1}$$

3.4.1 Serial Correlation Test

With $U_1, U_2, \ldots$ random, the statistical correlation $\mathrm{Corr}(U_i, U_{i+1}) = 0$ exactly. With $U_i = x_i/m$ in (3.4.1), it has been shown that

$$\mathrm{Corr}(U_i, U_{i+1}) = \frac{1}{a}(1 - \frac{6c}{m} + 6(\frac{c}{m})^2 + \epsilon), \tag{3.4.2}$$

with

$$|\epsilon| \le \frac{a+6}{m}. \tag{3.4.3}$$

If a and m are relatively prime, it is possible to replace ϵ by an exact expression. This will not be given here but is available in routine TSTM1 of the TESTRAND package, with details given in Dudewicz and Ralley (1981, p. 515). This test is a rough screen: we should probably not use any generator for which **both** the right-hand side of (3.4.2) is larger than 0.01 **and** the right-hand side of (3.4.3) is smaller than 0.005.

As an example, if we use URN13, which has, as discussed in Section 3.2, $c = 0$, $a = 663,608,941$, and $m = 2^{32}$, then we have estimated correlation

$$1/a = 0.0000000015$$

with error bound

$$(a+6)/m = 0.1545.$$

Thus, this generator passes our criteria for serial correlation. Since a and m are relatively prime, the exact serial correlation can be obtained by using TSTM1.

3.4.2   Interplanar Distance (or Spectral) Test

This test is essentially based on the result (see Marsaglia (1968)) that the n-tuples $(U_1, \ldots, U_n)$, $(U_2, \ldots, U_{n+1})$, $(U_3, \ldots, U_{n+2})$, $\ldots$ obtained from the generator (3.2.1) lie in a finite and small number of parallel equally spaced hyperplanes. That they lie in a finite number of hyperplanes is trivial: a finite number of points must be able to be captured by a finite number of equally spaced parallel hyperplanes. That the number is small is the surprising part (with true randomness, one might suspect that each plane could catch little better than 3 of the points in 3-space, since 3 points determine a plane). That small number is at most $(n!\,m)^{1/n}$. Later, Marsaglia gave a heuristic argument for thinking that the bound might be able to be shown to be in fact $\sqrt{n}m^{1/n}$; however, whether this is a valid bound still seems to be an open question. The $(n!\,m)^{1/n}$ bound and $\sqrt{n}m^{1/n}$ (the latter in parentheses), are shown below for some combinations of m and n.

$(n!\,m)^{1/n}$ for a Variety of n and m.

n	2	3	4	5	6	7	8	9
$m = 2^{16}$	362	73	35	23	19	16	15	14
	(362)	(69)	(32)	(20)	(15)	(12)	(11)	(10)
$m = 2^{32}$	92681	2953	566	220	120	80	60	48
	(92681)	(2815)	(512)	(188)	(98)	(62)	(45)	(35)

In many cases, there are several sets of parallel equally spaced hyperplanes that could be used to capture all of the points. **Consider the set where the spacing between the hyperplanes is the largest** (the coarsest set that still captures all the points); **its interplanar distance d_n in n-dimensions is the criterion for the interplanar distance test.** If this distance is large, the generator has n-space patterns that are too pronounced for it to be considered random.

An algorithm for determining d_n for cases with $c \neq 0$ and period m, or $c = 0$ and period $m - 1$, is given in TESTRAND by routine TSTM4. This routine uses an infinite-precision arithmetic package also incorporated in TESTRAND, so that no matter what word-length computer one is working on the size of a, m, c is no barrier to the computation.

For evaluation, clearly (for fixed n) the smaller d_n, the better the generator, comparatively. For evaluation of one generator, a criterion proposed in the literature is that $d_n \leq 2^{-30/n}$. For $n = 2, 3, \ldots, 9$ this rule of thumb evaluates to:

$2^{-30/n}$ **as a function of** n.

$n = 2$	$n = 3$	$n = 4$	$n = 5$	$n = 6$	$n = 7$	$n = 8$	$n = 9$
.00003	.001	.005	.016	.03	.05	.07	.10

For example, it is known (from Dudewicz and Ralley (1981), p. 138 and our own computations) that for the generator known as RANDU, which has $c = 0$, $a = 2^{16} + 3 = 65539$, $m = 2^{32}$, we have

$$d_2 = .00002, \quad d_3 = .092, \quad d_4 = d_5 = d_6 = d_7 = d_8 = d_9 = .093.$$

Thus, RANDU passes the interplanar distance test **only** in 2 dimensions, and is rejected (resoundingly) in 3 or more dimensions. As a **second example,** the generator known as LLRANDOM has $c = 0$, $a = 7^5 = 16807$, $m = 2^{31}-1$, and for it we have

$$d_2 = .00006, \quad d_3 = .002, \quad d_4 = .007, \quad d_5 = .015,$$
$$d_6 = .03, \quad d_7 = .06, \quad d_8 = .08, \quad d_9 = .11.$$

We see that, while this generator also fails the test, it comes much closer to passing.

This test is widely referred to as the **spectral test**, since early work on testing generators with hints of spectral analysis (time series analysis) developed something called the n-dimensional wave number ν_n, for which large values were good (indicative of randomness in a heuristic sense). It is now known that $\nu_n = 1/d_n$, so that in fact a basis now exists for the spectral test, which was previously a heuristic procedure with unclear theory to back it up.

All vectors of numbers $(U_i, \ldots, U_{i+n-1})$ lie in the n-dimensional unit cube. The smallest distance between two vertices of that cube is equal to the distance from the origin $(0, 0, \ldots, 0, 0)$ to $(0, 0, \ldots, 0, 1)$, namely

$$d_L(n) = ((0-0)^2 + (0-0)^2 + \ldots + (0-0)^2 + (0-1)^2)^{0.5} = 1,$$

and the largest is that from $(0, 0, \ldots, 0, 0)$ to $(1, 1, \ldots, 1, 1)$, namely

$$d_U(n) = ((0-1)^2 + (0-1)^2 + \ldots + (0-1)^2 + (0-1)^2)^{0.5} = \sqrt{n}.$$

Therefore, since a distance of length l can be equally divided by r planes into $r - 1$ equal lengths of length $l/(r - 1)$ each, it follows that a generator with number of planes approaching $(n!\,m)^{1/n}$, which cannot necessarily be attained in n dimensions, will have interplanar distance between the numbers

$$1/((n!\,m)^{1/n} - 1) \ \text{ and } \ \sqrt{n}/((n!\,m)^{1/n} - 1).$$

The former is the best possible interplanar distance in n-dimensions for a generator with modulus m; the latter results if the generator has a rotation of its numbers. If we compute the former for various n and m, we obtain the following.

Best Interplanar Distance as a Function of n and m.

n	2	3	4	5	6	7	8	9
$m = 2^{16}$	.0028	.014	.029	.044	.056	.065	.071	.076
$m = 2^{32}$	.00001	.0003	.0018	.005	.008	.013	.017	.021

We can regard this as **an absolute standard of goodness of a generator in terms of the interplanar distances $d_2, \ldots, d_9$ for a given** m. If we divide the d_ns of a generator by the "best" interplanar distance for its m, then we have **a ratio that is at least 1, and values closer to 1 are better. For example,** for RANDU the ratios are:

$$r_2 = .00002/.00001 = 2, \ r_3 = 307, \ r_4 = 52,$$
$$r_5 = 19, \ r_6 = 12, \ r_7 = 7, \ r_8 = 5, \ r_9 = 4.$$

This shows especially bad behavior in 3 and 4 dimensions. If more digits of accuracy are used for the d_i of RANDU and for the $m = 2^{32}$ best interplanar distances, we find

$$r_2 = 1.41, \ r_3 = 271.82, \ r_4 = 52.52, \ r_5 = 20.33,$$
$$r_6 = 11.11, \ r_7 = 7.37, \ r_8 = 5.5, \ r_9 = 4.43.$$

On the other hand, for generator LLRANDOM we find (rounding more accurate values to the nearest integer) the ratios

$$r_2 = 4, \ r_3 = 4, \ r_4 = 3, \ r_5 = 3, \ r_6 = 4, \ r_7 = 4, \ r_8 = 4, \ r_9 = 5,$$

which indicates a much better test result.

From the interplanar distance we can bound the number of planes needed. For example, RANDU has a 3-dimensional interplanar distance of $d_3 = .092$. Therefore, its number of planes must be between the number required to split a distance of 1 into parts .092 apart, and that required for a distance of $\sqrt{3}$, namely between

$$1/.092 + 1 = 11.8 \ \text{ and } \ \sqrt{3}/.092 + 1 = 19.8.$$

On the other hand, LLRANDOM requires, in 3 dimensions, between

$$1/.002 + 1 = 501 \text{ and } \sqrt{3}/.002 + 1 = 867$$

planes. Recall that the maximum number that could possibly be required in 3 dimensions, for a generator with $m = 2^{32}$ like RANDU, is 2953; while for a generator with $m = 2^{31} - 1$ like LLRANDOM the number is 2344. The 3-dimensional superiority of LLRANDOM over RANDU is thus clear and pronounced. Of course, LLRANDOM (URN01 in TESTRAND) was rejected by Dudewicz and Ralley (1981, p. 134), in part for its bad performance in 3 dimensions. So, we should note that its competitors include generators such as that with $c = 0$, $a = 69069$, $m = 2^{32}$ (URN03, described in the next section), which attains $d_3 = .00069$, and hence uses between

$$1/.00069 + 1 = 1450 \text{ and } \sqrt{3}/.00069 + 1 = 2511.$$

planes.

Example. Let us consider the generator

$$x_{i+1} = 7x_i \bmod 11, \quad x_0 = 1$$

using the tests of this section. First, from (3.4.2) we know that for the sequence $U_i = x_i/m$ (with $m = 11$) we have **serial correlation**

$$\mathrm{Corr}(U_{i+1}, U_i) = \frac{1}{7} + \epsilon = .14 + \epsilon,$$

and, from (3.4.3), $|\epsilon| \leq 13/11 = 1.18$. Thus, the generator does not fail this test probably only due to the lack of a good error bound. More exact computation using TESTRAND test TSTM1 is explored in the problems at the end of this chapter.

For the **spectral test** in $n = 2$ dimensions, first note that the sequence x_i produced is 1,7,5,2,3,10,4,6,9,8,1, which has period $m - 1 = 10$. Thus, 10 pairs (U_i, U_{i+1}) are available, namely the pairs obtained by dividing each number by $m = 11$ in the vectors (1,7), (7,5), (5,2), (2,3), (3,10), (10,4), (4,6), (6,9), (9,8), (8,1). The 10 pairs (U_i, U_{i+1}) are plotted in Figure 3.4–1.

The spectral test looks at parallel equispaced planes that contain all of the pairs. In Figure 3.4–2, we show a set of 10 planes with an interplanar distance of $1/11 = .0909$ that contain all of the pairs. However, we know that we can contain all the pairs from any multiplicative congruential generator with $m = 11$ in at most $(2!\,11)^{1/2} = 4$ planes. In Figure 3.4–3, we show a set of 4 planes, parallel with spacing $((3/13)^2 + (22/143)^2)^{.5}$, which contain all of

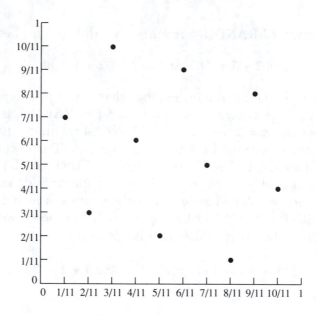

Figure 3.4–1. 10 pairs (U_i, U_{i+1}) available from the generator
$U_i = x_i/m$, $x_{i+1} = 7x_i \bmod 11$, $x_0 = 1$.

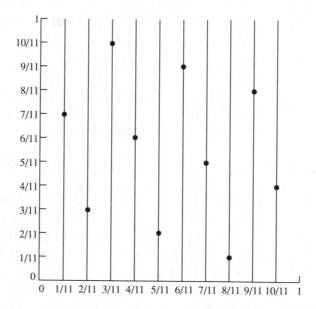

Figure 3.4–2. A set of 10 parallel planes at spacing of 1/11 that
contain all pairs ($U_i = b$ for b of 1/11,..., 10/11).

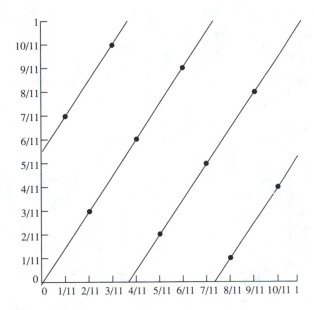

Figure 3.4–3. A set of 4 parallel planes at spacing of 0.2774 containing all pairs; $U_{i+1} = 1.5U_i + b$ for $b = 5.5/11$, 0, $-5.5/11$, $-11/11$.

the pairs. This, of course, is a worse result, and the fewer planes we can use, the worse the generator looks to us. In Figure 3.4–4, we see that in fact 3 planes can contain all of the pairs. These farthest-spaced equispaced parallel planes that can accomplish this have an interplanar distance of 0.3162, hence $d_2 = 0.3162$.

The best interplanar distance with $m = 11$ is, in 2-space, $1/(4 - 1) = 0.3333$, so this generator has a ratio of $0.3162/0.3333 = 0.95$. Thus, it is (of generators with $m = 11$) a fairly good generator in 2-space. However, if one is working on a computer where $m = 2^{32}$ can easily be used, which is true for most computers today, then a more reasonable comparison is with the best $m = 2^{32}$ 2-space distance, namely 0.00001. Then we find the ratio $0.3162/0.00001 = 31620$, which is very poor **hence this generator should not be used.** Note that the ratio we found of $0.3162/0.3333 = 0.95$ should be surprising, since our analysis yielding the 0.3333 called it the "best possible." How can any generator have a smaller value? This can happen since the analysis used its available planes to equally split the length l with r planes into $r - 1$ equal lengths of $1/(r - 1)$ each. However, one may not need planes at the ends of the interval unless there are numbers there, so the r

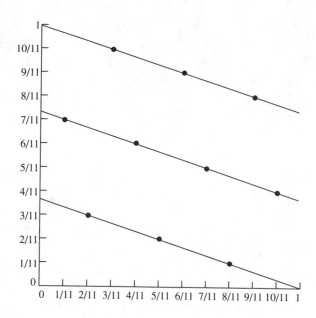

Figure 3.4–4. A set of 3 parallel planes at spacing $\sqrt{0.10}$ containing all pairs; $U_{i+1} = -U_i/3 + b$ for $b = 11/11$, 2/3, 1/3.

planes could possibly split the interval into smaller parts in its center, with larger lengths on each "end." This indicates a defect in the spectral test: the numbers may be "bunched" on a large number of close planes, with large regions at the ends uncovered, so that the generator will look deceptively good by the d_n criterion.

3.5   Test Results on Specific Random Number Generators

As may be inferred from the detailed considerations we have given to random number generation and testing, this subject is extremely important. Some important points are:

1. The use of a bad generator can ruin a simulation study (e.g., see pp. 28–41 of Dudewicz and Karian (1985) for examples).

2. Over 75% of the generators in widespread use fail tests of randomness

(e.g., see Dudewicz and Ralley (1981) and the results given below in this section).

3. Hence, any study which does not specify its random number generator (e.g., that of Zheng (1994)) must remain suspect until its results are corroborated using a documented random number generator.

Therefore, our recommendations that were given at the end of Section 3.1 bear repeating:

1. Choose a generator with speed and memory needs you can tolerate, and with a period of at least 1,000,000,000.

2. Verify that the generator has passed the TESTRAND tests; if not, test it yourself.

3. Do not trust any generator, ever, without seeing its test results on the TESTRAND tests.

4. Generators are noted below that have passed the TESTRAND tests; using them will satisfy the above recommendations. Generators that have failed are also noted below; do not use any of them for any reason. We will comment on a set of generators below and for each, we will give its TESTRAND name (if it has one), the source of the generator, and some notes on its method.

3.5.1 Random Number Generators

URN01. This generator, popularized by Learmonth and Lewis's (1973) package called LLRANDOM, is due to Lewis, Goodman, and Miller (1969), and is described by

$$x_{i+1} = 16807x_i \bmod (2^{31} - 1). \tag{3.5.1}$$

Note that $16807 = 7^5$. In LLRANDOM this is called generator RANDOM.

URN02. This generator was given by Marsaglia and Bray (1968). The three sequences

$$l_i = 65539l_{i-1} \bmod 2^{32}$$
$$m_i = 33554433m_{i-1} \bmod 2^{32} \tag{3.5.2}$$
$$k_i = 362436069k_{i-1} \bmod 2^{32}$$

are used in a fashion which is called a **composite** generator. Note that $65529 = 2^{16} + 3$, $33554433 = 2^{25} + 1$, $16777216 = 2^{24}$. They start with 128 initial odd integers stored in the locations $N(1)$, $N(2)$, $\ldots$, $N(128)$. The sequences l_i and m_i are used to choose one of the 128 storage locations. The number generated is based on the number in the location selected, and also on l_i and m_i. If location $N(J)$ is selected, it is then refilled from sequence k_i. Thus, except for the initial 128 numbers $N(1)$, $\ldots$, $N(128)$, this vector contains numbers from the sequence k_i.

URN03. This generator has been popularized by the SUPER–DUPER package due to Marsaglia, Ananthanarayanan, and Paul (1973), where it is called UNI. The results of the sequence

$$x_i = 69069 x_{i-1} \bmod 2^{32} \qquad (3.5.3)$$

are combined with results of a so-called feedback shift generator (with a right shift 15 and a left shift 17, i.e., based on the primitive trinomial $x^{17} + x^{15} + 1$), which also yields a number y_i with 32 bits. These x_i and y_i are added by **exclusive or** addition bitwise (in this addition, $1 \oplus 1 = 0$, $1 \oplus 0 = 0 \oplus 1 = 1$, and $0 \oplus 0 = 0$) to produce the sequence used, say z_i. For example,

$$x_i = 01001 \ldots 01010$$
$$y_i = 00101 \ldots 10110$$
$$z_i = x_i \oplus y_i = 01100 \ldots 11100. \qquad (3.5.4)$$

Note that $69069 = (3)(7)(11)(13)(23)$.

URN04. This is URN01 with a shuffling of its numbers. It goes by the name SRAND in the LLRANDOM package.

URN05. This is a generalization of the feedback-shift type of generator, due to Lewis and Payne (1973). Such generators were highly touted for a while, since they have astronomical periods (however, we have already seen very bad generators with large periods, like the generator $x_{i+1} = x_i + 1 \bmod 2^{500}$, $x_0 = 0$), and they can also repeat numbers without repeating the sequence. This is like sampling with replacement, which random sampling of numbers on the interval (0,1) should be.

URN06. This is a modification of URN02 of Marsaglia and Bray (1968). The modification, suggested by Lurie and Mason (1973), is also a composite generator. It forms

$$v_i = 65539 v_{i-1} \bmod 2^{32} \qquad (3.5.5)$$

$$w_i = 262147 w_{i-1} \bmod 2^{32}$$

and then sets

$$x_i = (v_i + w_i) \bmod 2^{31}. \tag{3.5.6}$$

The x_i sequence is used. Note that $65539 = 2^{16} + 3$, and $262147 = 2^{18} + 3$.

URN07, URN08, URN09. These are slightly different codings of the generator called RANDU first proposed by IBM (1970). URN07 is a fast version in assembler, URN08 is the original version, and URN09 is the version used in the statistical package SPSS, as noted in Nie, Bent, and Hull (1973). The generator is

$$x_i = 65539 x_{i-1} \bmod 2^{31}. \tag{3.5.7}$$

Note that $2147483647 + 1 = 2^{31}$.

URN10. This generator, given initially by Kruskal (1969), was touted because it could produce the same results on many computers. It was incorporated as the generator of the OMNITAB II statistical package (see Hogben, Peavy, and Varner (1971)). The generator sets

$$x_i = 5^3 x_{i-1} \bmod 2^{13}. \tag{3.5.8}$$

Note that $2^{13} = 8192$.

URN11. This is a portable congruential generator (ready for the user to specify the a, c, m to be used) provided in TESTRAND. It was tested with the choices $c = 0$, $a = 452807053$, $m = 2^{31}$, which were used in the CUPL language (see Walker (1967)). Note that the a used is $5^{15} = 30517578125$ taken mod 2^{31}.

URN12. This is a double-precision version of URN11. It was tested with the same c, a, m as used for URN11.

URN13. This generator, due to Ahrens and Dieter (1974), sets

$$x_i = 663608941 x_{i-1} \bmod 2^{32}. \tag{3.5.9}$$

URN14. This generator, due to Zarling (1971), uses a vector of integers, $N(1), \ldots, N(64)$. The sequence

$$y_{i+1} = 266245 y_i + 453816693 \bmod 2^{31} \tag{3.5.10}$$

is generated. However, this sequence is not used directly. Rather, it is shuffled by filling the table $N(\cdot)$ with it. An index into the table is selected by using another sequence,

$$\begin{cases} z_{i+1} = 10924 z_i + 6925 \\ z_{i+1} = z_{i+1} - 32769(z_{i+1}/32769). \end{cases} \tag{3.5.11}$$

If $z_{i+1} = 32768$, one iterates to the next term of the sequence. This will place z_i between 0 and 32767; then $z_{i+1}/512 + 1$ is between 1 and 64 (note that $64 \times 512 = 32768$) and is taken as the index. That entry of $N(\cdot)$ is used to form the next random number; the entry used is then refilled from sequence (3.5.10). Note that $266245 = (5)(7)(7607)$, $453816693 = (3^2)(11)(4584007)$, $10924 = (2^2)(2731)$, and $6925 = (5^2)(277)$.

URN15, URN16, URN17, URN18, URN19, URN20. These six generators with $c = 0$ and $m = 2^{31} - 1$ did well in the spectral test as applied by Hoaglin (1976). They use multipliers

$$a_{15} = 764261123$$
$$a_{16} = 1323257245$$
$$a_{17} = 1078318381 \qquad\qquad (3.5.12)$$
$$a_{18} = 1203248318$$
$$a_{19} = 397204094$$
$$a_{20} = 2027812808.$$

The full computer code for **generators URN01, URN02, ..., URN20** is contained in Dudewicz and Ralley (1981), and in the version of the TES-TRAND package that was available from The Ohio State University for IBM 360/370 computers and their look-alikes (copies could be obtained from Instruction and Research Computer Center, The Ohio State University, 1971 Neil Avenue, Columbus, OH 43210). In the version of TESTRAND for VAX–11 computers that was available from Denison University (copies could be obtained from Computer Center, Denison University, Granville, OH 43023), additional generators **URN21, ..., URN30** were included. (See the end of Section 3.1 for current TESTRAND availability.) Note that in practice one does not need all of URN01 through URN35. Many of these generators failed testing and should not be used; moreover, in most studies one only needs one or perhaps two good generators—recommendations and code are given later in this section.

We now briefly describe URN21 through URN30 and follow this by a discussion of URN31 through URN35, and other generators that have been studied more recently.

URN21. This generator has been popularized in chemical engineering by its authors, Swain and Swain (1980). Test results given later (see Dudewicz and Karian (1985), pp. 42–44) show very poor performance by this generator. The generator is in some ways similar to the additive congruences discussed in Section 3.1. It uses 3 initial numbers as seeds and sets successive terms by

$$\begin{cases} x_{i+1} = x_i + x_{i-1} + x_{i-2} \\ x_{i+1} = x_{i+1} + 1357 & \text{if } x_{i-1} < 50,000,000 \\ x_{i+1} = x_{i+1} - 100,000,000 & \text{if } x_{i+1} \geq 100,000,000 \\ x_{i+1} = x_{i+1} - 100,000,000 & \text{if } x_{i+1} \geq 100,000,000. \end{cases} \tag{3.5.13}$$

If each of the initial numbers x_0, x_1, x_2 is on the range 1 to $99,999,999$, then x_3 will be no larger than $299,999,997$. If in fact the x_1 was less than $50,000,000$, then 1357 is added to "jog" the least significant digits. We know from Section 3.1 that the least significant digits have marked patterns in a congruential generator, which of course URN21 is not, and the resulting x_3 will still be no larger than $299,999,997$, due to the small x_1 in this case. The next two subtractions then put x_3 onto the range 0 to $10^8 - 1$. This procedure is replicated in an obvious way for successive terms. Of course the "random" numbers are taken as $x_{i+1}/10^8$.

URN22. This generator is the library routine on VAX–11/780 computers under the VMS operating system, where it is called MTH\$RANDOM. It sets

$$x_{i+1} = 69069x_i + 1 \bmod 2^{32}, \tag{3.5.14}$$

which (except for $c = 1$ instead of $c = 0$) is sequence (3.5.3), which forms part of URN03. For further reference details, see card 594, frames L11 to L12 of the microfiche of operating system code for release 4.2 of VMS provided by the Digital Equipment Corporation.

URN23. This generator, proposed by Brody (1984), uses sequence (3.5.14) and a table, $T(1), \ldots, T(N)$. While (3.5.14) is used to generate x_{i+1} from x_i, another sequence is generated by exclusive or addition of x_{i+1} and one of the entries of $T(\cdot)$. (See (3.5.4) for the definition of exclusive or addition.) Bits are "pulled off" of the tabled number used, to find an index into the table, and that entry is used for the next random number and is subsequently replaced by the newly generated number.

URN24. This generator was formulated by Thesen (1985) especially for microcomputers. However, as previously noted for microcomputers, one may use methods described in Dudewicz, Karian, and Marshall (1985) to implement good generators that need larger numbers than are typically available on microcomputers. This generator does not produce random numbers by dividing integers by a number m. Instead, it produces bits byte-wise, which is in some ways similar to the methods used in Dudewicz and Ralley (1981) to produce multiple-precision random numbers. For further details, see Thesen (1985).

URN25. Gait (1977) suggested use of the Data Encryption Standard (DES) as a random number generator. For details and references on this complex bitwise generator, see Mathews (1986). In particular, Mathews (1986, p. 71) found that this generator takes 500 times as much computer time as does, e.g., generator URN22; in view of other testing results, given below, this method does not warrant further detail here.

URN26. This generator was an attempt by Mathews (1986) to improve on URN24; however, due to testing results that we will see later, we will not give further detail here. Mathews (1986, p. 69) calls this a generator that "...may seem like it should work reasonably well, but which does not." This underscores again the problems of "roll your own" (or self-created, home-brewed) generators.

URN27. This is the generator included in release 1 of the GPSS/H version of the GPSS simulation language (see Henriksen and Crane (1983)). It is a feedback-shift generator (see URN03 for notes on such generators) based on the trinomial $x^{31} + x^3 + 1$. For further details, see Whittlesey (1968).

URN28. This is a portable version of URN01, that uses the same algorithm, namely (3.5.1). However, its portability is achieved by coding in Pascal. Portable versions of URN01 were first suggested by Schrage (1979).

URN29. This generator, which arose in computer code written for calculating nuclear cross sections by Monte Carlo integration methods, uses the same sequence as does URN01, namely

$$x_{i+1} = 16807x_i \bmod (2^{31} - 1). \qquad (3.5.15)$$

However, it then takes $U_{i+1} = x_{i+1}/2147483711$, instead of $x_{i+1}/(2^{31} - 1)$ which would be $x_{i+1}/2147483647$. This almost seems like a data entry error. The results, as we see below, are disastrous, since now there cannot be any numbers between $(2^{31} - 2)/2147483711 = .9999999697$ and 1.

URN30. This generator, which is used in the SIMSCRIPT simulation language (see West and Johnson (1984)) uses the Lehmer (see Payne, Rabung, and Bogyo (1969)) sequence

$$x_{i+1} = 630360016x_i \bmod (2^{31} - 1). \qquad (3.5.16)$$

Note that 630360016 is $14^{29} = (2^{29})(7^{29}) \bmod (2^{31} - 1)$, and 630360016 = $(2^4)(11)(13)(137)(2011)$.

URN31, URN32, URN33. Miyazaki (1987) studied these generators and claimed that they are all "... good for practical use in simulation." The generator **URN31** sets

$$x_{i+1} = 2456949x_i \bmod 2^{24} \qquad (3.5.17)$$

with $x_0 = 1234567$. It has no initial segment and a period of $d = 4,194,304$.

URN32. Sets

$$x_{i+1} = 3513383x_i \bmod 2^{25} \qquad (3.5.18)$$

with $x_0 = 1234567$. It has no initial segment and a period of $d = 8,388,608$.

URN33. Sets

$$x_i = y_{2i+2} \qquad (3.5.19)$$

where

$$y_{2i+1} = 11257y_{2i} \bmod (10^8 + 7), \qquad (3.5.20)$$

and

$$y_{2i+2} = 12553y_{2i+1} \bmod (10^8 + 37), \qquad (3.5.21)$$

with $y_0 = 12345678$. This has an initial segment of 2,476,661 terms and period $d = 19,103,417$.

URN34. This generator was proposed by Sezgin (1991) for use on micro-computers. It was stated that the generator "... stood up quite well to all statistical tests." The method uses three sequences:

$$\begin{aligned} w_i &= 44w_{i-1} \bmod 2039, \\ y_i &= 45y_{i-1} \bmod 2037, \\ z_i &= 41z_{i-1} \bmod 2003, \end{aligned} \qquad (3.5.22)$$

with $w_0 = y_0 = z_0 = 1$, and sets

$$U_i = w_i/P_1 - y_i/(P_1P_2) + z_i/(P_1P_2P_3) \qquad (3.5.23)$$

where $P_1 = 2039$, $P_2 = 2027$, and $P_3 = 2003$. However, instead of direct use of (3.5.23), a permutation from a table is used to attempt to destroy any lattice structure that may be present. The code recommended, which is in FORTRAN, is given in Figure 3.5–1, where the values read in are KW=KY=KZ=1 and XR=.453. The value XR is initially used to calculate the indices of the tables and hence needs 3 digits.

URN35. This generator was proposed by L'Ecuyer (1987) especially for use on microcomputers. The generator is stated to be "... a high-quality pseudo-random number generator for 16-bit computers" and to have been

```
       DIMENSION TW(10),TY(10),TZ(10)
       READ*, XR,KW,KY,KZ
       DO 1 I=1,10
       KW=MOD(44*KW,2039)
       TW(I)=KW
       KY=MOD(45*KY ,2027)
       TY(I)=KY
       KZ=MOD(41*KZ ,2003)
1      TZ(I)=KZ
2      KW=MOD(44*KW,2039)
       KY=MOD(45*KY,2027)
       KZ=MOD(41*KZ,2003)
       IU=1000*XR
       I1=IU/100
       I=IU-I1*100
       I2=I/10
       I3=I-10*I2+1
       I1=I1+1
       I2=I2+1
       IW=TW(I1)
       IY=TY(I2)
       IZ=TZ(I3)
       TW(I1)=KW
       TY(I2)=KY
       TZ(I3)=KZ
       NMBR=MOD(IW,6)+1
       GO TO(10,20,30,40,50,60),NMBR
10     XR=FLOAT(IW)/2038.0-FLOAT(IY)/4128988.0+FLOAT(IZ)/8266233976.0
       GO TO 100
20     XR=FLOAT(IW)/2038.0-FLOAT(IZ)/4080076.0+FLOAT(IY)/8266233976.0
       GO TO 100
30     XR=FLOAT(IY)/2026.0-FLOAT(IZ)/4056052.0+FLOAT(IW)/8266233976.0
       GO TO 100
40     XR=FLOAT(IY)/2026.0-FLOAT(IW)/4128988.0+FLOAT(IZ)/8266233976.0
       GO TO 100
50     XR=FLOAT(IZ)/2002.0-FLOAT(IW)/4080076.0+FLOAT(IY)/8266233976.0
       GO TO 100
60     XR=FLOAT(IZ)/2002.0-FLOAT(IY)/4056052.0+FLOAT(IW)/8266233976.0
100    CONTINUE
       STOP
       END
```

Figure 3.5–1. The FORTRAN code for URN34.

"... submitted to extensive statistical testing." This generator, like URN34, combines several simple generators:

$$w_i = 157w_{i-1} \bmod 32363$$
$$y_i = 146y_{i-1} \bmod 31727$$
$$z_i = 142z_{i-1} \bmod 31657 \qquad (3.5.24)$$
$$x_i = (w_i + y_i + z_i - 3) \bmod 32362$$
$$U_i = (x_i + 1)/32363.$$

Here $w_0 = y_0 = z_0 = 1$ may be taken as seeds. The period is stated as being $(32362)(31726)(31656)/4 = 8.12543685 \times 10^{12}$.

URN36, URN37, and URN38 are generators for microcomputers. In Section 1.2.3, we discussed briefly how random number generators that have good characteristics (and often utilize large numbers in their algorithms) may be implemented on microcomputers. For a method of handling multiplications of large numbers when a microcomputer has a small word size, see equation (1.2.3) and the discussion that follows it. Three generators (coded in FORTRAN), appropriate for computers with limited word sizes, were tested by Dudewicz, Karian, and Marshall (1985). The details are as follows.

URN36. This generator, recommended by Jennergren (1984), can easily be implemented on computers with word sizes of 36 bits or greater. It uses the relationship

$$x_i = 5^{13}x_{i-1} \bmod 2^{35}$$
$$= 1,220,703,125x_{i-1} \bmod 8,589,934,592 * 4. \qquad (3.5.25)$$

URN37. This generator was proposed by Dudewicz, Karian, and Marshall (1985). It is a modification of URN36 obtained by adjusting the modulus to make the computations associated with (3.5.25) easily manageable on computers that use 32-bit words. URN37 uses the relationship

$$x_i = 5^{13}x_{i-1} \bmod 2^{31}$$
$$= 5^{13}x_{i-1} \bmod 2,147,483,648. \qquad (3.5.26)$$

URN38. This generator is another variation of URN36 for implementation on computers with 16-bit word size:

$$x_i = 5^{13}x_{i-1} \bmod 2^{15}$$
$$= 5^{13}x_{i-1} \bmod 32,768. \qquad (3.5.27)$$

URN38 was the method used in Problem 1.1 (not necessarily because it was good, but because it could be implemented on microcomputers that use small word length, with the methods of Section 1.2.3).

Detailed testing results are given in Dudewicz, Karian, and Marshall (1985) and are summarized in Table 3.5–2.

URN39. (A "perfect" random number generator?) Rey (1990) gave the following "perfect" random number generator.

$$x_0 = 2/(1 + \sqrt{5}),$$
$$x_i = (x_{i-1} + (i + x_{i-1})\sin(i)) \bmod 1. \tag{3.5.28}$$

This generator is still under study (Dudewicz and Bernhofen (1998)) and the results to date are encouraging except for problems of accuracy associated with the computation of the sine function with large arguments.

URN40 and URN41. (Random number generators with very large periods.) In recent years there has been considerable interest in random number generators with very large periods. As was pointed out in the first paragraph of this chapter, some computers can process in excess of 10,000,000 random numbers per second. Hence, it is understandable that generators with periods in excess of 10^7 would be desired. As we noted in the beginning of Section 3.5 (see the first point of the recommendations), any generator being considered for use should have a period of at least 10^9. We have also warned (see note at the end of Section 3.2) that a large period is not sufficient. For example, the generator (3.2.1) with $a = 1$, $c = 1$, $x_0 = 0$, and $m = 2^{500}$ will have period $2^{500} \approx 3 \times 10^{150}$, but its numbers are so non-random as to be useless. Thus, careful consideration of generators that boast "astronomic periods" is necessary lest we succumb to the latest snake-oil remedy. Here we consider two recently proposed generators that have periods of about $2^{95} \approx 4 \times 10^{28}$ and $2^{62} \approx 5 \times 10^{18}$.

URN40. This generator, called "KISS" (for Keep It Simple, Stupid) in a preprint by Marsaglia and Zaman, is claimed to have "impeccable randomness" and to have " ... passed extensive tests of randomness, including all those in our battery of tests DIEHARD." The generator uses the three integer sequences

$$w_i = 69069w_{i-1} + 23606797$$
$$y_i = y_i(w_i + R^{17})(w_i + L^{15}) \tag{3.5.29}$$
$$z_i = z_i(w_i + L^{18})(w_i + R^{13})$$

where y_i and z_i are kept as binary vectors and R and L are shift operators of the type used in URN03 (see 3.5.4)). The integer sequence produced is

$$x_i = w_i + y_i + z_i \bmod 2^{32}. \tag{3.5.30}$$

As will be seen below, our tests indicate this generator is rejected by the TESTRAND tests (it fails both serial pairs tests). This suggests that caution be used until truly extensive testing, of the TESTRAND type, has been performed on the "very large period" generators now being proposed. We plan to publish results on this aspect of a number of such generators in our forthcoming book (Karian and Dudewicz (1999)).

URN41. This is a linear congruential generator (described in (3.2.1)) with

$$m = 2^{64}, \quad a = 39468974943, \quad x_0 = 123456789, \quad c = 0.$$

It will be shown below that this generator passes the TESTRAND tests. The results of the spectral test for this generator are:

$$r_2 = 1.85, \quad r_3 = 2.23, \quad r_4 = 3.69, \quad r_5 = 2.94,$$
$$r_6 = 3.92, \quad r_7 = 3.41, \quad r_8 = 4.19, \quad r_9 = 4.94.$$

By our criterion

$$r_i \leq 5, \text{ for } i = 2, 3, \dots 9,$$

this generator shows very good spectral performance.

URN41 is of particular value to users of 64-bit computers (these include most current workstations from Sun Microsystems and the Digital Equipment Corporation) because it can take advantage of the 64-bit architecture of these computers and deliver very fast execution times. The next desktop computer chip from the Intel Corporation, scheduled for late 1998 or 1999 will have a 64-bit design. In a few years 64-bit desktop computers will be commonplace, making URN41 universally accessible through standard programming languages. The code for URN41, along with that of several other high-quality generators, is given at the end of this section.

3.5.2 Tests of Random Number Generators and Their Results

High-quality random number generators are essential in modern computer-based scientific and technological investigations. For example, such generators are required in simulation and Monte Carlo studies, statistical bootstrapping, many expert systems and artificial intelligence systems, and computational physics, as well as in many other settings. Hundreds of random number generators have been proposed in a wide-ranging literature;

of these, 41 have been tested using TESTRAND, a software package containing the most extensive series of random number quality tests available. Of the 41 evaluated generators, 13 passed the tests. These generators are: URN02, URN03, URN12, URN13, URN14, URN15, URN22, URN30, URN35, URN36, URN37, URN39 and URN41.

What we have referred to as "the TESTRAND tests" are the chi-square on chi-square tests in the format originally given by Dudewicz and Ralley (1981, p. 141). That is, a number of tests are performed, each of which results in an approximately chi-square distributed statistic under the hypothesis of randomness of the pseudo-random number sequence being tested. Each test is replicated 1000 times and a chi-square test on the 1000 chi-square values is performed, as detailed above at the end of Section 3.3. **The tests used are 19 in number**, all described in Section 3.3. We now list them for ease of reference, along with their parameter choices:

1.	UDISTB	11.	POKER (HAND 4, PART. 4)
2.	COUPON D=5	12.	POKER (HAND 5, PART. 6)
3.	COUPON D=5, FR(100*R)	13.	POKER (HAND 5, PART. 4)
4.	COUPON D=10	14.	RUNS UP R
5.	GAP BELOW MEAN	15.	RUNS UP FR(10*R)
6.	GAP ABOVE MEAN	16.	RUNS UP FR(100*R)
7.	GAP (.333,.667)	17.	SERIAL PAIRS 3×3
8.	PERMUTATION 3's	18.	SERIAL PAIRS 10×10
9.	PERMUTATION 4's	19.	SERIAL PAIRS 20×20.
10.	PERMUTATION 5's		

For each test, the CSCS test statistic is at a certain percentage point of the chi-square distribution with 99 degrees of freedom (which is the appropriate distribution since the hypothesis that the 1000 values are chi-square is tested with 100 cells, with boundaries for the cells having the appropriate degrees of freedom). **In evaluating a generator, we reject the generator if any of the 19 tests yields a CSCS statistic at or beyond the 99% point**. In addition, **the number of CSCS statistics at or beyond the 95% point is listed; a generator is rejected if this number is 2 or more**. Finally, for those generators where it is appropriate, the spectral test (discussed in Section 3.4) is used and **generators that fail the spectral test are also eliminated**. Table 3.5–2 summarizes the results of these tests. Detailed results for URN01 through URN20 are given in Dudewicz and Ralley (1981, pp. 141–142); for URN21 through URN30 in Mathews (1986, pp. 89–91); in Table 3.5–3 for URN31 through URN35; in Dudewicz, Karian and Marshall (1985) for URN36 through URN38; in Dudewicz and Bernhofen (1998) for URN39; and in Table 3.5–3 for URN40 and URN41.

The CSCS test results are based on 1000 chi-square values for a batch of 10,000 random numbers each. The numbers reported in Table 3.5–3 are the probabilities below the CSCS test statistic for each test.

Table 3.5–2. CSCS tests on 1000 batches of 10,000 numbers.

	URN										
	01	02	03	04	05	06	07	08	09	10	11
No. $\geq$ .99	0	0	0	1	3	2	3	3	3	19	1
No. $\geq$ .95	0	0	1	1	6	4	5	4	4	19	3
Passes?	Y[†]	Y[‡]	Y	N	N	N	N	N	N	N	N

	URN										
	12	13	14	15	16	17	18	19	20	21	22
No. $\geq$ .99	0	0	0	0	0	1	1	0	0	13	0
No. $\geq$ .95	0	1	0	1	2	1	1	2	2	13	1
Passes?	Y	Y	Y	Y	N	N	N	N	N	N	Y

	URN										
	23	24	25	26	27	28	29	30	31	32	33
No. $\geq$ .99	1	8	1	19	11	0	19	0	4	0	0
No. $\geq$ .95	3	10	3	19	14	0	19	0	5	1	3
Passes?	N	N	N	N	N	Y[†]	N	Y	N	Y[†]	N

	URN							
	34[††]	35	36	37	38[‡‡]	39	40	41
No. $\geq$ .99	5	0	0	0		0	1	0
No. $\geq$ .95	5	1	0	1		1	2	1
Passes?	N	Y	Y	Y	N	Y	N	Y

[†]Fails the spectral test.

[‡]Fails the entropy-uniformity test.

[††]Tests on 500 batches of 10,000 numbers, due to the long generation times for URN34.

[‡‡]Due to very poor performance on both the Coupon D=5 tests and on the Runs-Up FR(100*R) test, and large computer time requirement, the full tests were not run on URN38 (as it was clear they would be failed).

The time required to generate numbers is also a consideration, though this is relative, since a slow generator may be speeded up through better programming. The passing generators are ranked according to their execution times. While differences among computers make absolute timings of little direct interest, unless we are using the same computer, relative timings are not usually affected markedly. For the generators that passed the above testing (as noted in Table 3.5–2), in Table 3.5–4 we give timings relative to the fastest generator in the timing group for each passing generator. The entries in Table 3.5–4 are the ratios of the (time per random number with i-th passing generator) to the (time per random number for the fastest generator in the passing group). Since URN41 is the only generator of quality in its category of large period generators, its timing data is not included in Table 3.5–4. The CPU times for generating 10^7 numbers with URN41 range from 1.8 seconds (on a Digital Equipment Corporation Alpha workstation) to 4.9 seconds (on a 233 MhZ Pentium II computer).

Table 3.5–3. CSCS test results for URN31 through URN35.

Test	URN				
	31	32	33	34[†]	35
1. UDISTB	.72	.92	.66	.99	.62
2. COUPON D=5	.36	.61	.61	.57	.10
3. COUPON D=5, FR(100*R)	.99	.57	.42	.39	.13
4. COUPON D=10	.28	.30	.71	.53	.34
5. GAP BELOW MEAN	.93	.87	.22	.47	.12
6. GAP ABOVE MEAN	.77	.94	.12	.45	.93
7. GAP (.333,.667)	.84	.96	.97	.73	.74
8. PERMUTATION 3's	.93	.82	.51	.21	.73
9. PERMUTATION 4's	.76	.23	.46	.61	.62
10. PERMUTATION 5's	.60	.80	.95	.50	.27
11. POKER (HAND 4, PART. 4)	.77	.79	.15	.10	.79
12. POKER (HAND 5, PART. 6)	.51	.36	.61	.40	.97
13. POKER (HAND 5, PART. 4)	.32	.76	.40	.52	.58
14. RUNS UP R	.99	.14	.18	.59	.20
15. RUNS UP FR(10*R)	.75	.03	.93	.81	.33
16. RUNS UP FR(100*R)	.99	.94	.46	.99	.91
17. SERIAL PAIRS 3x3	.05	.61	.05	.99	.77
18. SERIAL PAIRS 10x10	.98	.66	.97	.99	.94
19. SERIAL PAIRS 20x20	.99	.83	.53	.99	.66

[†]Tests on 500 batches of 10,000 numbers, due to the long generation times of URN34.

As a consequence of an important result of Dudewicz, van der Meulen, SriRam, and Teoh (1995), URN02 fails the combination chi-square entropy-uniformity test (see Section 3.3.9) because it lacks the proper distribution

Table 3.5–4. Relative timings of some generators chosen in Table 3.5–2.

	02	03	12	13	14	15
URN01–20	2.8	1.1	4.4	1.0	3.7	2.3
	22	30				
URN21–30	1.0	3.2				
	35					
URN31–35	1.0					

of the $H_{m,n}$ statistic based on samples of size $n = 10,000$ with $m = 4$. In addition, Bernhofen, Dudewicz, Levendovszky, and van der Meulen (1996) found that URN35, which passed the TESTRAND tests and claims to have a period of

$$32362 \times 31726 \times 31656/4 = 8.12544 \times 10^{12},$$

repeats values 1,434 times in the first 10,000 numbers generated. No periodicity seems evident as the repetitions were traced through a sequence of random numbers and seem to occur randomly. While (with up to 8 decimal places of computation) any generator could repeat even though there may not be repetitions with larger number of digits, the large frequency of repetitions raises serious doubts about the quality of URN35. This generator was originally developed for use on 16-bit computers; therefore, its coding is tailored for computers with limited memory and perhaps it should only be used in the setting for which it was intended. For this reason we have not included it on our list of recommended generators below.

Thus, **based on the quality of numbers, generators 03, 12, 13, 14, 15, 22, 30, 36, 37, 39, and 41 appear fine for use.** Of these **the fastest include URNs 03, 13, and 22**, though reprogramming could substantially speed up some of the others. As a debugging aid of those who may program or reprogram their own version of one of these generators, and as a source for a small quantity of good random numbers, 100 random numbers (with the input used in their generation) are given for each of these generators in Appendix G. URN39 has been excluded because it has problems of breakdowns of its sequence when many numbers are generated. A suitable code for implementing generator URN39, without extended-precision arithmetic, is still under development—final results will be in Dudewicz and Bernhofen (1998). Note that URN35 was very slow and would have a time ratio of nearly 50.0 in comparison with other passing generators.

Many users have told us that having code for some of the best generators would be of considerable value and that in some cases they have continued to use rejected generators due to problems associated with developing programs for good generators. Therefore, below we give the code in C and/or FORTRAN for some of the best generators. Minor modifications may be necessary to conform these programs to compilers that are being used. If properly implemented, these programs should produce the first 100 numbers given in Appendix G, except possibly at the least significant digit or two because of differences in floating-point arithmetic. Figure 3.5–5 shows a C program for the generator URN03 and Figures 3.5–6 and 3.5–7 give FORTRAN programs for URN36 and URN37, respectively. To make sure that readers have access to a generator with a long period, we give, in Figures 3.5–8 and 3.5–9, respectively, the FORTRAN and C codes for URN41.

```c
void rstrt(int *ix, int *jx)
{    if (*ix != 0) *ix = (*ix | 1);
     if (*jx != 0) *jx = (*jx & 0x000007ff) | 1;     }

void urn03(int *ix, int *jx,double *x, int nbatch)
{    int r2,r4,r5,r6,r7; double r8;
    r2  =  nbatch;   r4  =  0;   r5  =  *ix;   r6  =  *jx;
    while (r2 != 0)
       {  r7  =  r6;
          r7  =  r7 >> 15;
          r7  =  r7 & 0x0001ffff; /*convert arith. shift to logical*/
          r6  =  r6 ^ r7; /*r6 = r6 XOR r7*/
          r7  =  r6;
          r7  =  r7 << 17;
          r6  =  r7 ^ r6;
          r5  =  (69069 * r5);
          r7  =  r5 ^ r6;
          r8  =  (double) r7; /*convert to double prec float pt. */
          if (r8 < 0.0)
              r8 = r8 + 4294967296.0;
          r8 = r8 / 4294967296.0;
          *(x+ nbatch-r2)  =  r8;
          r2-=1;
       }
    *ix  =  r5;   *jx  =  r6;   /*save for future use*/
}
```

Figure 3.5–5. C code for URN03; procedure `rstrt`
must be executed, followed by `urn03`.

```
NBATCH = 100
MULT=1220703125
INTIN(1) = 32767
INTIN(2) = INTIN(1) /32768
    J = INTIN(1) * 65536
    IF (J .LT. 0) J = J + 2147483647 + 1
INTIN(3) = J /65536
INTIN(4) = MULT /32768
    J = MULT * 65536
    IF(J.LT.0) J = J + 2147483647 + 1
INTIN(5) = J/65536


SUBROUTINE URNWN(X,NBATCH,INTIN(1),INTIN(2),INTIN(3),
        INTIN(4),INTIN(5))
INTEGER JEN, L16CB2,R20SUM,R5B1C1,B1,B2,C1,C2
DIMENSION X(1)
INTEGER JEN, L16CB2,R20SUM,R5B1C1,B1,B2,C1
C1 = INTIN(4)
C2 = INTIN(5)
B1 = INTIN(2)
B2 = INTIN(3)
DO 10 I=1,NBATCH
    J2 = (B2*C1 + B1*C2) * 2048
    IF (J2.LT.0) J2 = J2 + 2147483647 + 1
    R20SUM = J2/2048
    J3 = B1*C1*67108864
    IF(J3.LT.0) J3 = J3 + 2147483647 + 1
    R5B1C1 = J3/67108864
    L16CB2 = (C2*B2)/32768
    J4 = ((L16CB2+R20SUM+R5B1C1*32768) * 2048)
    IF(J4.LT.0) J4 = J4 + 2147483647 + 1
    B1 = J4/2048
    JEN = B1/32
    J5 = C2*B2*65536
    IF(J5.LT.0) J5 = J5 + 2147483647 + 1
    B2 = J5/65536
    X(I) = FLOAT(JEN)/32768.0
10   CONTINUE
INTIN(2) = B1
INTIN(3) = B2
RETURN
```

Figure 3.5–6. FORTRAN code for URN36.

```
          NBATCH = 100
          IX0 = 32767

          SUBROUTINE URNWN(X,NBATCH,IX0)
          DIMENSION X(1)
          INTEGER J
          J = IX0
          DO 10 I =1, NBATCH
               J= J* 1220703125
               IF (J.LT.0) J = J + 2147483647 + 1
               X(I) = FLOAT(J) * .4656613E-9
   10     CONTINUE
          IX0 = J
          RETURN
```

Figure 3.5–7. FORTRAN code for URN37.

```
          PROGRAM URN41
          INTEGER*8 X,A,M,I
          REAL R,MM, SUM
          X=123456789
          A=39468974943
          M=9223372036854775807
          MM=2.0*M+2
          DO 10 I=1,100
          X=A*X
          IF (X .LT. 0) THEN
               R=1+X/MM
          ELSE
               R=X/MM
          END IF
          PRINT *,R
   10     CONTINUE
          END
```

Figure 3.5–8. FORTRAN code for URN41.

```
void URN41(long long NumNumbers)
{
long long x,a,m,i;
double r, mm, sum=0.0;

    x=123456789;
    a=39468974943;
    m=9223372036854775807;
    mm=2.0*(double)m+2.0;
    for (i=0; i<NumNumbers;i++)
    {    x=a*x;
         if (x < 0)
                { r=1.0+(double)x/mm; }
         else
                { r=(double)x/mm; }
         printf("%11.10f ",r);

    }
}
```

Figure 3.5–9. C program for URN41.

Some remarks on shuffling. There is a myth in simulation that, if a generator produces numbers $U_1, U_2, \ldots$ and those numbers are "shuffled," then the resulting sequence will be better (i.e., more random). That this is wrong can be seen from our testing results: URN01 passed the CSCS tests (though it failed the spectral test and is thus not recommended for use), but its shuffled version, URN04, failed; URN22 passed the CSCS tests, but its version with shuffling, URN23, failed.

This does not mean that all generators which incorporate shuffling are undesirable, for in fact URN14 uses shuffling and passed the testing. Its unshuffled version was not tested; it would be an interesting project to test and compare these versions. However that turns out, we know that URN14 is suitable for use. In fact, it is clear theoretically that a good shuffling method could have such benefits as:

1. Numbers could be repeated without repeating the whole sequence (which is desirable, since true random numbers would, with low but nonzero probability, have $U_{i+1} = U_i$ occasionally).

2. Results on the interplanar distance test would be markedly improved.

To illustrate the second point, suppose that we have available a shuffle so effective that when one takes the successive numbers (U_i, U_{i+1}) from the shuffled sequence, all pairs (U_i, U_j) are possible. Then for the generator studied in Figure 3.4–4, which previously generated only 10 possible pairs (which could all be contained in just 3 parallel equidistant planes at spacing of 0.3162), we now find there are 100 possible pairs (to contain all of which one needs 10 parallel equidistant planes at spacing $1/11 = 0.0909$). This is shown in Figure 3.5–10. Note that before 3 planes contained all 10 points, whereas now 10 planes will be needed to contain the 100 possible pairs (U_i, U_{i+1}). To achieve results this good **with an** m **large enough for practical use** may require a large table to shuffle in; however, with current computer speed and memory, we expect that this will be possible in the years ahead, resulting in speed- and memory-feasible generators with an astronomical period **and** the ability to pass statistical and spectral testing without qualification. This will be especially important in applications where vectors of numbers $(U_1, U_2, \ldots, U_n)$ are used for a relatively high n, which is where the spectral characteristics of current generators are often deficient.

Recall that our **summary recommendations on generators** (recommended, and not recommended) are given immediately before "Some remarks on shuffling."

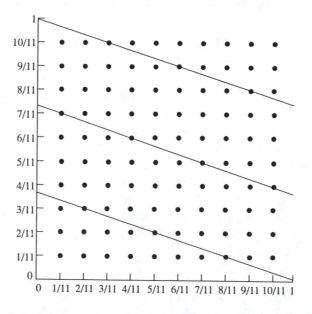

Figure 3.5–10. Results of an effective shuffle on the spectral structure of the generator in Figure 3.4–4.

3.6   Notes on the Use of **TESTRAND**

On pages 267 and following, Dudewicz and Ralley (1981) give sample TES-TRAND programs with their full output. (On pages 268–270 is a program with full JCL for IBM 360/370 and look-alike computers. With a "cataloged procedure" (EXEC TESTRAND) to invoke TESTRAND, the user need not give most of the JCL, and the program on page 271 will accomplish the same functions.) Those programs were appropriate to the installation at The Ohio State University.

To allow access from terminals, at Syracuse University one could in the past submit "batch mode" TESTRAND programs by

1. First, creating a file; e.g., one named FILE: TEST1 in the first example given in Figure 3.6–1.

2. Next, saying SUBMIT FILE MVS.

The examples illustrated in Figures 3.6–1, 3.6–2, and 3.6–3 show programs with the following characteristics and functions:

FILE: TEST1. This program generates and prints 100 random numbers from each of the 20 random number generators URN01 ... URN20 built into the TESTRAND package.

FILE: TEST2. This program generates 1000 numbers from random number generator URN20 and prints the first 100 of them. It then performs the chi-square on chi-squares test (TST01) on 1000 samples of 1000 random numbers each. This is NOT an appropriate program to use for practice in getting a TESTRAND program to run, as it takes a moderate amount of computer time. Program TEST1 above should be used instead.

FILE: TEST4. This program performs a complete test of a microcomputer random number generator, and provides results that correspond to those in Dudewicz, Karian, and Marshall (1985) for their generator numbered (2). This program shows how to enter a generator that is **not** in the TESTRAND package and then test it. This is the most costly of the three programs to run, and should not be used as a test. It is shown for reader ease in constructing a program to test *other* random number generators. Most computer centers and packages have at least

```
FILE: TEST1 MVS A S.U. ACADEMIC COMPUTING SERVICES VM/SP
//TEST1 JOB (           ),MJRAC,REGION=1500K
/*ROUTE PRINT SUVM.TESTRND
//    EXEC TESTRAND
//FORT.SYSIN    DD    *
      CALL TESTER
      STOP
      END
//GO.SYSIN DD *
URN01              100NO    100
URN02              100NO    100
URN03              100NO    100
URN04              100NO    100
URN05              100NO    100
URN06              100NO    100
URN07              100NO    100
URN08              100NO    100
URN09              100NO    100
URN10              100NO    100
//
```

Figure 3.6–1. A program to generate and print random
numbers from URN01, ..., URN10.

```
FILE: TEST2 MVS A S.U. ACADEMIC COMPUTING SERVICES VM/SP
//TEST2 JOB (    ,    ),MJRAA620,REGION=1500K
/*ROUTE PRINT SUVM.MJRAA620
//    EXEC TESTRAND
//FORT.SYSIN    DD    *
      CALL TESTER
      STOP
      END
//GO.SYSIN DD *
URN03              100YES   100
TST01    10000     1                    CSCS  1
01T02
//
```

Figure 3.6–2. A program to generate 1000 numbers from
URN03, print 100, and use TST01.

```
//TEST4 JOB (    ,     , 120,40,0),TESTRND,REGION=2000K
/*ROUTE PRINT SUVM.TESTRND
// EXEC TESTRAND,TIME=120
//FORT.SYSIN DD *
      COMMON/OWNPRM/INTIN(200),REALIN(200),IA,IR
      IA=1
      IR=0
      INTIN(1)=32767
      CALL TESTER
      STOP
      END
      SUBROUTINE URNWN(X,NBATCH)
      COMMON/OWNPRM/INTIN(200),REALIN(200),IA,IR
      DIMENSION X(1)
      INTEGER J
      J=INTIN(1)
      DO 10 I=1,NBATCH
         J=J * 1220703125
         IF (J .LT. 0) J = J + 2147483647 + 1
         X(I) = FLOAT(J) * 0.4656613E-9
10    CONTINUE
      INTIN(1)=J
      RETURN
      END
//GO.SYSIN DD *
URNWN            10000YES100                  2
TST03   10000
TSTM4   5**13       2**35           2,3,4,5,6,7,8,9,     Q
TST01   1000        03              CS2KS211111110       11
01T02
01T04   1           10000
01T05   333         666             1       10000
01T06   10000       001             1
01T07   1           10000
01T08   1           10000
01T09   1           10000
01T10   1           10000
01T11   1           2000            01      1
//
```

Figure 3.6–3. A program to perform a complete test for generator (2) of Dudewicz, Karian, and Marshall (1985). (This is a good generator—it passes the TESTRAND tests.)

one generator, most of which are awful and should be tested (and, if their awfulness is confirmed, discarded and replaced with one that has passed the TESTRAND tests).

Note that the columns in which the entries appeared were crucial to the correct operation of the program.

State-of-the-art TESTRAND programs (as opposed to the above, described for historical reasons so the literature of TESTRAND test results will be clear to the reader) are described in the last paragraph of Section 3.1.

Problems for Chapter 3

3.1 (Section 3.1) Flip a coin repeatedly. Each time a "Head" occurs, record a "1," and each time a "Tail" occurs record a "0." Record the results of 100 flips, in the order in which they occurred. (For example, if flip number 1 results in Heads, flip no. 2 in Tails, flip no. 3 Tails, ..., flip no. 100 is Heads, then your record will be: 100...1.)

3.2 (Section 3.2) By direct enumeration, find the period of the following mixed congruential generators:

   a. $x_{i+1} = 5x_i \bmod 7, \quad x_0 = 1.$

   b. $x_{i+1} = 11x_i \bmod 32, \quad x_0 = 1.$

   c. $x_{i+1} = 121x_i + 567 \pmod{1}000, \quad x_0 = 0.$

3.3 (Section 3.2) Use the results in Section 3.2 to study the periods of the generators in Problem 3.2.

3.4 (Section 3.2) The TSTM2 routine of the TESTRAND package calculates the period of generators of the form (3.2.1). If this package is available at your computer center, use it to find the periods of the generators in Problem 3.2.

Note: For this problem the items to be submitted include: the input and call for execution, and the output (labeled to clearly show the periods desired).

3.5 (Section 3.3) For the 100 random bits you generated by coin flipping in Problem 3.1, perform the following tests of randomness:

a. The Uniformity of Distribution Test, with the two categories "Number is 0" and "Number is 1" (which have probability .50 each under the null hypothesis).

b. The Coupon Collector's Test with $M = 2$. Note that here your categories are $Q = 2$, $Q = 3, \ldots$, and $Q \geq L$, with L chosen so that the chi-square approximation to the test statistic T is reasonable.

c. The Gap Test for "runs below the mean," and also for "runs above the mean." (If you find GN gaps in the 100 numbers, then use all categories j as individual categories, as long as $GN * p_j \geq 4$. Collect the other categories together into one category for the test.)

d. Successive pairs of the bits are (under randomness) equally likely to take on the values 00, 01, 10, 11. Use this fact to test randomness with a test similar to the Permutation Test with $T = 2$ (which cannot be directly used here since ties are possible).

Similarly, successive triples of the bits may take on any of the values 000, 001, 010, 100, 011, 101, 110, 111. Use these categories to test randomness with a test like the Permutation Test for $T = 3$. In each case, show details of your calculations, including the 100 numbers you are starting with, the derived numbers, the category counts, the chi-square contributions (i.e., the terms in the sum that yields T), etc.

3.6[1] (Section 3.4) For the generator (3.4.1) with $c = 0$, $a = 6$, and $m = 11$, perform the serial correlation test using formulas (3.4.2) and (3.4.3). What is the estimated correlation, and what is the error bound on it? Does the generator pass this test?

3.7[1] (Section 3.4) For the generator of Problem 3.6, use routine TSTM1 of the TESTRAND package to perform the serial correlation test.

3.8[1] (Section 3.4) For URN13, which has $c = 0$, $a = 663,608,941$, and $m = 2^{32}$, we found an estimated correlation of .0000000015 with an error bound of .1545. Use TSTM1 of the TESTRAND package to perform the serial correlation test.

3.9[1] (Section 3.4) For the generator $x_{i+1} = 6x_i \bmod 11, x_0 = 1$,

a. Perform the serial correlation test, using (3.4.2) and (3.4.3).

---

[1]Give input and output, as well as answers, whenever a TESTRAND program is used.

b. List all pairs (U_i, U_{i+1}) that this generator can provide. How many such pairs are there?

c. Plot the pairs of Part b on graph paper.

d. Show (on a copy of the graph of Part b) all sets of parallel equally spaced planes that contain all of the points. For each such set, find the interplanar distance. What is d_2 for this generator? What is the ratio r_2 in comparison with other generators that have modulus $m = 11$?

3.10[1] (Section 3.4) Use routine TSTM4 of the TESTRAND package to

a. Test the generator with $c = 0$, $a = 7$, $m = 11$ in 2 through 9 dimensions.

b. Test the generator with $c = 0$, $a = 6$, $m = 11$ in 2 through 9 dimensions.

c. Test the generator with $c = 0$, $a = 663,608,941$, $m = 2^{32}$ (i.e., URN13) in 2 through 9 dimensions. Tell how many iterations were performed (I), and how many would be needed for an exhaustive search (E), in each case.

Note: In the specifications for the required columns for use of TESTRAND test TSTM1 on p. 225 of Dudewicz and Ralley (1981), under "REMARKS" for use when one or more of L, C, ZZ exceeds your machine's allowable maximum integer value, it should state the columns as

$$
\begin{array}{ll}
1 - 12 & \text{for IPARM(1)} \\
13 - 24 & \text{for IPARM(2)} \\
25 - 36 & \text{for IPARM(3)}
\end{array}
$$

3.11[1] (Section 3.4)

a. For the generator $x_{i+1} = 6x_i \bmod 10$, with $x_0 = 5$, use routine TSTM4 to test the generator in 2 dimensions. Interpret a graph comparable to that in Figure 3.4–4.

---

[1]Give input and output, as well as answers, whenever a TESTRAND program is used.

b. As in Part a, but for the generator $x_{i+1} = 4x_i \bmod 5$. (Hint: If all pairs are contained in one line, intuitively we should have $d_2 = +\infty$ in order to denote this as an undesirable situation, since small values of d_2 are associated with good generators.)

3.12[1] (Section 3.4) A defect of the spectral test was noted at the end of Section 3.4. Namely, it is conceptually possible that for some a,c,m all of the pairs will be contained on L planes $y = c_1 + c_0x$, $y = c_2 + c_0x$, $\ldots$, $y = c_L + c_0x$ such that the interplanar distance d_2 is very small, such as $d_2 = 0.00001$, but L will be relatively small (so that these planes will be "bunched up" in the unit square, and the generator will thus be terrible, even though its d_2 would lead us to think that it was a good generator). Find a,c,m such that this occurs. Hint: A theoretical analysis will use the fact that $x_{i+1} = ax_i + c \bmod m$, while an example may be sought by manipulating a,c for small m and seeking the desired characteristics.

3.13 (Section 3.5) New generators are constantly appearing in the literature, usually with a minimal amount of testing that does not satisfy the standards of the TESTRAND testing approach. From recent journals, find one such generator (in addition to the generators URNxx for which results are given in this section), and test it using the TESTRAND tests. Be sure to submit all of

a. The input and output of the TESTRAND package.

b. A summary, with results, of the 19 CSCS tests, and an evaluation of them.

c. Considerations of speed, memory requirements, and period of the generator.

d. Your overall evaluation of the generator.

Of course, give the full reference in which the generator first appeared, and give particular attention to comparing your results and recommendations with those of the author of that reference.

3.14 (Section 3.5) Find the period for the generator URN36.

3.15 (Section 3.5) Find the period for the generator URN37.

---

[1]Give input and output, as well as answers, whenever a TESTRAND program is used.

3.16 (Section 3.5) Find the period for the generator URN38. Perform the full TESTRAND tests and use them to complete Table 3.5–2.

3.17 (Section 3.5) Apply the spectral test to the generators URN36, URN37, and URN38. Discuss the changes (if any) needed in the recommendations regarding these generators based on the results of the spectral test.

Chapter 4

Random Variable Generation

In Chapter 3, we saw how to produce a sequence of random variables $U_1, U_2, U_3, \ldots$ that are independent and each of which has a uniform probability density function on the interval of real numbers $(0, 1)$, i.e., what is called a sequence of **random numbers**. (We also say that $U_1, U_2, U_3, \ldots$ are uniform random variables or that they have a uniform distribution.) In this chapter we deal with the question of **how to obtain random variables from nonuniform distributions**. We need to do this because **the random variables required to run a simulation** (such as the interarrival times in Section 1.2, or service times, etc.) **typically do not follow the uniform distribution**. However, in this chapter we will see that from $U_1, U_2, U_3, \ldots$ we can often obtain random variables with the nonuniform distributions we need. A simple example of this was given in the introduction to Chapter 3 where we obtained exponential random variables with mean 2 by letting $X_i = -2\ln(U_i)$.

In Section 4.1 we consider the question of **how to select the needed (nonuniform) distribution for the random variables**. Sections 4.2 through 4.6 consider both general and special methods of providing random variables from the distribution we have selected.

The information provided in this chapter will be very useful for those who are writing their own routines for generating nonuniform random variables. For those using a special simulation language, the language itself usually provides such routines (but may allow the use of a user-provided routine). Since there are many "bad" algorithms, including some that we discuss in this chapter and label as such, we caution the reader as follows: before using a package's routine for a nonuniform variable, identify its algorithm. If it is a bad one, replace it with a well-tested, user-provided routine for a good algorithm that we provide.

4.1   Selection of a Distribution

If we need $X_1, X_2, \ldots$ which are independent and identically distributed random variables for some aspect of a simulation, we first must know **what statistical distribution these random variables are to have.** A random variable X has a probability of having a value no larger than 3.2—that is, $P(X \leq 3.2)$—and this probability is the value of the distribution function of X at 3.2. There is such a probability for every number x we might specify, so we have in general

$$F_X(x) = P(X \leq x), \qquad -\infty < x < +\infty, \qquad (4.1.1)$$

which is called the **distribution function of** X. **The question we need to answer can therefore be restated as "what distribution function are these random variables to have?"**

One way to deal with this question is to **avoid the question by using real-world data sequences (historical data) to run the simulation.** For example, if we wish to simulate the operation of a banking system, we can gather information on interarrival times of customers, say $X_1, X_2, \ldots, X_N$, and then use these to run our simulation. **The problems with this approach include the following:**

1. The number of random variables N needed to run a simulation is typically very large; for example, to run a simulation of one year's operation 10 times, we will need 10 years of customer information. Typically, we cannot wait the 10 years needed to gather this information.

2. If we did have 10 years of records on file for the needed information, often there would have been changes in the patterns over such a long time span, so that much of the data would no longer be representative of the current experience; hence, the data would not be appropriate for use.

3. Even if we did have large enough N from our records and the data distribution had not changed over the time period in which it was recorded, we would often desire to look at the effect of future changes in the data stream (such as more frequent arrivals to the system, to see what workload it could handle). Historical data does not allow us to accomplish this.

4. Even if N were large enough, and the distribution had not changed, and we did not want to look at future changes, our simulation would still allow us to look only at the one stream of data, although we know that in the future we will see data with other peaks and valleys and we will need to know how the system will respond to that random environment. For example, when designing a dam we need to allow for the "100-year flood," and if we have 50 years of data on rainfall, that 100-year flood may well not be represented in the data.

A second answer is to **take the data available, and from it develop an approximation to** $F_X(x)$. One way to do this is to fit a histogram to the data, then sample from the histogram. Some problems with this include:

1. We will never generate a value larger than the largest in the data set upon which the histogram was based.

2. A histogram is "rough," while often we know that the real distribution is "smooth."

For these reasons, often we would want to at least smooth the histogram (for some methods, see Section 4.8 of Dudewicz and Mishra (1988), in particular the "continuous-empiric p.d.f."). **The method** that has been found very appropriate in many simulations, and that **we recommend, is to fit a generalized lambda distribution using the available data; this is treated in detail in Section 4.10 for both univariate and (in Section 4.10.4) bivariate cases.** For full details, see Karian and Dudewicz (1999).

A currently popular **third answer** is to use what is called a "bootstrap" approach: if one has historical data $x_1, \ldots, x_N$, when the next X is needed choose one of $x_1, \ldots, x_N$ at random and use that value for X. With this approach one will never generate any value not in the historical data. For this reason, we strongly recommend against using this approach. Instead, use the Generalized Bootstrap (see Section 6.6 and Karian and Dudewicz (1999)).

The **fourth answer**, and the most frequently used approach, is to select a distribution on theoretical grounds and generate values from that distribution. The justification for this approach is that one may be able to show that, under certain assumptions, a particular distribution will be appropriate for the variables needed. Also, the literature of the application may use a particular distribution. Some problems with this approach include:

1. The theoretical analysis may rest on assumptions that are not met in the application being considered.

2. The literature may just be following the first few papers in the area, which often make arbitrary assumptions to get a "first approximation" (but are afterward often viewed almost as dogma in the area).

One can ameliorate these drawbacks by using not only the specific theoretical distribution to run the simulation, but also others that will allow one to see if the results of the simulation are sensitive to the distribution being assumed. Our recommendations are:

1. Use a theoretical distribution only for (a) theoretical studies (such as a study attempting to determine what characteristics the system will have under the particular distribution), and (b) studies where the theoretical distribution is a good fit to the real data $X_1, \ldots, X_N$.

2. Otherwise, use a distribution fitted to the data (the generalized lambda distribution will often be appropriate).

3. Do not use the historical data directly and do not sample from the histogram; do use the generalized bootstrap method (see Section 6.6), but not the naive bootstrap method.

4.2   Generating Univariate Random Variables: Inverse Distribution Function Method

Suppose that **we have selected a distribution function** $F_X(\cdot)$ for the random variable of interest **and now wish to generate independent random variables** $X_1, X_2, \ldots$ **each of which has this distribution function**. If $F_X(\cdot)$ is a continuous and increasing function then it has an inverse function, often called the inverse distribution function, or percentile function, denoted by $F_X^{-1}(\cdot)$.

A general continuous and increasing distribution function $F_X(\cdot)$ might look like the function shown in Figure 4.2–1. For each value of the argument x, this function tells us the probability that X takes on a value no larger than x; e.g., $F_X(3.2) = P[X \leq 3.2]$. In the case shown, $F_X(3.2) = .65$. On the other hand, the **inverse distribution function** $F_X^{-1}(y)$ tells us at what value x we will have $F_X(x) = y$; e.g., $F_X^{-1}(.65)$ is the value of x such that $F_X(x) = .65$. In the case shown in Figure 4.2–2, $F_X^{-1}(.65) = 3.2$.

$F_X^{-1}(y)$ is, in fact, a function of y—say, $G(y)$—such that $F_X(G(y)) = y$. To find $G(y)$, turn the page so that Figure 4.2–2 lies on it sideways and

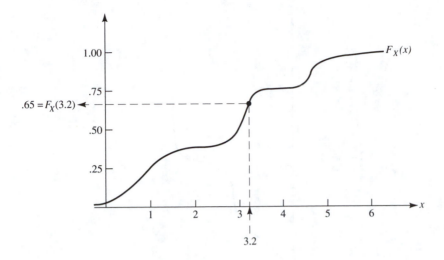

Figure 4.2–1. A typical continuous, increasing distribution function; at $x = 3.2$, $F_X(x) = .65$.

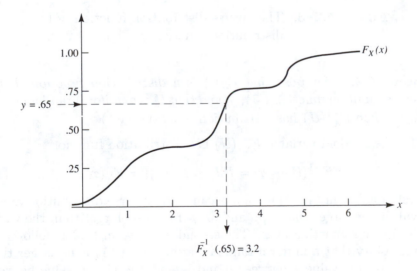

Figure 4.2–2. A typical continuous, increasing distribution for $y = .65$, $x = F_X^{-1}(y) = 3.2$.

look through the page from the other side. You will see the function shown in Figure 4.2–3. The following theorem allows us to generate the random variables $X_1, X_2, \ldots$ when the inverse distribution function is available.

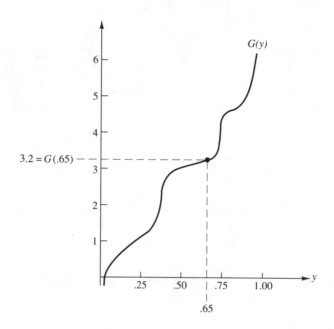

Figure 4.2–3. The inverse distribution function of the
distribution function in Figure 4.2–2.

Theorem 4.2.1. *Suppose that $F_X(\cdot)$ is a distribution function that has
inverse distribution function $F_X^{-1}(\cdot)$, and that U is a uniform random variable
on (0,1). Then $F_X^{-1}(U)$ has distribution function $F_X(\cdot)$.*

Proof: The random variable $F_X^{-1}(U)$ has distribution function

$$P[F_X^{-1}(U) \le x] = P[U \le F_X(x)] = F_X(x). \qquad (4.2.2)$$

The first equality in (4.2.2) follows because the inverse distribution function
at U will be no larger than x, if and only if U is no larger than the value of
the distribution function at x. The second equality in (4.2.2) follows since
the probability that a uniform random variable on (0,1) is no larger than z
equals z (for any value z between 0 and 1, and $F_X(x)$ is a value between 0
and 1).

Corollary 4.2.3. *If $U_1, U_2, \ldots$ are independent uniform random variables
on $(0,1)$, then*

$$X_1 = F_X^{-1}(U_1), X_2 = F_X^{-1}(U_2), \ldots \qquad (4.2.4)$$

are independent random variables each with distribution function $F_X(\cdot)$.

4.2.1 The Exponential Distribution

Suppose we wish to generate random variables $X_1, X_2, \ldots$ which are independent and each have the **exponential distribution with mean** λ. The exponential distribution with mean λ has probability density function

$$f(x) = \begin{cases} \frac{1}{\lambda} e^{-x/\lambda} & \text{if } 0 \leq x \\ 0 & \text{otherwise,} \end{cases} \tag{4.2.5}$$

and hence its distribution function is

$$F(x) = \int_{-\infty}^{x} f(z)\, dz = \begin{cases} 1 - e^{-x/\lambda} & \text{if } 0 \leq x \\ 0 & \text{otherwise.} \end{cases} \tag{4.2.6}$$

To find the inverse distribution function, for any y (between 0 and 1) we need to solve the equation $y = F(x)$ for x in terms of y. To solve

$$y = 1 - e^{-x/\lambda} \tag{4.2.7}$$

we observe that

$$\begin{aligned} e^{-x/\lambda} &= 1 - y \\ -x/\lambda &= \ln(1 - y) \\ x &= -\lambda \ln(1 - y). \end{aligned} \tag{4.2.8}$$

Therefore (for y between 0 and 1)

$$F^{-1}(y) = -\lambda \ln(1 - y). \tag{4.2.9}$$

It then follows from Corollary 4.2.3 that if $U_1, U_2, \ldots$ are independent uniform random variables on $(0, 1)$, **then**

$$X_1 = -\lambda \ln(1 - U_1), \quad X_2 = -\lambda \ln(1 - U_2), \ldots$$

are independent exponential random variables with mean λ.

In the literature one will often find this result stated with $X_i = -\lambda \ln(U_i)$. Since when U_i is uniform on $(0, 1)$ it is also true that $1 - U_i$ is uniform on $(0, 1)$, this result is also valid. This modification, using U_i instead of $1 - U_i$, has been in use because of its slight improvement on speed due to the elimination of a subtraction. **We recommend that this shortcut not be used, for several reasons.** First, as we will see in Section 6.7.3, to increase the precision of comparisons in a simulation, one often attempts to induce correlation. Positive correlation is desired in simple settings where the same input needs to be made to different system configurations. In more complex systems, use of the same random numbers $U_1, U_2, \ldots$ to generate the different inputs may be used to achieve this. However, if one uses U_i instead of $1 - U_i$, the correlation may end up negative instead of positive, decreasing the precision of comparisons in a simulation. Second, the use of shortcuts, particularly when several are combined, can make the computer code unintelligible. Third, by relying on the statement that the code uses "the inverse distribution function method," a user may perform calculations that are not valid when a shortcut has been used in the code in an attempt to speed it up.

4.2.2   The Bernoulli Distribution

Suppose that we wish to generate random variables $X_1, X_2, \ldots$ which are independent and each X_i has the **Bernoulli distribution with probability of success** p $(0 \leq p \leq 1)$. A random variable X has the Bernoulli distribution if it takes on values 0 and 1, and

$$P[X = 0] = 1 - p, \quad P[X = 1] = p \qquad (4.2.10)$$

(and $P[X = x] = 0$ for all x other than 0 and 1). For this discrete distribution the distribution function is

$$F(x) = P[X \leq x] = \begin{cases} 1 & \text{if } x \geq 1 \\ 1 - p & \text{if } 0 \leq x < 1 \\ 0 & \text{if } x < 0, \end{cases} \qquad (4.2.11)$$

which is plotted in Figure 4.2–4. From that plot, we see that **there is no inverse function**. That is, while there is a distribution function since at any value x there is a probability $F(x)$ that a value no larger than x will be obtained, at each y when we come across at height y to try to hit the curve we have a problem: either we do not hit the curve at all (y is "jumped over"), or we hit the curve at a continuum of points (y corresponds to a "flat" part of the curve).

Although there is no inverse function, we can construct a function $G(y)$ such that $P[G(U) \leq x] = F(x)$, by letting

$$G(y) = \begin{cases} 1 & \text{if } y = 1 \\ 1 & \text{if } 1 - p < y < 1 \\ 0 & \text{if } y = 1 - p \\ 0 & \text{if } 0 \leq y < 1 - p. \end{cases} \qquad (4.2.12)$$

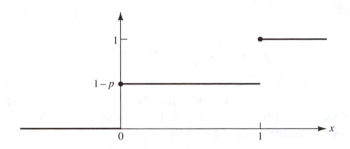

Figure 4.2–4. The Bernoulli distribution function, which is not increasing and is not continuous for all x.

Note that $P[G(U) \leq x] = 0$ if $x < 0$, and $P[G(U) \leq x] = 1 - p$ if $0 \leq x < 1$, and $P[G(U) \leq x] = 1$ if $x \geq 1$; hence, $P[G(U) \leq x] = F(x)$, as claimed. Therefore, $G(U_1), G(U_2), \ldots$ **are independent Bernoulli random variables with success probability** p.

Thus, though there was no inverse function, we constructed a function with the crucial property, namely $P[G(U) \leq x] = F(x)$. We did this by defining $G(y)$ to be the smallest x for which $F(x) = y$ in those cases where y has many xs with $F(x) = y$, such as $y = 1$ and $y = 1 - p$ in Figure 4.2–4. And, for y that has no x with $F(x) = y$, we defined $G(y)$ to be the x value at which $F(x)$ "jumped over" height y, i.e., $F(x^*) < y$ for $x^* < x$ and $F(x) > y$. For further details, see Dudewicz and Ralley (1981, pp. 54–55).

Note that $G(U_i)$ is a very simple function, which can be expressed as

$$G(U_i) = \begin{cases} 0 & \text{if } U_i \leq 1 - p \\ 1 & \text{if } 1 - p < U_i. \end{cases} \qquad (4.2.13)$$

Thus, in implementing this method of generating Bernoulli random variables, one sets X_i to 0 if $U_i \leq 1 - p$, and to 1 otherwise.

4.2.3   The Normal Distribution

Suppose we wish to generate random variables $X_1, X_2, \ldots$ which are independent and each have the **normal distribution with mean** μ **and variance** σ^2, often denoted by $N(\mu, \sigma^2)$. A random variable X has the $N(\mu, \sigma^2)$ distribution if its distribution function is

$$F_X(x) = \int_{-\infty}^{x} f_X(z)\, dz \qquad (4.2.14)$$

where

$$f_X(z) = \frac{1}{\sqrt{2\pi}\sigma} e^{-0.5(z-\mu)^2/\sigma^2}. \qquad (4.2.15)$$

Since it is known that there is no simple closed-form expression for $F_X(\cdot)$, we may correctly assume that it will not be possible to solve $y = F_X(x)$ for $x = F_X^{-1}(y)$ in a usable simple formula.

There are, however, many approximations to $F_X(x)$, such as that of Shah (see Dudewicz and Mishra (1988, p. 144)), which states that, for $\mu = 0$ and $\sigma^2 = 1$,

$$F_X(x) = x(4.4 - x)/10 + 0.5 \qquad (4.2.16)$$

with an error that is at most 0.005 for x between 0 and 2.2. In fact, there are much more accurate approximations to $F_X(x)$, such as that of Milton and Hotchkiss (see p. 161 of Dudewicz and Ralley (1981) for its code in FORTRAN, and pp. 155–160 for a study of its accuracy in comparison with other methods) which has an error of at most 10^{-10} for x between -8 and $+8$. We might therefore conjecture that there would be many approximations to $F_X^{-1}(\cdot)$ for the normal distribution (this is true), and that the best of them may be very accurate.

In fact, an approximation of Milton and Hotchkiss has been shown (see Dudewicz and Ralley (1981, p. 164)) to find $F_X^{-1}(y)$ to within 10^{-5} when y is between .00000030 and .99999970. This implies that if U_i would lead to an X_i between -5 and $+5$ with the exact inverse, it will lead to a value within 10^{-5} of X_i using this method. The accuracy is not as good, though, in the tails of the distribution. For example, if U_i would lead to an X_i near -6, then the approximation yields a value within 10^{-3} of that X_i; and, around -7, one does not even obtain an accuracy of 10^{-1}. **Since we take the view that when one states that normal variables will be used, exact normal variables should be used,** unless no method can produce them, **we do not recommend the above method.** In Section 4.6.2 we will see how to produce exact normal variables.

From the developments of Sections 4.2.1, 4.2.2, and 4.2.3, we can conclude that for some distributions, such as the exponential distribution, the inverse function method works well. With some extensions, it also works in discrete cases such as the Bernoulli distribution. However, **in many important cases $F_X^{-1}(\cdot)$ is not available,** and an overemphasis on the inverse function method may lead us to use approximations when exact methods are available, **as in the normal distribution case.**

Algorithms that generate $N(0,1)$ random variables can be modified to become general $N(\mu, \sigma^2)$ generators. If X is $N(0,1)$, we simply compute

$$Y = \mu + \sigma X \qquad (4.2.17)$$

to obtain a random variable with the $N(\mu, \sigma^2)$ distribution.

4.2.4 The Uniform Distribution

Suppose we wish to generate random variables $X_1, X_2, \ldots$ which are independent and each have the **uniform distribution on the range** (a, b), with $a < b$. Here the distribution function of the random variables is

$$F_X(x) = \begin{cases} 1 & \text{if } x > b \\ \dfrac{x-a}{b-a} & \text{if } a \le x \le b \\ 0 & \text{if } x < a. \end{cases} \qquad (4.2.18)$$

Solving $y = F_X(x)$, we find

$$x = F_X^{-1}(y) = a + y(b - a). \qquad (4.2.19)$$

It then follows from Corollary 4.2.3 that if $U_1, U_2, \ldots$ are independent uniform random variables on $(0, 1)$, **then**

$$X_1 = a + U_1(b - a), \quad X_2 = a + U_2(b - a), \ldots \qquad (4.2.20)$$

are independent uniform random variables on (a, b).

4.2.5 The Binomial Distribution

Suppose we wish to generate random variables $X_1, X_2, \ldots$ which are independent and each have the **binomial distribution with parameters** n (number of trials) **and** p (success probability). A random variable X has the binomial distribution if it takes on values $0, 1, 2, \ldots, n$ and

$$P[X = i] = \binom{n}{i} p^i (1 - p)^{n-i}, \quad i = 0, 1, 2, \ldots, n \qquad (4.2.21)$$

and $P[X = x] = 0$ for all x other than $0, 1, 2, \ldots, n$. For this discrete distribution the distribution function is

$$
F_X(x) = \begin{cases}
1 & \text{if } x > n \\
P[X = 0] + \cdots + P[X = n-1] & \text{if } n-1 \le x < n \\
\vdots & \\
P[X = 0] + P[X = 1] + P[X = 2] & \text{if } 2 \le x < 3 \\
P[X = 0] + P[X = 1] & \text{if } 0 \le x < 1 \\
P[X = 0] & \text{if } 0 \le x < 1 \\
0 & \text{if } x < 0.
\end{cases} \qquad (4.2.22)
$$

From a plot of $F_X(\cdot)$, which looks like Figure 4.2–4 but has jumps at values of x of $0, 1, 2, \ldots, n$, we see that there is no inverse function. However, using ideas from Section 4.2.2 we can find a function $G(y)$ such that $P[G(U) \le x] = F_X(x)$. That $G(y)$ is defined by

$$
G(y) = \begin{cases}
n & \text{if } F_X(n-1) < y \\
n-1 & \text{if } F_X(n-2) < y \le F_X(n-1) \\
n-2 & \text{if } F_X(n-3) < y \le F_X(n-2) \\
\vdots & \\
3 & \text{if } F_X(2) < y \le F_X(3) \\
2 & \text{if } F_X(1) < y \le F_X(2) \\
1 & \text{if } F_X(0) < y \le F_X(1) \\
0 & \text{if } y \le F_X(0).
\end{cases} \qquad (4.2.23)
$$

It is easy to see that $P[G(U) \le x] = F_X(x)$, hence $G(U_1), G(U_2), \ldots$ **are independent binomial random variables with parameters n and p.**

To implement this method, suppose that we create a vector $P(1)$, $P(2), \ldots, P(n+1)$ containing $F_X(0), F_X(1), \ldots, F_X(n)$. To generate X_i we first compute U_i and then search through $P(\cdot)$, an increasing sequence of numbers, until we find the first time that $P(I) < U_i \le P(I+1)$, and set $X_i = I$. Thus, we compare U_i with $P(1)$, if $U_i \le P(1)$ we set $X_i = 0$; otherwise we compare U_i with $P(2)$, if $U_i \le P(2)$ we set $X_i = 1$; ...; we compare U_i with $P(n-1)$, if $U_i \le P(n-1)$ we set $X_i = n-2$; otherwise we compare U_i with $P(n)$, if $U_i \le P(n)$ we set $X_i = n-1$; otherwise we must have $U_i \le P(n+1) = 1$ and we set $X_i = n$.

This method requires creation of a table that will fill $n+1$ **locations in memory**, and a search that requires $X_i + 1$ comparisons. The **mean number of comparisons will be $np+1$,** since a binomial random variable with parameters n and p has expected value np. **Although the method**

yields exact binomial random variables, it will not be suitable for large n, due to the need to create the table, the substantial storage required, and the fact that on the average $np + 1$ comparisons are needed to find each X_i. For example, if a process produces items that are defective with probability $p = .005$ (on the average 1 defective in each 200), we may wish to simulate quality control inspection of lots of n items. Suppose that we are talking of shipping lots of size $n = 100,000$ items (4000 boxes each containing 25 computer diskettes would yield this quantity). Then we will, in generating the number X_i of defectives in the shipment, need $n + 1 = 100,001$ entries in $P(\cdot)$. Even if the memory required were feasible, we would need an average of $np + 1 = 501$ comparisons to generate each X_i. An additional problem is that many of the probabilities needed in the $P(\cdot)$ vector will be very small, requiring slow high-precision arithmetic. Due to these problems, we will study other methods for generating binomial random variables, in Sections 4.3 and 4.6.1.

4.3 Generating Univariate Random Variables: Discrete Distributions

In Section 4.2.2, we studied the Bernoulli distribution with parameter p, and by a slight extension of the inverse distribution function method we showed that taking $X_i = 0$ if $U_i \leq 1 - p$ and $X_i = 1$ if $1 - p < U_i$ yields a Bernoulli random variable X_i. As usual, U_i is assumed to be uniform on $(0, 1)$.

In Section 4.2.5, we studied the binomial distribution with parameters n and p, and by the same slight generalization of the inverse distribution function method we showed that taking

$$X_i = I \ \text{ if } \ P(I) < U_i \leq P(I + 1), \tag{4.3.1}$$

where $P(I)$ is the probability that a binomial random variable is $\leq I - 1$, results in X_i being a binomial random variable. For example,

$$P[X_i = 0] = P[P(0) < U_i \leq P(1)] = P[0 < U_i \leq p_0]$$
$$= p_0 - 0 = p_0 \tag{4.3.2}$$

where p_0 is the probability that a binomial random variable takes value 0. Now **suppose we wish to generate random variables $X_1, X_2, \ldots$ which** are independent and **such that values $0, 1, 2, \ldots, n$ are taken on with probabilities $p_0, p_1, p_2, \ldots, p_n$** (where $n \geq 0$ is a nonnegative integer, and

$p_i \geq 0$ for all i with $p_0 + p_1 + p_2 + \ldots + p_n = 1$). While **this is the general discrete distribution on the values** $0, 1, 2, \ldots, n$, the only difference from the binomial case is that, instead of the specific values (4.2.21) for the probabilities, we now have general probabilities $p_0, p_1, p_2, \ldots, p_n$. Thus, the generalization of the inverse distribution function method leads us to generate X_i as follows:

$$
\begin{array}{llll}
X_i = 0 \text{ if } & 0 & < U_i \leq & p_0 \\
X_i = 1 \text{ if } & p_0 & < U_i \leq & p_0 + p_1 \\
X_i = 2 \text{ if } & p_0 + p_1 & < U_i \leq & p_0 + p_1 + p_2 \\
& \vdots & & \\
X_i = I \text{ if } & p_0 + \cdots + p_{I-1} & < U_i \leq & p_0 + \cdots + p_i \\
& \vdots & & \\
X_i = n \text{ if } & p_0 + \cdots + p_{n-1} & < U_i \leq & p_0 + \cdots + p_n = 1.
\end{array}
\tag{4.3.3}
$$

This means that $X_i = I$ with probability

$$
\begin{aligned}
P[X_i = I] &= P[p_0 + \cdots + p_{I-1} < U_i \leq p_0 + \cdots + p_{I-1} + p_i] \\
&= (p_0 + \cdots + p_{I-1} + p_I) - (p_0 + \cdots + p_{I-1}) = p_I, \quad (4.3.4)
\end{aligned}
$$

which is exactly as desired. **This method may be implemented by tabling** $P(1) = p_0, P(2) = p_0 + p_1, P(3) = p_0 + p_1 + p_2, \ldots, P(n+1) = p_0 + p_1 + p_2 + \cdots + p_n$, **then searching through** $P(\cdot)$ **to find the first time that** $P(I) < U_i \leq P(I+1)$, **then setting** $X_i = I$. This is precisely what was recommended in the binomial case in Section 4.2.5.

If, instead of the values $0, 1, 2, \ldots, n$ **we wish to generate some** $n+1$ **values** $a_0, a_1, a_2, \ldots, a_n$, **we could simply** modify (4.3.3) by using value a_I instead of I (for all I):

$$
\begin{array}{llll}
X_i = a_0 \text{ if } & 0 & < U_i \leq & p_0 \\
X_i = a_1 \text{ if } & p_0 & < U_i \leq & p_0 + p_1 \\
& \vdots & & \\
X_i = a_I \text{ if } & p_0 + \cdots + p_{I-1} & < U_i \leq & p_0 + \cdots + p_i \\
& \vdots & & \\
X_i = a_n \text{ if } & p_0 + \cdots + p_{n-1} & < U_i \leq & p_0 + \cdots + p_n = 1.
\end{array}
\tag{4.3.5}
$$

To implement this, we now need, in addition to $P(\cdot)$, a vector $A(\cdot)$ containing the values $a_0, a_1, \ldots, a_n$ in $A(1), A(2), \ldots, A(n+1)$. We can now state the general results as a theorem.

Theorem 4.3.6. *Let $a_0, a_1, \ldots, a_n$ be any $n+1$ numbers, and $p_0, p_1, \ldots,$ p_n be any $n+1$ nonnegative numbers that sum to 1. To generate a random variable X that takes value a_I with probability p_I (all I), we may set*

$$X = a_I \quad if \quad p_0 + \cdots + p_{I-1} < U \leq p_0 + \cdots + p_{I-1} + p_I \qquad (4.3.7)$$

where U is uniform on $(0, 1)$.

Corollary 4.3.8. *If $U_1, U_2, U_3, \ldots$ are independent uniform random variables on $(0, 1)$, and if $X_1, X_2, X_3, \ldots$ are the results of using (4.3.7) successively with $U = U_1, U = U_2, U = U_3, \ldots$, then $X_1, X_2, X_3, \ldots$ are independent random variables that take values $a_0, a_1, a_2, \ldots, a_n$ with probabilities $p_0, p_1, p_2, \ldots, p_n$, respectively.*

Note that in Theorem 4.3.6 (and hence in Corollary 4.3.8) no specific order of the values $a_0, a_1, \ldots, a_n$ was required; whatever order we chose, the probabilities needed to be the ones corresponding to the values. Thus, we want to know the **order of values that makes the algorithm most efficient.** One measure of efficiency is the number of comparisons we will need to perform in generating the value X. We will need one comparison if $U_i \leq p_0$, two if $p_0 < U_i \leq p_0 + p_1$, and so on, up to $n+1$ if $p_0 + \cdots + p_{n-1} < U_i \leq p_0 + \cdots + p_{n-1} + p_n$. In other words, we will need one comparison if $X = a_0$, two if $X = a_1, \ldots, n+1$ if $X = a_n$. Thus, the expected number of comparisons will be

$$1 \cdot P[X = a_0] + 2P[X = a_1] + \cdots + (n+1)P[X = a_n]$$
$$= 1 \cdot p_0 + 2p_1 + \cdots + (n+1)p_n. \qquad (4.3.9)$$

It is then clear that we should associate the smallest values (coefficients) with the largest probabilities in order to minimize (4.3.9), and this is done by arranging the coefficients in order of decreasing probabilities. Lemma 4.3.10 summarizes this idea.

Lemma 4.3.10. *The method of Theorem 4.3.6 and Corollary 4.3.8 is most efficient when the values $a_0, a_1, \ldots, a_n$ are put in the order where their respective probabilities $p_0, p_1, \ldots, p_n$ satisfy $p_0 \geq p_1 \geq p_2 \geq \ldots \geq p_n$.*

Example 4.3.11. Suppose that we wish to generate a binomial random variable with $n = 5$ and $p = .90$. Then the possible values and their probabilities are

$$P[0 \text{ successes}] = \binom{5}{0}(.9)^0(.1)^5 = (1)(.00001) = .00001,$$

$$P[1 \text{ success}] = \binom{5}{1}(.9)^1(.1)^4 = (5)(.00009) = .00045,$$

$$P[2 \text{ successes}] = \binom{5}{2}(.9)^2(.1)^3 = (10)(.00018) = .00810, \quad (4.3.12)$$

$$P[3 \text{ successes}] = \binom{5}{3}(.9)^3(.1)^2 = (10)(.00729) = .07290,$$

$$P[4 \text{ successes}] = \binom{5}{4}(.9)^4(.1)^1 = (5)(.06561) = .32805,$$

$$P[5 \text{ successes}] = \binom{5}{5}(.9)^0(.1)^0 = (1)(.59049) = .59049.$$

If we use Corollary 4.3.8 with

$$\begin{cases} a_0 = 0, & p_0 = .00001, & a_1 = 1, & p_1 = .00045, \\ a_2 = 2, & p_2 = .00810, & a_3 = 3, & p_3 = .07290, \\ a_4 = 4, & p_4 = .32805, & a_5 = 5, & p_5 = .59049 \end{cases} \quad (4.3.13)$$

we will obtain the discrete distribution (4.3.12). The average number of comparisons needed to generate each X_i will be

$$p_0 + 2p_1 + 3p_2 + 4p_3 + 5p_4 + 6p_5 = 5.5. \quad (4.3.14)$$

However, if we use the ordering recommended in Lemma 4.3.10, we will take

$$\begin{cases} a_0' = 5, & p_0' = .59049, & a_1' = 4, & p_1' = .32805, \\ a_2' = 3, & p_2' = .07290, & a_3' = 2, & p_3' = .00810, \\ a_4' = 1, & p_4' = .00045, & a_5' = 0, & p_5' = .00001. \end{cases} \quad (4.3.15)$$

Again we will obtain distribution (4.3.12) for the random variables we generate. Now, however, the average number of comparisons needed to generate each X_i will be

$$p_0' + 2p_1' + 3p_2' + 4p_3' + 5p_4' + 6p_5' = 1.5. \quad (4.3.16)$$

The ratio of (4.3.16) to (4.3.14) is $1.5/5.5 = 0.27$. Therefore, we expect that use of the ordering given in (4.3.15) will save approximately $100(1 - 0.27) = 73\%$ of the comparison time used by the ordering in (4.3.13) to generate the random variables.

Remark 4.3.17. The methods of this section are sometimes used **to generate random variables from continuous distributions**. To accomplish this, one replaces the desired range of values (a, b) with a set of $n+1$ discrete values $a_0, a_1, \ldots, a_n$, and assigns probabilities $p_0, p_1, \ldots, p_n$ to these values in such a way that the continuous distribution function $F_1(\cdot)$ which is desired is in some sense approximated by the distribution function $F_2(\cdot)$ of the discrete random variable which is generated. One method often used in such cases is that of **equal percentiles**. Then one chooses $a_0, a_1, \ldots, a_n$ subject to

$$F_1(a_i) = i/100 \text{ for } i = 0, 1, \ldots, 100. \tag{4.3.17}$$

Since $F_1(b) = 1$, $a_{100} = b$ is taken. If the distribution is on an infinite range, x_{100} is set to some relatively arbitrary value. Choice of the 100 values in (4.3.18), with corresponding probabilities $p_0 = \cdots = p_{99} = .01$, will result in the discrete distribution function $F_2(\cdot)$, which approximates $F_1(\cdot)$, in fact equaling $F_1(\cdot)$ at $x_1, \ldots, x_{100}$:

$$F_1(a_i) = F_2(a_i) \text{ for } i = 0, 1, \ldots, 100. \tag{4.3.18}$$

A random variable, X, with distribution function $F_2(\cdot)$ can now be generated by

$$X = A(INT(100 * (U + .01))), \tag{4.3.19}$$

where $INT(\cdot)$ is the function that takes the integer part of its argument (e.g., $INT(3.68) = 3$). The comparisons of the search for the value U are avoided. However, **we do not recommend this method. Although it is a good way to compress the continuous distribution to 100 values, and it can be implemented efficiently on a computer via (4.3.20), one has replaced a random variable that should be able to take on any value between** a **and** b**, with a random variable that can take on only 100 possible values.**

Remark 4.3.21. The methods of this section are sometimes used to generate **univariate discrete distributions that take on infinitely many values,** such as the Poisson distribution, which can take on any nonnegative integer value. This is done by truncating the values so that only a finite number are involved, in which case use of the methods of this section can be applied. In the Poisson case, one may truncate at some n such that the cumulative probability of the values $n, n+1, n+2, \ldots$ is no larger than .01 and all of that probability is then assigned to the value n. **Since the random variable generated does not have the exact distribution desired, we do not recommend this method.** Some important cases of discrete random variables with infinitely many possible values are considered in Section 4.4.

4.4 Generating Poisson and Geometric Random Variables (Discrete Univariate Distributions that Take on Infinitely Many Values)

Remark 4.3.21 of Section 4.3 describes a method that provides approximations to random variables which are discrete but that take on an infinite number of possible values. In this section, we provide **exact** methods for generating some of the most important such random variables.

4.4.1 The Poisson Distribution

A random variable X has the **Poisson distribution with mean** $\mu > 0$ if its possible values are $0, 1, 2, \ldots$ with probabilities

$$P[X = i] = e^{-\mu}\mu^i/i! \quad (i = 0, 1, 2, \ldots). \tag{4.4.1}$$

The following theorem and corollary allow us to generate exact Poisson random variables.

Theorem 4.4.2. *Let $U_1, U_2, \ldots$ be independent uniform random variables on $(0,1)$. For a fixed $\mu > 0$, define a variable X as follows:*

$$\begin{cases} \text{if } U_1 < e^{-\mu}, & & \text{let } X = 0; \\ \text{if } U_1 \geq e^{-\mu}, & \text{and } U_1U_2 < e^{-\mu}, & \text{let } X = 1; \\ \text{if } U_1U_2 \geq e^{-\mu}, & \text{and } U_1U_2U_3 < e^{-\mu}, & \text{let } X = 2; \\ \vdots \\ \text{if } U_1 \cdots U_I \geq e^{-\mu}, & \text{and } U_1 \cdots U_{I+1} < e^{-\mu}, & \text{let } X = I; \\ \vdots \\ . \end{cases} \tag{4.4.3}$$

Then X has the Poisson distribution with mean μ.

Corollary 4.4.4. *Let $X_1, X_2, \ldots$ be the results of using (4.4.3) successively, with disjoint sets of uniform random variables ($U_1, U_2, \ldots, U_{X_1+1}$ are used to generate X_1; $U_{X_1+2}, \ldots, U_{X_1+X_2+2}$ are used to generate X_2; etc.). Then $X_1, X_2, \ldots$ are independent Poisson random variables with mean $\mu > 0$.*

We will prove Theorem 4.4.2, after which Corollary 4.4.4 will be obvious. In fact, we give two proofs of Theorem 4.4.2. The first proof is conceptually simpler, but uses advanced facts about the nature of a Poisson process. The second uses straightforward calculations covered in the usual calculus-based first course on mathematical statistics and probability.

First Proof of Theorem 4.4.2. If events occur in time in accord with a Poisson process with parameter ν, then the probability of i occurrences in the time interval $(0, T)$ is $e^{-\nu T}(\nu T)^i/i!$ $(i = 0, 1, \ldots)$, and the time lapses between successive occurrences are independent exponential random variables with mean $1/\nu$. (E.g., see pp. 135, 174 of Parzen (1962).) Take $\nu = 1$ and $T = \mu$. Then the number X of occurrences in $(0, \mu)$ **is** the desired Poisson random variable. However, this is $X = k$ where k is determined by

$$E_1 + E_2 + \cdots + E_k \leq \mu < E_1 + E_2 + \cdots + E_k + E_{k+1} \qquad (4.4.5)$$

and $E_1, E_2, \ldots$ are independent exponential random variables with mean $1/\nu = 1$. Since such $E_1, E_2, \ldots$ are (see the paragraph following (4.2.8)) $-\ln(U_1), -\ln(U_2), \ldots$, we have $X = k$ for the k such that

$$-\ln(U_1) - \cdots - \ln(U_k) \leq \mu < -\ln(U_1) - \cdots - \ln(U_k) - \ln(U_{k+1}),$$

or

$$-\ln(U_1 \cdots U_k) \leq \mu < -\ln(U_1 \cdots U_k U_{k+1}),$$

or

$$\ln(U_1 \cdots U_k) \geq -\mu > \ln(U_1 \cdots U_k U_{k+1}),$$

or

$$U_1 \cdots U_k \geq e^{-\mu} > U_1 \cdots U_k U_{k+1},$$

which proves the theorem.

Second Proof of Theorem 4.4.2. By the stated algorithm, we need to find

$$P[X = k] = P[U_1 \cdots U_k \geq e^{-\mu}, \ U_1 \cdots U_k U_{k+1} < e^{-\mu}]$$

$$= P[U_{k+1} < \frac{e^{-\mu}}{U_1 \cdots U_k}, \ U_1 \cdots U_k \geq e^{-\mu}] \qquad (4.4.6)$$

$$= \int_{e^{-\mu}}^{1} P[U_{k+1} < e^{-\mu}/x] \ f_{U_1 \cdots U_k}(x) \, dx,$$

where $f_{U_1 \cdots U_k}(\cdot)$ is the probability density function of the product $U_1 \cdots U_k$.

$$\begin{aligned} F_{U_1 \cdots U_k}(x) &= P[U_1 \cdots U_k \leq x] \\ &= P[\ln(U_1) + \cdots + \ln(U_k) \leq \ln(x)] \\ &= P[-\ln(U_1) - \cdots - \ln(U_k) \geq -\ln(x)] \qquad (4.4.7) \\ &= P[Y_1 + \cdots + Y_k \geq -\ln(x)] \\ &= 1 - P[Y_1 + \cdots + Y_k \leq -\ln(x)] \end{aligned}$$

where (see the paragraph following (4.2.8)) $Y_1, \ldots, Y_k$ are independent exponential random variables with mean 1. Differentiating (4.4.7) to find the probability density function of $U_1, \cdots, U_k$, we have

$$f_{U_1\cdots U_k}(x) = \frac{d}{dx} F_{U_1\cdots U_k}(x) = \frac{1}{x} f_{Y_1+\cdots+Y_k}(-\ln(x)). \qquad (4.4.8)$$

However, it is well known (e.g., see Dudewicz and Mishra (1988), p. 277) that a sum of independent exponential random variables each with the same mean has a gamma distribution, and when the mean of each exponential is 1 that gamma distribution is

$$f_{Y_1+\cdots+Y_k}(z) = \frac{1}{(k-1)!} z^{k-1} e^{-z} \qquad (z > 0). \qquad (4.4.9)$$

Using (4.4.9) in (4.4.8), we have

$$f_{U_1\cdots U_k}(x) = \frac{1}{(k-1)!} \left(\ln \frac{1}{x}\right)^{k-1}. \qquad (4.4.10)$$

Substituting (4.4.10) into the integral of (4.4.6) and using the substitution $y = \ln(1/x)$, we now find that

$$\begin{aligned}
P[X = k] &= \int_{e^{-\mu}}^{1} \frac{e^{-\mu}}{x} \frac{1}{(k-1)!} \left(\ln \frac{1}{x}\right)^{k-1} dx \\
&= \int_{0}^{\mu} \frac{e^{-\mu}}{(k-1)!} y^{k-1} \, dy = \frac{e^{-\mu}}{k!} y^k \Big|_0^{\mu} \\
&= e^{-\mu} \mu^k / k!
\end{aligned}$$

as was to be shown.

The method of Theorem 4.4.2 is a reasonable one, and is simple to implement, when μ is not large. For more efficient methods when μ is large, and for reference to other methods, see pp. 210–211 of Tadikamalla (1984). Note that **the average number of uniform random variables needed to generate one Poisson via Theorem 4.4.2 is** $E(X+1) = E(X)+1 = \mu+1$.

4.4.2   The Geometric Distribution

The possible values of a geometric random variable are $1, 2, \ldots$ with probabilities

$$P[X = i] = p(1-p)^{i-1}, \qquad i = 1, 2, \ldots, \qquad (4.4.11)$$

where p is a number with $0 < p < 1$. **Such a random variable X arises as the number of trials needed to obtain the first success in a sequence of independent trials with probability p of success on each trial.** This interpretation provides us with a first method of generating a geometric random variable, by taking successive uniform random variables U_i on $(0, 1)$ until we find the first one which is less than or equal to p; X is the number of uniform random variables used. This is more formally stated in Theorem 4.4.12.

Theorem 4.4.12. *If $U_1, U_2, \ldots$ are independent uniform random variables on $(0, 1)$, for a fixed p $(0 < p < 1)$, define X by*

$$
\begin{cases}
\text{if } U_1 \leq p, & \text{let } X = 1; \\
\text{if } U_1 > p, & \text{and } U_2 \leq p, \text{ let } X = 2; \\
\text{if } U_1 > p,\, U_2 > p, & \text{and } U_3 \leq p, \text{ let } X = 3; \\
\quad\vdots & \\
\text{if } U_i > p \;\; (i = 1, 2, \ldots, I - 1), & \text{and } U_i \leq p, \text{ let } X = I; \\
\quad\vdots & \\
\quad.
\end{cases}
\tag{4.4.13}
$$

Then X has the geometric distribution with probability p.

Corollary 4.4.14. *Let $X_1, X_2, \ldots$ be the results of using (4.4.13) successively with disjoint sets of uniform random variables. Then $X_1, X_2, \ldots$ are independent geometric random variables with success probability p. (Note that $U_1, \ldots, U_{X_1}$ are used to generate X_1; $U_{X_1+1}, \ldots, U_{X_1+X_2}$ are used to generate X_2; etc.)*

While the method of Theorem 4.4.12 is intuitively appealing, **the number of uniform random variables needed to generate one geometric random variable is X, which has mean $E(X) = 1/p$** (see Dudewicz and Mishra (1988), Problem 5.2.11 with $r = 1$), and for small values of p this is very inefficient. Since studies with small values of p are important in many applications of simulation, we will develop a method that is more efficient for small p.

If we attempted to use the method of Theorem 4.3.6, we would first define the numbers $a_0 = 1$, $a_1 = 2$, $a_2 = 3, \ldots$ with corresponding probabilities $p_0 = p$, $p_1 = p(1 - p)$, $p_2 = p(1 - p)^2, \ldots$. To generate a random variable X which takes value a_I with probability p_I (all I) we would set

$$
X = a_I = I + 1 \quad \text{if} \quad p_0 + \cdots + p_{I-1} < U \leq p_0 + \cdots + p_I \tag{4.4.15}
$$

where U is uniform on $(0,1)$. Now this is equivalent to

$$X = I + 1 \text{ if}$$
$$p + p(1-p) + (p(1-p)^2 + \cdots + p(1-p)^{I-1} \qquad (4.4.16)$$
$$< U \leq p + p(1-p) + p(1-p)^2 + \cdots + p(1-p)^I, $$

or

$$X = I + 1 \text{ if}$$
$$p \sum_{i=0}^{I-1} (1-p)^i < U \leq p \sum_{i=0}^{I} (1-p)^i. \qquad (4.4.17)$$

Since for $|x| < 1$ we have the well-known summation result

$$1 + x + \cdots + x^n = \frac{1 - x^{n+1}}{1 - x}, \qquad (4.4.18)$$

from (4.4.17) and (4.4.18) we obtain

$$X = I + 1 \text{ if } p\frac{1 - (1-p)^I}{1 - (1-p)} < U \leq p\frac{1 - (1-p)^{I+1}}{1 - (1-p)}$$
$$\text{if } 1 - (1-p)^I < U \leq 1 - (1-p)^{I+1}$$
$$\text{if } (1-p)^I > 1 - U \geq (1-p)^{I+1} \qquad (4.4.19)$$
$$\text{if } I \ln(1-p) > \ln(1-U) \geq (I+1) \ln(1-p)$$
$$\text{if } I < \frac{\ln(1-U)}{\ln(1-p)} \leq I + 1.$$

In other words, X can be taken to be $\ln(1-U)/\ln(1-p)$ rounded up to an integer value. We formalize this in the following theorem.

Theorem 4.4.20. *If $U_1, U_2, \ldots$ are independent uniform random variables on $(0,1)$, and if $X_1, X_2, \ldots$ are generated as*

$$X_i = (\text{The number } \ln(1-U_i)/\ln(1-p) \text{ rounded up to an integer}),$$

then $X_1, X_2, \ldots$ are independent geometric random variables with probability p.

While the method of Corollary 4.4.14 took on the average $1/p$ uniform random variables to produce one geometric random variable, **the method of Theorem 4.4.20 requires only one uniform random variable to produce one geometric random variable.**

Remark 4.4.21. The method of Theorem 4.3.6 is useful in cases where the number of possible values of the discrete random variable is finite. We have used it here for **a case (the geometric) where the number of possible values is infinite with values** $1, 2, 3, \ldots$. This was possible because the sum $p_0 + p_1 + \cdots + p_I$ has a simple closed-form expression, so that we were able to derive the simple expression (4.4.19) for when (4.4.15) holds. In other cases, such as the Poisson, there is no simple expression for $p_0 + p_1 + \cdots + p_I$, so in such cases this use of Theorem 4.3.6 is not possible.

4.5   Generating Bivariate and Multivariate Discrete Distributions

Thus far our discussion has been limited to generating **univariate random variables** X, which are single numbers with a distribution function $F_X(x)$. This suffices when we have one random quantity to generate, or when we have several but they are statistically independent (in which case we may generate each by the appropriate univariate technique, irrespective of the value of the other). In other settings, however, we encounter **bivariate random variables** (X_1, X_2), which are pairs of numbers with a distribution function

$$F_{X_1,X_2}(x_1, x_2) = P[X_1 \leq x_1, X_2 \leq x_2] \tag{4.5.1}$$

which gives probabilities such as $F_{X_1,X_2}(3.2, 1.8)$, the probability that X_1 does not exceed 3.2 and simultaneously X_2 does not exceed 1.8. In the **discrete case**, there are pairs of numbers (x_1, x_2) that are "possible" (for which $P[X_1 = x_1, X_2 = x_2] > 0$), while for all other pairs the probability is zero. If there are a **finite number of pairs with a positive probability**, then we may use the method of Theorem 4.3.6 to generate values of the random variable (X_1, X_2) by simply replacing $a_0, a_1, \ldots, a_n$ with the possible pairs.

Example 4.5.2. A manufactured item is composed of two parts (for example, a water tank may have a top half and a bottom half; a table may have a top assembly and a leg assembly; etc.). Let X_1 denote whether, when inspected, the first part of the assembly is found to be free of defects ($X_1 = 0$), in need of reworking ($X_1 = 1$), or fit only for being scrapped ($X_1 = 2$). Similarly, let X_2 denote the results of inspection of the second part of the assembly: $X_2 = 0$ (free of defects), $X_2 = 1$ (in need of reworking), or $X_2 = 2$ (to be scrapped). Suppose that the parts of the assembly are not independent, but rather have a bivariate distribution with the following positive probabilities on the nine possible pairs of points:

$$
\begin{aligned}
&P[X_1 = 0, X_2 = 0] = 2/9, &\quad &P[X_1 = 0, X_2 = 1] = 1/9, \\
&P[X_1 = 0, X_2 = 2] = 0, &\quad &P[X_1 = 1, X_2 = 0] = 1/18, \\
&P[X_1 = 1, X_2 = 1] = 2/9, &\quad &P[X_1 = 1, X_2 = 2] = 1/18, \qquad (4.5.3)\\
&P[X_1 = 2, X_2 = 0] = 0, &\quad &P[X_1 = 2, X_2 = 1] = 1/9, \\
&P[X_1 = 2, X_2 = 2] = 2/9. & &
\end{aligned}
$$

The probabilities of all the possible pairs (x_1, x_2) add to 1.00, and all other pairs have probability zero, as is the case for every discrete bivariate probability distribution. To generate a random variable (X_1, X_2) with the bivariate discrete probability distribution given in (4.5.3), in the efficient manner described in Lemma 4.3.10, we list the pairs in order of decreasing probability, omitting pairs with probability zero.

$$
\begin{cases}
a_0 : & (0,0) \text{ with probability } 2/9 & = p_0 \\
a_1 : & (1,1) \text{ with probability } 2/9 & = p_1 \\
a_2 : & (2,2) \text{ with probability } 2/9 & = p_2 \\
a_3 : & (0,1) \text{ with probability } 1/9 & = p_3 \qquad (4.5.4)\\
a_4 : & (2,1) \text{ with probability } 1/9 & = p_4 \\
a_5 : & (1,0) \text{ with probability } 1/18 & = p_5 \\
a_6 : & (1,2) \text{ with probability } 1/18 & = p_6.
\end{cases}
$$

We then take a uniform random variable U and generate

$$
\begin{cases}
a_0, & (0,0), & \text{if } U \le 2/9 \\
a_1, & (1,1), & \text{if } 2/9 < U \le 4/9 \\
a_2, & (2,2), & \text{if } 4/9 < U \le 6/9 \\
a_3, & (0,1), & \text{if } 6/9 < U \le 7/9 \qquad (4.5.5)\\
a_4, & (2,1), & \text{if } 7/9 < U \le 8/9 \\
a_5, & (1,0), & \text{if } 8/9 < U \le 17/18 \\
a_6, & (1,2), & \text{if } 17/18 < U \le 1.
\end{cases}
$$

Thus, if our first random number is $U = .44683$, we will generate $(X_1, X_2) = (2,2)$ since $4/9 = .44444 < .44683 \le .66667 = 6/9$.

The same ideas are used in the **multivariate case**, i.e., where the random variable $(X_1, X_2, \ldots, X_p)$ of interest has some number $p \ge 2$ of components, and the variable is **discrete with a finite number of p-tuples of numbers that have positive probability**.

Theorem 4.5.6. *Let $a_0, a_1, \ldots, a_n$ denote the $n+1$ possible values, each a p-tuple of numbers, of a p-dimensional random variable $(X_1, X_2, \ldots, X_p)$, with respective probabilities $p_0, p_1, \ldots, p_n$ (for which $p_0 + p_1 + \cdots + p_n = 1$). To generate a value of the random variable $(X_1, X_2, \ldots, X_p)$ that takes a_I with probability p_I $(I = 0, 1, \ldots, n)$ we may set*

$$(X_1, X_2, \ldots, X_p) = a_I \text{ if } p_0 + \cdots + p_{I-1} < U \leq p_0 + \cdots + p_I \quad (4.5.7)$$

where U is uniform on $(0, 1)$. The generation algorithm will be most efficient if we order the a's so that $p_0 \geq p_1 \geq \ldots \geq p_n$.

4.6 Generating Specific Univariate Distributions

After dealing with **selection of a distribution** (Section 4.1), we studied general methods for generating random variables through the **inverse distribution function method** of Section 4.2, and **its generalization for discrete cases** in Section 4.3. The discrete distributions and **Poisson and geometric distributions** were considered in Section 4.4, and general, **bivariate and multivariate** distributions were considered in Section 4.5. We now consider specially developed methods for some of the most commonly arising distributions. General procedures will be considered in later sections; here we concentrate on methods particular to the specific distribution under consideration.

4.6.1 The Binomial Distribution

Method 1: Use of Tables. The binomial distribution is specified by (4.2.21) for its probability of each value, and by (4.2.22) for its distribution function. In Section 4.2.5 we showed how to use a table, essentially a generalization of the inverse distribution function approach, to generate this distribution. Since $n+1$ storage locations, and (on the average) $np+1$ comparisons, are needed with this procedure, it is desirable to have other exact methods for large n and/or large p.

Method 2: Via Bernoulli Trials. If $B_1, B_2, \ldots$ are independent Bernoulli random variables with success probability p (as defined in (4.2.10) and 4.2.11)), then it is known that $B = B_1 + B_2 + \cdots + B_n$ is a binomial random variable with n trials and success probability p. Thus, we can also generate binomial random variables by generating sequences of n Bernoulli random variables and adding them up. This does not require large storage, does not require tabulation of the binomial probability distribution for large n (which can be difficult), and it is quite feasible with only minimal programming effort. It does, however, require generation of n Bernoulli random variables for each binomial generated. This is its main disadvantage, since it may take too much computer time, even though the method, (4.2.13), for generating Bernoulli random variables is very simple.

Method 3: Via Geometric Intersuccess Times. In a sequence of Bernoulli trials, it can be shown that the number of trials needed to obtain the first success (including that success) has the geometric distribution of (4.4.11). It follows that if $G_1, G_2, \ldots$ are independent geometric random variables, then

$$B = (\text{Largest } i \text{ such that } G_1 + G_2 + \cdots + G_i \leq n) \qquad (4.6.1)$$

has the binomial distribution with n trials and success probability p. It takes G_1 trials to get the first success, G_2 trials to get a second success,..., G_i trials to get an i-th success, and the number of trials to get an $i + 1^{\text{st}}$ success would take us beyond n trials, hence in the first n trials there are i successes. Since geometric random variables can be efficiently generated via Theorem 4.4.20, (4.6.1) can be used to generate binomial random variables. Since the number of geometric random variables used is the number of successes, it therefore follows that the number of geometric random variables used has a binomial distribution with mean np. Therefore, this method will be particularly efficient when p is very small. Similar efficiency can be obtained in the case of p large by generating B' with n trials and success probability $1 - p$, since then $B = n - b'$ has the desired distribution.

Method 4: Via the Beta Integral. Recall that, in Method 1 above, we used (4.2.23) to set the value of the binomial random variable B, i.e., we set

$$B = i \text{ such that } \sum_{k=0}^{i-1} P[B = k] < U \leq \sum_{k=0}^{i} P[B = k], \qquad (4.6.2)$$

where U is a uniform random variable on $(0, 1)$. Since

$$\sum_{k=0}^{n} P[B = k] = 1,$$

this is equivalent to

$$B = i \text{ such that } 1 - \sum_{k=i}^{n} P[B = k] < U \leq 1 - \sum_{k=i+1}^{n} P[B = k]$$

which simplifies to

$$\sum_{k=i+1}^{n} P[B = k] \leq 1 - U < \sum_{k=i}^{n} P[B = k]$$

and

$$\sum_{x=i+1}^{n} \binom{n}{x} p^x (1-p)^{n-x} \leq 1 - U < \sum_{x=i}^{n} \binom{n}{x} p^x (1-p)^{n-x}, \qquad (4.6.3)$$

where we used expression (4.2.21) for the probabilities. It is known (see p. 154 of Dudewicz and Mishra (1988), for example) that

$$\sum_{x=a}^{n} \binom{n}{x} p^x (1-p)^{n-x} = \frac{\Gamma(n+1)}{\Gamma(a)\Gamma(n-a+1)} \int_0^p x^{a-1} (1-x)^{n-a} \, dx$$

$$= I_p(a, n - a + 1) \qquad (4.6.4)$$

where $I_p(\cdot, \cdot)$ is called the **incomplete beta function**, defined in general by

$$I_p(n_1, n_2) = \frac{\Gamma(n_1 + n_2)}{\Gamma(n_1)\Gamma(n_2)} \int_0^p x^{n_1 - 1} (1-x)^{n_2 - 1} \, dx. \qquad (4.6.5)$$

From (4.6.3) and (4.6.4), we see that a binomial random variable is obtained from

$$B = i \text{ such that } I_p(i + 1, n - i) \leq 1 - U < I_p(i, n - i + 1). \qquad (4.6.6)$$

In (4.6.5), $\Gamma(\cdot)$ is the **gamma function** defined by

$$\Gamma(x) = \int_0^\infty t^{x-1} e^{-t} \, dt, \quad x > 0 \qquad (4.6.7)$$

which has the property that

$$\Gamma(x) = (x - 1)\Gamma(x - 1). \qquad (4.6.8)$$

Hence for positive integers x we have $\Gamma(x) = x!$.

The drawback of this method has been that accurate computation of the incomplete beta function is generally very time consuming, especially for large n_1 and n_2 in (4.6.5). Inaccurate computation would put this method in the same undesirable class as the methods in Section 4.2.3 for the normal distribution. However, today many software systems have high-quality algorithms for computing the incomplete beta function as well as the integral in (4.6.7) with limits of 0 and p (called the **incomplete gamma function**). Thus, this method may be feasible in such settings and in others where a high-accuracy algorithm for the incomplete beta and gamma functions could be added to the available software.

4.6.2   The Normal Distribution

Method 1: Inverse Distribution Function. The normal distribution $N(\mu, \sigma^2)$, specified by (4.2.14), can be generated via $\mu + \sigma F^{-1}(U)$, as discussed in Section 4.2.3. This approach has problems with accuracy, especially in the tails of the distribution.

Method 2: Central Limit Theorem Approach. In one of its simplest classical forms, the Central Limit Theorem (e.g., see p. 315 of Dudewicz and Mishra (1988)) can be stated as follows.

Theorem 4.6.9. *If $X_1, X_2, \ldots$ are independent and identically distributed random variables with $E(X_1) = \mu$ and Var $(X_1) = \sigma^2 > 0$ (both finite), then (for all z, $-\infty < z < +\infty$) as $n \to \infty$,*

$$P\left(\frac{(X_1 - \mu) + \cdots + (X_n - \mu)}{\sqrt{n}\,\sigma} < z\right) \to \frac{1}{\sqrt{2\pi}} \int_{-\infty}^{z} e^{-0.5y^2}\, dy. \quad (4.6.10)$$

From Theorem 4.6.9 we can conclude that for large n the distribution function of

$$\frac{(X_1 - \mu) + \cdots + (X_n - \mu)}{\sqrt{n}\,\sigma} = \frac{X_1 + \cdots + X_n - n\mu}{\sqrt{n}\,\sigma} \qquad (4.6.11)$$

is close to that of the standard normal distribution.

It is known (see Dudewicz and Mishra (1988), p. 222) that if U is uniform on (a, b) with $a < b$, then

$$E(U) = (a + b)/2, \quad \text{and} \quad \text{Var } (U) = (b - a)^2/12. \qquad (4.6.12)$$

Thus, if $a = 0$ and $b = 1$, as they do for uniform random variables on $(0, 1)$, we will have

$$E(U) = 1/2, \quad \text{and} \quad \text{Var } (U) = 1/12. \qquad (4.6.13)$$

The rules for dealing with means and variances of sums of independent random variables state that for independent random variables $X_1, X_2, \ldots, X_n$,

$$E(X_1 + \cdots + X_n) = E(X_1) + \cdots + E(X_n) \qquad (4.6.14)$$

and

$$\text{Var}(X_1 + \cdots + X_n) = \text{Var}(X_1) + \cdots + \text{Var}(X_n). \qquad (4.6.15)$$

If the variables have the same mean and variance, then these become

$$E(X_1 + \cdots + X_n) = nE(X_1), \quad \text{Var}(X_1 + \cdots + X_n) = n\text{Var}(X_1). \quad (4.6.16)$$

Since $E(cX) = cE(X)$ and $\text{Var}(cX) = c^2\text{Var}(X)$ for any random variable X, we have the following result.

Theorem 4.6.17. *If $U_1, U_2, \ldots, U_k$ are independent uniform random variables on $(0, 1)$, then*

$$X = \frac{U_1 + \cdots + U_k - \frac{k}{2}}{\sqrt{k/12}} = \sqrt{12/k}\,\left(U_1 + \cdots + U_k - \frac{k}{2}\right)$$

has mean zero and variance 1. The distribution function of X is close to that of the standard normal distribution for large k.

With this "Central Limit Theorem" approach to generation of standard normal random variables, one sums k uniforms, subtracts $k/2$, and multiplies by the square root of $12/k$. Usually, when this method is used, k is taken to be 12, since then the method reduces to generating

$$X = U_1 + \cdots + U_{12} - 6, \tag{4.6.18}$$

which is faster since the square root and multiplication are not needed. This approach was once fairly common, due to the need for enough speed to make simulation feasible, but today we specifically recommend that this method **not** be used, because it is not exact. In particular, the distribution generated has **no** tails beyond the range $(-6, 6)$. In a true normal distribution $N(0, 1)$, the probability outside the range $(-6, 6)$ is 1.973175×10^{-9}. While this is small, note that:

1. The probability of obtaining at least one value outside the range $(-6, 6)$ in 10^8 trials is .28, which is not small.

2. A true normal distribution has positive probability of all ranges of values.

For example, if one has a simulation involving nuclear safety and there are $20,000$ nuclear weapons/nuclear plants under consideration, and if each day represents a new trial for each, then in a year there will be about $7,300,000$ trials. A total of $100,000,000$ represents the number of trials in a simulation involving 14 years, or in one simulation involving one year replicated 14 times. Since replication, often for many thousands of times, underlies the accuracy of simulation, the seemingly rare event outside a restricted range may well occur. In fact, it may be the crux of the simulation (as is the highest flood, the most severe earthquake, the highest wind speed, etc.).

The use of $k = 48$ in 4.6.17 addresses some, but not all, of these concerns. This choice of k would yield $X = ((U_1 + \cdots + U_{48})/2) - 12$, which is reasonably economical to compute. The range of possible values would now be expanded to $(-12, 12)$, eliminating some of the concerns about range restriction.

Method 3: Box-Muller Transformation. In 1958 Box and Muller suggested that the following result from mathematical statistics be used to provide normal random variables (for a proof, see Dudewicz and Mishra (1988), p. 186).

Theorem 4.6.19. *If U_1 and U_2 are independent uniform random variables on $(0,1)$, then*

$$X_1 = \sqrt{-2\ln(U_1)}\, \sin(2\pi U_2) \tag{4.6.20}$$

is exactly $N(0,1)$.

Using this method, **one takes a sequence of independent $N(\mu, \sigma^2)$ random variables to be**

$$\begin{cases} \mu + \sigma X_1, & X_1 = \sqrt{-2\ln(U_1)}\, \sin(2\pi U_2), \\ \mu + \sigma X_2, & X_2 = \sqrt{-2\ln(U_3)}\, \sin(2\pi U_4), \\ \mu + \sigma X_3, & X_3 = \sqrt{-2\ln(U_5)}\, \sin(2\pi U_6), \\ \vdots \\ \cdot \end{cases} \tag{4.6.21}$$

This method is relatively fast, and provides exact normal random variables, hence it is **recommended** for use.

One should not use overlapping U_i, such as $\sqrt{-2\ln(U_2)}\, \sin(2\pi U_3)$, since then independence would not hold. Note that (4.6.21) yields one normal random variable for each two uniform ones used. It can be shown that the variables calculated exactly as in (4.6.21) but with $\sin(\cdot)$ replaced by $\cos(\cdot)$ are also normal **and** independent of the random variables in (4.6.21). Thus, it might be tempting to use both transformations to produce two normals from each two uniforms. This is **not** recommended, even though it is widely done, because defects in the uniform random number generator will produce a lack of independence in the supposed normals that result.

Method 4: Ratio-of-Uniforms. The Ratio-of-Uniforms method of Theorem 4.6.42 may also be used to generate normal random variable.

Remark 4.6.22. There are, in fact, literally hundreds of proposed methods for generating normal random variables (e.g., see Tadikamalla (1984), pp. 206–207 for some references and notes on which are exact and fast). There are also many generators for other distributions, and the literature is active and growing, often with good new techniques. (For example, Dagpunar (1988) gives FORTRAN code for 15 selected distributions as well as a general discussion; those consulting Dagpunar's book should also consult its review by Stadlober (1989) for a discussion of more recent developments on discrete distributions. Devroye (1986) is encyclopedic in nature, but more theoretical and lacks algorithms.) Our choice has been motivated by the following considerations:

1. Exactness: One should not, with today's fast computers, use an approximation algorithm, unless perhaps one's choice of distribution is only a rough approximation itself.

2. Speed: An exact algorithm should not be so slow as to seriously slow the simulation program.

3. Memory requirements: The algorithm should not require excessively large amount of storage.

4. Ease of implementation: Even the best algorithms will remain unused if major effort is needed on the part of the professionals involved in the simulation.

In addition, we have covered methods that are in wide use but are seriously flawed, such as the "Central Limit Theorem" approach to generating normal random variables. The widespread use of defective algorithms needs to be pointed out, lest new simulationists be misled into using them.

Remark 4.6.23. The Box-Muller transformation method for generating normal random variables is suggestive of methods we will now use for a number of distributions. If it is known from mathematical statistics that the distribution we wish to generate **can be obtained through a transformation of random variables that we already know how to generate**, this immediately furnishes an algorithm for generating the new distribution.

Remark 4.6.24. It has been shown in the literature that the Box-Muller method has defective performance if used in conjunction with a multiplicative congruential generator with a small period (which would contravene our recommendations on the period of a generator, but is not rare in generators embedded in the libraries of computer centers or in software packages). The problem became apparent because of **a practice of making a histogram of the first several hundred values generated by each random variable generator in a simulation**. It was observed that one tail of the distribution was twice as high as the other. We recommend making such a histogram for several reasons:

1. Defects of the algorithm, such as the interaction noted of the Box-Muller transformation with certain types of random number generators, may be revealed.

2. Defects of coding and/or understanding of the algorithm may be revealed.

The Kolmogorov–Smirnov One-Sample Test may be used to test the hypothesis that the generated $X_1, X_2, \ldots, X_n$ are in fact a random sample from the desired distribution function $F_0(x)$. This is a valuable supplement to the visual impression conveyed by the histogram, which should be plotted with a superimposed rendition of the distribution being sought for maximum visual information. To perform this test, one computes the discrepancy statistic

$$D_n = \sup_x |F_n(x) - F_0(x)|, \tag{4.6.25}$$

where $F_n(x)$ is the **empirical distribution function** of the generated random variables $X_1, X_2, \ldots, X_n$, i.e.

$$F_n(x) = (\text{Number of } X_1, X_2, \ldots, X_n \text{ that are } \leq x)/n. \tag{4.6.26}$$

Since D_n is the maximum vertical distance between the functions $F_0(x)$ and $F_n(x)$, and since $F_n(x)$ is a step function which takes a jump at each value $X_1, X_2, \ldots, X_n$, it follows that D_n **occurs at or just before a jump point of $F_n(x)$**, simplifying the calculation of (4.2.26). It can be shown (e.g., see Dudewicz and Mishra (1988), p. 670) that **for large n one rejects the hypothesis at level of significance .01 if and only if**

$$D_n > 1.63/\sqrt{n}. \tag{4.6.27}$$

For level of significance .05, the hypothesis is rejected if $D_n > 1.36/\sqrt{n}$, and for level .10 it is rejected if $D_n > 1.22/\sqrt{n}$. For small n ($n \leq 40$) exact values of the rejection criterion are available in Gibbons (1997), p. 446; these values differ little from (4.6.27) unless n is very small.

Remark 4.6.28. The numbers 1.63 for level .01, 1.36 for level .05, and 1.22 for level .10, each divided by $\sqrt{n}$, are computed from the result that under the hypothesis, for all $t > 0$,

$$\lim_{n \to \infty} P(\sqrt{n}\, D_n \leq t) = 1 - 2 \sum_{i=1}^{\infty} (-1)^{i-1} e^{-2i^2 t^2}. \tag{4.6.29}$$

For example, choosing $t = 1.63$, we find that the right-hand side of (4.6.29) is equal to

$$1 - 2 \sum_{i=1}^{\infty} (-1)^{i-1} e^{-5.3138 i^2}$$

$$= 1 - 2(e^{-5.3138} - e^{-(5.3138)(4)} + e^{-(5.3138)(9)} - e^{-(5.3138)(16)} + \cdots)$$
$$= 1 - 2(.0049231803 - .50 \times 10^{-9} + 1.70 \times 10^{-21} - 1.19 \times 10^{-37} + \cdots)$$
$$= .99015.$$

Thus, for large n,

$$P(D_n \leq 1.63/\sqrt{n}) = P(\sqrt{n}\, D_n \leq 1.63) \approx .99105 \approx .99, \quad (4.6.30)$$

so that we have level of significance approximately $1 - .99 = .01$. When $n = 40$, $P(D_{40} \geq 1.63/\sqrt{40}) = P(D_{40} \geq .2577) \approx .01$. The exact result (see Gibbons (1985), p. 400) is $P(D_{40} \geq .252) = .01$.

Remark 4.6.31. A defect of the Box-Muller transformation **if** both the $\sin(\cdot)$ and $\cos(\cdot)$ parts are used has been noted by Bratley, Fox, and Schrage (1983), pp. 210–211. With a multiplicative congruential generator the pairs

$$(X_i, X_{i+1}) = (\sqrt{-2\ln(U_i)}\, \sin(2\pi U_{i+1}), \ \sqrt{-2\ln(U_i)}\, \cos(2\pi U_{i+1})) \quad (4.6.32)$$

with $i = 1, 2, \ldots$ will all lie on a spiral, due to the periodic nature of the $\sin(\cdot)$ and $\cos(\cdot)$ functions. For instance, with the generator used as an example in Section 2.4, namely

$$y_{i+1} = 7y_i \bmod 11, \quad y_0 = 8 \quad (4.6.33)$$

taking $U_i = y_i/11$, we calculate as in Table 4.6–1. The results are plotted in Figure 4.6–2.

Table 4.6–1. Pairs from $\sin(\cdot)$ and $\cos(\cdot)$ Box-Muller transformation.

i	U_i, U_{i+1} from (4.6.30)	(X_i, X_{i+1}) from (4.6.29)
1	(1/11, 7/11)	(−1.66, −1.43)
3	(5/11, 2/11)	(1.14, .52)
5	(3/11, 10/11)	(−.87, 1.36)
7	(4/11, 6/11)	(−.40, −1.36)
9	(9/11, 8/11)	(−.63, −.09)

Note that (4.6.32) with (4.6.33) yields, using the periodicity of the $\sin(\cdot)$ and $\cos(\cdot)$ functions,

$$(X_i, X_{i+1}) = (\sqrt{-2\ln(U_i)}\, \sin(14\pi U_i), \ \sqrt{-2\ln(U_i)}\, \cos(14\pi U_i)), \quad (4.6.34)$$

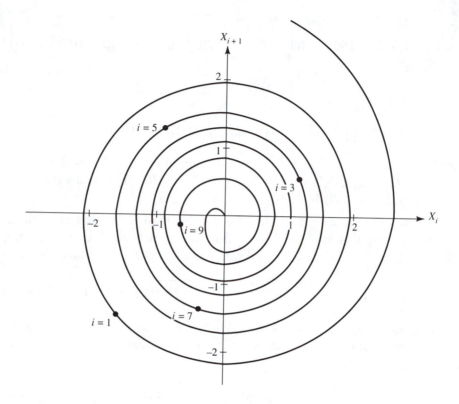

Figure 4.6–2. Pairs (X_i, X_{i+1}) from (3.6.29) if both sin and
cos are used (i odd, generator (3.6.30)).

so that all points will lie on the curve

$$\left(\sqrt{-2\ln(t)}\, \sin\left(14\pi t\right),\ \sqrt{-2\ln(t)}\, \cos\left(14\pi t\right)\right),\ 0 \le t \le 1, \quad (4.6.35)$$

which is plotted in Figure 4.6–2. As t increases from 0, the curve spirals
clockwise with a monotone decreasing distance from the origin. One cycle is
completed between $t = 0$ and $t = 1/7$, with an additional cycle each change
of $1/7$ in the values of t. The last spiral ends at the origin.

This "defect" of the Box-Muller transformation method, combined with a multiplicative congruential generator, is visually im-
pressive. However, we consider this less of a problem than the fact
that in n-dimensions, the vectors from such a generator lie on a (small) finite number of hyperplanes (see Section 3.4). Whether pairs lie on planes, or
spirals, or some other figure in two dimensions does not matter; what does

matter is how well that set of points covers the two-dimensional space. If the multiplier of the generator is large, the spiral will have so many loops that it will "cover" the plane "well" (though that does not answer the question of how many of the "loops" are needed to cover all of the pairs generated). We basically consider the spiral result of this remark an oddity, rather than an item of strong fundamental concern.

4.6.3 The Chi-Square Distribution

The chi-square distribution with n degrees of freedom, denoted by $\chi_n^2(0)$ where n is a positive integer, has probability density function

$$f(x) = \begin{cases} \dfrac{1}{2^{n/2}\Gamma(n/2)} x^{(n/2)-1} e^{-x/2}, & 0 \le x < \infty \\ 0, & \text{otherwise.} \end{cases} \tag{4.6.36}$$

Note that the gamma function defined in (4.6.7) arises in this distribution.

Method 1: Via Normals. It is known that X has the $\chi_n^2(0)$ distribution if (see Dudewicz and Mishra (1988), pp. 141, 267)

$$X = Y_1^2 + Y_2^2 + \cdots + Y_n^2 \tag{4.6.37}$$

where $Y_1, Y_2, \ldots, Y_n$ are independent $N(0,1)$ random variables. Since we already have good methods of generating $N(0,1)$ random variables which are independent, we can generate such an X from n such Y_is by summing their squares.

The above method may be improved when n is large by using the fact that the chi-square is a special case of the gamma distribution. A random variable G has the gamma distribution $G(\alpha, \beta)$ if its probability density function is

$$f_G(x) = \begin{cases} \dfrac{1}{\beta^{\alpha+1}\Gamma(\alpha+1)} x^\alpha e^{-x/\beta}, & 0 \le x < \infty \\ 0, & \text{otherwise} \end{cases} \tag{4.6.38}$$

where $\alpha > -1$ and $\beta > 0$. The $\chi_n^2(0)$ distribution arises if we choose $\alpha = (n/2) - 1$ and $\beta = 2$.

Method 2: Ratio-of-Uniforms. The Ratio-of-Uniforms method of Theorem 4.6.42 may also be used to generate chi-square random variables when the degrees of freedom is at least 2. This method can be quite efficient.

4.6.4   Student's *t*-Distribution

The Student's *t*-distribution family has been widely used in both theoretical and applied statistics since W.S. Gosset ("Student") published his original paper in 1908 (see Section 3.1), and a variety of methods have been developed to generate *t* random variables. The methods below have been included for their simplicity and exactness (Method 1), or for speed and exactness (Method 2). Thus, Method 1 below is easy to derive and exact. However, it can (especially if the degrees of freedom *n* is large) be relatively slow, which can be unsuitable for simulations where many *t* random variables are needed. On the other hand, Method 2 below (while not as simple to derive) is also exact, but much faster: for $n = 1$ Method 2 needs only about 50% of the time of Method 1; for $n = 2$, 25%; $n = 3$, 20%; $n = 10$, 6%; $n = 20$, 2%; $n = 70$, .2% (i.e., when $n = 70$, Method 1 takes about 510 times as long as does Method 2, so with Method 2 one can generate about 510 *t* random variables in the time needed for just one *t* random variable with Method 1). Clearly this sort of speed disparity can make the difference between time-feasible and time-infeasible computing when one has a simulation with large *n* which needs many replications for accuracy of the final results.

Method 1: Via Normal and Chi-Square. The Student's *t*-distribution with *n* degrees of freedom (*n* being a positive integer) has probability density function

$$f(x) = \frac{\Gamma((n+1)/2)}{\sqrt{n\pi}\,\Gamma(n/2)} \frac{1}{(1+x^2/n)^{(n+1)/2}}, \quad -\infty < x < \infty. \quad (4.6.39)$$

It is known (see Dudewicz and Mishra (1988), p. 481) that *X* has the Student's *t*-distribution with *n* degrees of freedom if

$$X = \frac{Y_1}{\sqrt{Y_2/n}} \quad (4.6.40)$$

where Y_1 is a $N(0,1)$ random variable and Y_2 is a $\chi_n^2(0)$ random variable independent of Y_1. We already have methods of generating Y_1 and Y_2 (which will be independent as long as disjoint uniforms are used in their generation), hence we can use (4.6.40) to generate Student's *t* random variables.

When $n = 1$, the distribution is also sometimes called the **Cauchy distribution**, and the probability density function is then

$$f(x) = \frac{1}{\pi(1+x^2)}, \quad -\infty < x < \infty. \quad (4.6.41)$$

In verifying this from (4.6.39), one uses the fact that $\Gamma(1/2) = \sqrt{\pi}$.

Method 2: Ratio-of-Uniforms. This method, due to Kinderman, Monahan, and Ramage (1977), is applicable to many distributions, not just to the t-distribution. It is based on the following theorem.

Theorem 4.6.42. *Let $f(x)$ be a bounded probability density function, for which $x^2 f(x)$ is also bounded. Then the set of points*

$$A = \{(u(x), v(x)) : u(x) = \sqrt{f(x)}, \; v(x) = x\sqrt{f(x)}\}, \qquad (4.6.43)$$

where x ranges over the real numbers, is contained in a rectangle.

While we will not give a formal proof of Theorem 4.6.42, note that the largest possible value of $u(x)$ is the square-root of the largest possible value of $f(x)$ (which exists since $f(x)$ was assumed to be bounded). Also, the largest possible value of $v^2(x)$ is the largest possible value of $x^2 f(x)$ (which also exists since $x^2 f(x)$ was assumed to be bounded).

Now suppose we generate a random point (U, V) in the region A. (Note that this point (U, V) is easy to generate. If the rectangle with base running from a to b, and vertical height running from c to d includes A, let S_1 be uniform on $[a, b]$ and let S_2 be uniform on $[c, d]$, with S_1 and S_2 independent random variables. If (S_1, S_2) is in A, let $(U, V) = (S_1, S_2)$. Otherwise, try again until a pair in A is generated.) Then we have

Theorem 4.6.44. *Let $X = V/U$. Then X has probability density function $f(\cdot)$.*

While we will not prove Theorem 4.6.44, note that we know the joint density of (U, V) is uniform over A. Thus, finding the distribution function of X reduces to finding the area of that part of A where $X = V/U$ is $\leq x$. Differentiating this yields the probability density function of X as $f(\cdot)$. In Figures 4.6–3, 4.6–4, and 4.6–5 we graph A for Student's t, chi-square, and normal p.d.f.s respectively. In Figure 4.6–3, the curves are labeled with numbers that indicate the degrees of freedom for the corresponding t distribution; the same is true in Figure 4.6–4. In Figure 4.6–5, the normal distributions, $N(\mu, \sigma^2)$, that correspond to the curves have $\mu = 0$ and σ^2 as shown in the figure. Note that the $\chi^2_1(0)$ p.d.f. does not satisfy the boundedness requirement; hence, A is not bounded in that case.

An algorithm implementing the above efficiently can be developed as follows (Kinderman and Monahan (1980)). Noting that the Student's t probability density function $f(x)$ in (4.6.39) can be written as

$$f(x) = ch(x) \qquad (4.6.45)$$

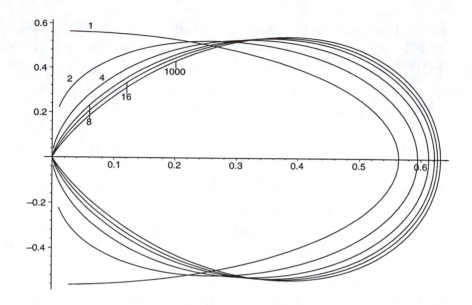

Figure 4.6–3. The region A of Theorem 4.6.42 for Student's t
p.d.f.s with various degrees of freedom.

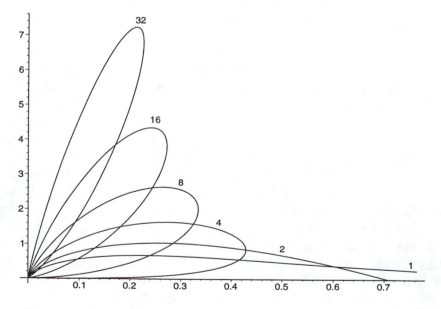

Figure 4.6–4. The region A of Theorem 4.6.42 for chi-square
p.d.f.s with various degrees of freedom.

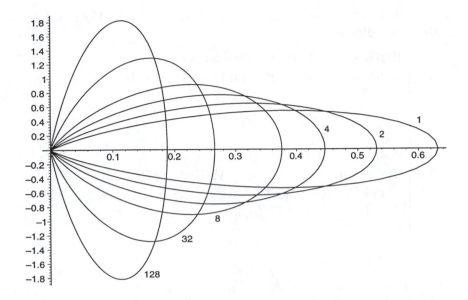

Figure 4.6–5. The region A of Theorem 4.6.42 for normal p.d.f.s with mean 0 and various variances.

where

$$c = \Gamma((n+1)/2)(n\pi)^{-0.5}/\Gamma(n/2), h(x) = (1 + x^2/n)^{-(n+1)/2} \quad (4.6.46)$$

and c and $h(x)$ are functions of n, we easily derive the following facts. First,

$$\frac{u(x)}{\sqrt{c}} = \sqrt{h(x)} = (1 + \frac{x^2}{n})^{-(n+1)/4} \quad (4.6.47)$$

is a decreasing function of x^2 whose maximum value of 1 occurs at $x = 0$. Since $v(x) = xu(x)$, the minimum and maximum values of $v(x)/\sqrt{c}$ (say m and M, respectively) can be shown to be

$$-m = M = (2n)^{0.5}(n+1)^{-(n+1)/4}(n-1)^{(n-1)/4} \quad (4.6.48)$$

for any positive integer n. (Note that this value in (4.6.48) equals 1 when $n = 1$.) Then we can show the following.

Theorem 4.6.49. *Let $U_1, U_2, \dots$ and $V_1, V_2, \dots$ be independent uniform random variables, with $U_1, U_2, \dots$ uniform on $[0, 1]$ and $V_1, V_2, \dots$ uniform* *on $[m, M]$. Calculate $p = -(n+1)/4$, $a = 4(1+1/n)^{-p}$, $e = 16/a$ and define a variable X as follows:*

$$
\begin{cases}
\text{if } aU_1 \leq 5 - (V_1/U_1)^2, \text{ then let } X = V_1/U_1; \\
\text{if } aU_1 > 5 - (V_1/U_1)^2, \text{ and } (V_1/U_1)^2 < e/U_1 - 3, \\
\quad \text{and } U_1 \leq (1 + (V_1/U_1)^2/n)^p, \\
\qquad \text{then let } X = V_1/U_1; \\
\vdots \\
\text{if } aU_i \leq 5 - (V_i/U_i)^2, \text{ then let } X = V_i/U_i; \qquad (4.6.50) \\
\text{if } aU_i > 5 - (V_i/U_i)^2, \text{ and } (V_i/U_i)^2 < e/U_i - 3, \\
\quad \text{and } U_i \leq (1 + (V_i/U_i)^2/n)^p, \\
\qquad \text{then let } X = V_i/U_i; \\
\vdots \\
.
\end{cases}
$$

If $n = 1$, 2, or 3, then the conditions $(V_i/U_i)^2 < e/U_i - 3$ (for $i = 1, 2, , \ldots$) are omitted from (4.6.50). Then X has probability density function (4.6.39), i.e., Student's t with n degrees of freedom.

4.6.5   The Erlang Distribution

Method 1: Via Exponentials. The Erlang distribution with parameters r (a positive integer) and $\alpha > 0$ has probability density function

$$
f(x) = \begin{cases} \dfrac{\alpha}{\Gamma(r)} (\alpha x)^{r-1} e^{-\alpha x}, & x \geq 0 \\ 0, & \text{otherwise.} \end{cases} \qquad (4.6.51)
$$

It is known (see Dudewicz and Mishra (1988), pp. 275–277) that if

$$
X = Y_1 + Y_2 + \cdots + Y_r \qquad (4.6.52)
$$

where $Y_1, \ldots, Y_r$ are independent exponential random variables with mean $1/\alpha$ (and hence have probability density function (4.2.5) with $\lambda = 1/\alpha$), then X has the Erlang(r, α) distribution. Since we know how to generate such exponential random variables, they can be used to generate the Erlang distribution.

Method 2: Via Gamma. From the expression for the probability density function of the gamma distribution with parameters α and β in (4.6.38), we see that by choosing

$$
\alpha = r - 1, \quad \beta = 1/\alpha_0, \qquad (4.6.53)
$$

we obtain the Erlang(r, α_0) density in (4.6.51). Hence, methods appropriate for generating gamma random variables can also be used to generate Erlang random variables.

4.6.6 Double-Exponential Distribution: Via Exponential

The double-exponential (or **Laplace**) distribution with parameter $\lambda > 0$ has probability density function (see Dudewicz and Mishra (1988), p. 364)

$$f(x) = \frac{1}{2\lambda} e^{-|x|/\lambda}, \qquad -\infty < x < \infty. \tag{4.6.54}$$

If Y and Z are independent random variables, Y is exponential with mean λ, and hence probability density function (4.2.5), and Z takes values ± 1 with probability 0.5 each, then $X = YZ$ has probability density function (4.6.54). Since we know how to generate exponential random variables, and when we set

$$Z = \begin{cases} 1 & \text{if } U \geq 0.5 \\ -1 & \text{if } U < 0.5 \end{cases} \tag{4.6.55}$$

we have $P[Z = 1] = P[Z = -1] = 0.5$, we are in a position to generate the desired Laplace distribution, which thus turns out to be an exponential with mean λ with a random sign attached to it.

The proof that YZ has the desired probability density function proceeds as follows. First, the distribution function of YZ is

$$F_{YZ}(x) = P[YZ \leq x] = \frac{1}{2}P[Y \leq x] + \frac{1}{2}P[-Y \leq x]$$

$$= \begin{cases} \frac{1}{2}\int_0^x \frac{1}{\lambda}e^{-y/\lambda}\,dy + \frac{1}{2} = \frac{1}{2}(1 - e^{-x/\lambda}) + \frac{1}{2} \\ \qquad = 1 - \frac{1}{2}e^{-x/\lambda}, \qquad\qquad\qquad \text{if } 0 \leq x \\[2mm] \frac{1}{2}\cdot 0 + \frac{1}{2}P[Y \geq -x] = \frac{1}{2}(1 - P[Y \leq -x]) \\ \qquad = \frac{1}{2} - \frac{1}{2}(1 - e^{x/\lambda}) = \frac{1}{2}e^{x/\lambda}, \ \text{if } x < 0. \end{cases} \tag{4.6.56}$$

Hence the probability density function of YZ is

$$f_{YZ}(x) = \begin{cases} \frac{1}{2}\cdot\frac{1}{\lambda}e^{-x/\lambda} & \text{if } 0 \leq x \\ \frac{1}{2}\cdot\frac{1}{\lambda}e^{x/\lambda} & \text{if } x < 0 \end{cases} = \frac{1}{2\lambda}e^{-|x|/\lambda} \tag{4.6.57}$$

as was shown.

4.6.7   The F Distribution: Via Chi-Squares

If X_1 is a chi-square random variable with n_1 degrees of freedom and X_2 is a chi-square random variable with n_2 degrees of freedom, and X_1 and X_2 are independent random variables, then the ratio

$$X = \frac{X_1/n_1}{X_2/n_2} \qquad (4.6.58)$$

has the F-distribution with n_1 and n_2 degrees of freedom. The probability density function of X is (see Dudewicz and Mishra (1988) pp. 145, 186)

$$f(x) = \frac{\Gamma((n_1+n_2)/2))(n_1/n_2)^{n_1/2}}{\Gamma(n_1/2)\Gamma(n_2/2)} \frac{x^{(1/2)(n_1-2)}}{(1+n_1 x/n_2)^{(1/2)(n_1+n_2)}} ,\, 0 < x \quad (4.6.59)$$

and $f(x) = 0$ if $x \leq 0$. Since we know from (4.6.37) how to generate chi-square random variables that are independent, F random variables can be generated via (4.6.58).

4.6.8   The Beta Distribution

Method 1: Via F (Special Parameter Values). The beta distribution with parameters $\alpha > -1$ and $\lambda > -1$ has probability density function

$$f(x) = \frac{1}{\beta(\alpha+1,\lambda+1)} x^\alpha (1-x)^\lambda, \quad 0 \leq x \leq 1 \qquad (4.6.60)$$

and $f(x) = 0$ for all other x, where the beta function is defined as

$$\beta(\alpha+1,\lambda+1) = \int_0^1 x^\alpha (1-x)^\lambda \, dx = \frac{\Gamma(\alpha+1)\Gamma(\lambda+1)}{\Gamma(\alpha+\lambda+2)}. \qquad (4.6.61)$$

For proof of the second equality, see Dudewicz and Mishra (1988), p. 134.

Suppose that $n_1 = 2(\lambda+1)$ and $n_2 = 2(\alpha+1)$ are integers and Y is a random variable with an F distribution with n_1 and n_2 degrees of freedom. Then it is known (see Dudewicz and Mishra (1988), p. 186) that

$$X = \frac{1}{1 + \dfrac{\lambda+1}{\alpha+1} Y} \qquad (4.6.62)$$

has a beta distribution with parameters α and λ. Since we know how to generate F random variables by (4.6.58), **we can generate beta random** variables via (4.6.62) whenever $2(\lambda + 1)$ and $2(\alpha + 1)$ are integers, i.e., whenever α and λ take on the values $-0.5, 0, 0.5, 1, 1.5, 2, 2.5, 3, \ldots$.

Method 2: Via Inverse Distribution Function. To use the inverse distribution function method of Section 4.2 to generate a beta random variable with parameters $\alpha > -1$ and $\lambda > -1$, one sets

$$X = F^{-1}(U), \tag{4.6.63}$$

so that $F(X) = U$ or, by (4.6.51), X is the solution of

$$\int_0^X \frac{1}{\beta(\alpha + 1, \lambda + 1)} x^\alpha (1 - x)^\lambda \, dx = U. \tag{4.6.64}$$

Using the incomplete beta function of (4.6.5), X is a solution of

$$I_X(\alpha + 1, \lambda + 1) = U. \tag{4.6.65}$$

If accurate computation of the incomplete beta function is available, then **Newton-Raphson iteration** can be used to converge on the root X of (4.6.65), using the sequence of successive approximations $x_0, x_1, x_2, \ldots$ where

$$x_{i+1} = x_i - \frac{F(x_i) - U}{f(x_i)}. \tag{4.6.66}$$

In (4.6.66) $F(\cdot)$ is the beta distribution function (4.6.64) (equivalently, the left side of (4.6.65)), and $f(\cdot)$ is the beta density function (4.6.60). x_0 is a first approximation to the root, and can be taken as $1/2$, or one could evaluate (4.6.64) at each of $x = .01, .02, \ldots, .99$ and take the one that is closest to value U as the first approximation. Newton's method of iteration has been employed in many recent papers to achieve speedy calculations of statistical tables; e.g., see Chen and Mithongtae (1986). Note that, in general, if one wishes a root of the function $g(x)$, the Newton-Raphson sequence of approximations uses

$$x_{i+1} = x_i - \frac{g(x_i)}{g'(x_i)}. \tag{4.6.67}$$

If Newton's method suggests a "root" which is not between the two best so far (e.g., the closest two of $x = .01, .02, \ldots, .99$ when one starts), or is perhaps not even in the interval $(0, 1)$, one should replace the Newton iterate with $x_M = (x_L + x_U)/2$, where x_L is the closest so far which is below the root (and x_U is the closest so far which is above the root).

Method 3: Ratio-of-Uniforms. The Ratio-of-Uniforms method of Theorem 4.6.42 may be used to generate beta random variables when the power parameters α and λ are positive, which guarantees that $f(x)$ is bounded.

See pp. 207–208 of Tadikamalla (1984) for references to some exact and fast methods of generating beta random variables.

4.6.9   The Weibull Distribution: Via the Exponential

The Weibull distribution with parameters $\alpha > 0$ and $\beta > 0$ has probability density function

$$f(x) = \alpha\beta x^{\beta-1} e^{-\alpha x^\beta}, \quad x \geq 0 \tag{4.6.68}$$

and $f(x) = 0$ for all other x. It is known (see Dudewicz and Mishra (1988), p. 175) that if Y is exponential with mean $1/\alpha$, then $Y^{1/\beta}$ is Weibull with probability density function (4.6.68). Since we know how to generate an exponential random variable with mean $1/\alpha$, we can generate a Weibull random variable by taking

$$X = Y^{1/\beta}. \tag{4.6.69}$$

4.6.10   The Lognormal Distribution: Via the Normal

The random variable X is said to have a lognormal distribution with parameters μ and σ^2 if $\ln(X)$ is $N(\mu, \sigma^2)$. Since we know how to generate such $N(\mu, \sigma^2)$ variables, we can generate the desired X by taking

$$X = e^Y \tag{4.6.70}$$

since then $\ln(X) = \ln(e^Y) = Y$ is $N(\mu, \sigma^2)$.

4.6.11   The Gamma Distribution

A random variable X is said to have the gamma distribution with parameters $\alpha > -1$ and $\beta > 0$ if its probability density function is

$$f(x) = \frac{x^\alpha e^{-x/\beta}}{\beta^{\alpha+1}\Gamma(\alpha+1)}, \quad 0 < x \tag{4.6.71}$$

and $f(x) = 0$ for all other x, where the gamma function is defined in (4.6.7).

Method 1: Via Exponentials (Special Parameter Values). We saw at (4.4.9) that a sum of k independent exponential random variables each with mean one has the gamma distribution with $\alpha = k - 1$ and $\beta = 1$. As Section 4.2.1 shows how to generate such exponential random variables, this method can be used for these special parameter values.

Method 2: Ratio-of-Uniforms. The Ratio-of-Uniforms method of Theorem 4.6.42 may be used to generate gamma random variables if α is positive.

Method 3: Mixture Method for $\alpha < 1$, Rejection Method for $\alpha > 1$. Tadikamalla and Johnson (1981) studied gamma variate generation extensively. They gave FORTRAN code for algorithms GS and G4PE (as well as for three other algorithms) on their pp. 227–233, and included ten test values that can be used to assure correct implementation of their code. (A correction note to Tadikamalla and Johnson (1981) appears in Tadikamalla and Johnson (1982), and has a one-line correction to the G4PE code; the rest of their 1981 paper was based on the correct version of G4PE.)

Algorithm GS uses what is called the "mixture method" where $f_1(x)$ is the power function density and $f_2(x)$ is the truncated exponential distribution. This method is based on representing the density f (such as the gamma) from which variates are to be generated as $f(x) = p_1 f_1(x) + p_2 f_2(x)$ where $p_1 + p_2 = 1$ (each non-negative) and f_1 and f_2 are each p.d.f.s (which is the case of a mixture of $k = 2$ densities). This algorithm handles cases where $\alpha < 1$, and is highly recommended for cases with $0 < \alpha < 1$.

Algorithm G4PE is a rejection method which uses several uniforms and an exponential (in the tail) as an "envelope" for the gamma. It handles cases where $\alpha \geq 1$, and is highly recommended for those cases (see their pp. 223–224). (For optimal rejection methods for the gamma, see pp. 127–129 of Dieter (1991).)

4.7 Generating p.d.f.s Bounded on a Closed Interval

Many **univariate** probability density functions $f(x)$ from which we may wish to generate random variables are **bounded on a closed interval**, i.e., there are numbers $a < b$ and $m > 0$ such that

$$f(x) = 0 \text{ for } x < a,$$
$$f(x) = 0 \text{ for } x > b, \qquad\qquad (4.7.1)$$
$$f(x) \leq m \text{ for } a \leq x \leq b.$$

For example, distributions fitted to a histogram (by methods discussed later in this chapter) commonly have this property. For those distributions that are not bounded on a closed interval (e.g., exponential, normal) we have other methods of generation. A typical p.d.f. $f(x)$ satisfying (4.7.1) is shown in Figure 4.7–1. Theorem 4.7.2 allows us to generate exact random variables with p.d.f. $f(x)$. The procedure is called a **rejection** method.

Theorem 4.7.2. *Let $U_1, U_2, \ldots$ be independent uniform random variables on $(0, 1)$. Let $f(\cdot)$ be a p.d.f. satisfying (4.7.1) for some numbers $a < b$, and $m > 0$. Define a variable X as follows:*

$$
\begin{cases}
\text{if } mU_2 \leq f(a + (b - a)U_1), \text{ then let } X = a + (b - a)U_1; \\
\text{if } mU_2 > f(a + (b - a)U_1), \text{ and } mU_4 \leq f(a + (b - a)U_3), \\
\quad \text{then let } X = a + (b - a)U_3; \\
\vdots \\
\text{if } mU_{i+1} > f(a + (b - a)U_i) \text{ for } i = 1, 3, \ldots, I - 2, I, \text{ and} \\
\quad mU_{I+3} \leq f(a + (b - a)U_{I+2}), \text{ then let } X = a + (b - a)U_{I+2}; \\
\vdots
\end{cases}
\tag{4.7.3}
$$

Then X has probability density function $f(\cdot)$.

The process described in Theorem 4.7.2 is illustrated in Figure 4.7–2. There, one selects numbers with a uniform distribution on $[a, b]$ until one finds the first such number, say S_1, for which $f(S_1)$ is not exceeded by a random number S_2 on $[0, m]$. Each pair (S_1, S_2) is independent of the previous pair, making each pair a random point on the rectangle with base $[a, b]$ and height $[0, m]$. Now one sets $X = S_1$. Intuitively, this process yields the "right" distribution of X since values on the x-axis occur in the "right" proportions: if $f(x_1)$ is larger than $f(x_2)$, then an $S_1 = x_1$ is proportionately more likely to be accepted than is an $S_1 = x_2$.

Proof of Theorem 4.7.2. If $F(\cdot)$ denotes the distribution function corresponding to probability density function $f(\cdot)$, then for any value x

$$
F(x) = \int_{-\infty}^{x} f(s) \, ds.
\tag{4.7.4}
$$

Let S_1 be uniform on $[a, b]$, and S_2 be uniform on $[0, m]$, with S_2 independent of S_1. Let X be the value produced by the algorithm of Theorem 4.7.2. Then

$$
P[X \leq t] = P[S_1 \leq t | S_2 \leq f(S_1)],
\tag{4.7.5}
$$

and since by the definition of conditional probability for any events A and B we have $P(A|B) = P(A \text{ and } B)/P(B)$, we have

$$
P[X \leq t] = \frac{P[S_1 \leq t \text{ and } S_2 \leq f(S_1)]}{P[S_2 \leq f(S_1)]}.
\tag{4.7.6}
$$

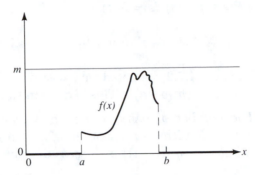

Figure 4.7–1. A typical p.d.f., $f(\cdot)$, bounded on a closed interval.

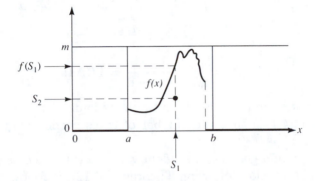

Figure 4.7–2. A pair (S_1, S_2) that leads to $X = S_1$ in Theorem 4.7.2.

The denominator of (4.7.6) is $1/(m(b-a))$, which we show later in detail. Therefore, we have

$$
\begin{aligned}
P[X \le t] &= P[S_1 \le t \text{ and } S_2 \le f(S_1)] \cdot m(b-a) \\
&= \int_{-\infty}^{\infty} P[S_1 \le t \text{ and } S_2 \le f(S_1)|S_1 = s] f_{S_1}(s)\, ds \cdot m(b-a) \\
&= \int_a^b P[s \le t \text{ and } S_2 \le f(s)]\frac{1}{b-a}\, ds \cdot m(b-a) \qquad (4.7.7) \\
&= \int_a^b P[s \le t]\frac{f(s)}{m}\frac{1}{b-a}\, ds \cdot m(b-a)
\end{aligned}
$$

$$= \int_a^b P[s \leq t] f(s) \, ds = F(t).$$

From this we see that X has distribution function (4.7.4), and thus p.d.f. $f(\cdot)$.

In the proof of Theorem 4.7.2, we noted that we would show later that the denominator of (4.7.6) is $1/(m(b-a))$. This is done in the following Lemma.

Lemma 4.7.8. *The number of pairs of random variables* $(U_i, U_{i+1}), i = 1, 3, 5, \ldots,$ *used by the algorithm of Theorem 4.7.2 is a geometric random variable with probability* $1/(m(b-a))$ *of success on each trial.*

Proof. The probability that a pair (U_i, U_{i+1}) terminates the algorithm is (see (4.7.6))

$$P[S_2 \leq f(S_1)] = \int_a^b P[S_2 \leq f(S_1)|S_1 = s] \frac{1}{b-a} \, ds$$

$$= \int_a^b P[S_2 \leq f(s)] \frac{1}{b-a} \, ds \qquad (4.7.9)$$

$$= \int_a^b \frac{f(s)}{m} \frac{1}{b-a} = 1/(m(b-a)).$$

Since pairs are tried independently until one is found for which $S_2 \leq f(S_1)$, it follows (see (4.4.11)) that the number of trials needed is geometric with $p = 1/(m(b-a))$.

Since the mean of a geometric random variable with success probability p is $1/p$ (see the discussion following Theorem 4.4.12), it follows immediately from Lemma 4.7.8 that we have the following.

Corollary 4.7.10. *The average number of uniform random variables required by the algorithm of Theorem 4.7.2 to produce one random variable with p.d.f.* $f(\cdot)$ *is* $2m(b-a)$.

From Corollary 4.7.10, we see that it is desirable to have m be the least upper bound on $f(x)$, since then the algorithm will be most efficient. Also, the algorithm will be most efficient for distributions that are "not far" from uniform.

The ideas above also extend to the case of a **multivariate** probability density function $f(x_1, \ldots, x_n)$ from which we may wish to generate random variables, as long as $f(x_1, \ldots, x_n)$ **is bounded on a closed rectangle**, i.e., there are numbers $a_1 < b_1, a_2 < b_2, \ldots, a_n < b_n$, and $m > 0$ such that

$$f(x_1, \ldots, x_n) \leq m \text{ for } a_1 \leq x_1 \leq b_1, \ldots, a_n \leq x_n \leq b_n,$$
$$f(x_1, \ldots, x_n) = 0 \text{ for all other } (x_1, \ldots, x_n). \qquad (4.7.11)$$

For example, multivariate distributions fitted to a data set will often be of this type (i.e., will often have bounded range and height). The following theorem allows us to generate exact random variables with p.d.f. $f(x_1, \ldots, x_n)$.

Theorem 4.7.12. *Let $U_1, U_2, \ldots$ be independent uniform random variables on $(0, 1)$. Let $f(x_1, \ldots, x_n)$ be an n-variate p.d.f. satisfying (4.7.11) for some numbers $a_i \leq b_i$ $(i = 1, \ldots, n)$ and $m > 0$. Define an n-variate variable $(X_1, \ldots, X_n)$ as:*

$$
\left\{
\begin{array}{l}
\text{if } mU_{n+1} \leq f(a_1 + (b_1 - a_1)U_1, \ldots, a_n + (b_n - a_n)U_n), \text{ then let} \\
\quad (X_1, \ldots X_n) = (a_1 + (b_1 - a_1)U_1, \ldots, a_n + (b_n - a_n)U_n); \\
\text{if } mU_{n+1} > f(a_1 + (b_1 - a_1)U_1, \ldots, a_n + (b_n - a_n)U_n), \text{ and} \\
\quad mU_{2n+2} \leq f(a_1 + (b_1 - a_1)U_{n+2}, \ldots, a_n + (b_n - a_n)U_{2n+1}), \text{ then let} \\
\quad (X_1, \ldots X_n) = (a_1 + (b_1 - a_1)U_{n+2}, \ldots, a_n + (b_n - a_n)U_{2n+1}); \\
\vdots \\
\text{if for } i = 1, \ldots I \; mU_{i(n+1)} > \\
\quad f(a_1 + (b_1 - a_1)U_{(i-1)(n+1)+1}, \ldots, a_n + (b_n - a_n)U_{(i-1)(n+1)+n}) \\
\quad \text{and } mU_{(I+1)(n+1)} \leq \\
\quad f(a_1 + (b_1 - a_1)U_{I(n+1)+1}, \ldots, a_n + (b_n - a_n)U_{I(n+1)+n}), \\
\quad \text{then let } (X_1, \ldots X_n) = \\
\quad (a_1 + (b_1 - a_1)U_{I(n+1)+1}, \ldots, a_n + (b_n - a_n)U_{I(n+1)+n}); \\
\vdots \\
.
\end{array}
\right.
$$

$$(4.7.13)$$

Then $(X_1, \ldots, X_n)$ has probability density function $f(\cdot, \ldots, \cdot)$.

The process described in Theorem 4.7.12 is illustrated similarly to Figure 4.7–1, except that there one takes x, a, and b to be, respectively, the vectors $(x_1, \ldots, x_n)$, $(a_1, \ldots, a_n)$, and $(b_1, \ldots, b_n)$. One selects vectors uniformly on the rectangle that contains $(x_1, \ldots, x_n)$ until a point $(S_1, \ldots, S_n)$ is found for which $f(S_1, \ldots, S_n)$ is not exceeded by a random number on $[0, m]$. Intuitively, this process yields the "right" distribution for $(X_1, \ldots, X_n)$ since values occur in the "right" proportion: if $f(x'_1, \ldots, x'_n)$ is larger than $f(x_1, \ldots, x_n)$ then an $(S_1, \ldots, S_n) = (x'_1, \ldots, x'_n)$ is more likely to be accepted than is an $(S_1, \ldots, S_n) = (x_1, \ldots, x_n)$. The proof of Theorem 4.7.12 is a minor generalization of that of Theorem 4.7.2, and will be omitted.

An important generalization is to the case where we have a p.d.f. $g(\cdot)$ and a number $a > 1$ such that

$$f(x) < ag(x) \text{ for all } x. \tag{4.7.14}$$

Note that in this case we do not require that $f(\cdot)$ be bounded on a closed interval; it is enough if it is bounded by a p.d.f. $g(\cdot)$ from which we know how to generate random variables. The result here (whose proof is similar to that of Theorem 4.7.2 and will therefore be omitted) is the following.

Theorem 4.7.15. *Let $U_1, U_2, \ldots$ be independent uniform random variables on $(0,1)$, and let $Y_1, Y_2, \ldots$ be independent random variables with p.d.f. $g(\cdot)$. Let $f(\cdot)$ be a p.d.f. satisfying (4.7.14) for some number $a > 1$. Define a variable X as follows:*

$$
\begin{cases}
\text{if } ag(Y_1)U_1 \le f(Y_1), & \text{then let } X = Y_1; \\
\text{if } ag(Y_1)U_1 > f(Y_1), \text{ and } ag(Y_2)U_2 \le f(Y_2), & \text{then let } X = Y_2; \\
\quad\vdots & \\
\text{if } ag(Y_i)U_i > f(Y_i) \text{ for } i = 1, \ldots, I, \text{ and} & \quad\quad (4.7.16) \\
\quad ag(Y_{I+1})U_{I+1} \le f(Y_{I+1}), & \text{then let } X = Y_{I+1}; \\
\quad\vdots & \\
\quad\cdot &
\end{cases}
$$

Then X has probability density function $f(\cdot)$.

Note that the process described in Theorem 4.7.15 takes a random variable Y that has p.d.f. $g(\cdot)$, and accepts that value for X if a uniform random variable on $[0, ag(Y)]$ is no larger than $f(Y)$. Thus, values with larger $f(Y)$ are more likely to be accepted as the final value for X.

4.8   Generating Multivariate Normal Random Variables

In the **multivariate case**, the random variable $(X_1, X_2, \ldots, X_p)$ of interest has some number $p \ge 2$ of components. How to generate such a random variable in the **discrete case** was dealt with in Section 4.5 (for the case where only a finite number of p-tuples of numbers have positive probability). In the **continuous case**, the distribution function

$$F_{X_1, \ldots, X_p}(x_1, \ldots, x_p) = P[X_1 \le x_1, \ldots, X_p \le x_p] \quad\quad (4.8.1)$$

can be written as the integral of a function $f(x_1, \ldots, x_p)$ that is nonnegative and integrates (with limits $-\infty$ to ∞ on each of $x_1, \ldots, x_p$) to 1 called the **probability density function** of $(X_1, \ldots, X_p)$:

$$F_{X_1, \ldots, X_p}(a_1, \ldots, a_p) = \int_{-\infty}^{a_p} \cdots \int_{-\infty}^{a_1} f(x_1, \ldots, x_p)\, dx_1 \ldots dx_p. \quad (4.8.2)$$

If $f(x_1, \ldots, x_p)$ is bounded on a closed rectangle (and zero otherwise), how to generate $(X_1, \ldots, X_p)$ was dealt with in Section 4.7. One of the most important multivariate continuous distributions (believed by some to be *the* most important such distribution) is **the multivariate normal distribution**; in this section we study how to generate random variables with that distribution.

The p-variate random variable $(X_1, \ldots, X_p)$ is said to have the **multivariate normal distribution** (abbreviated **MVN**) if and only if its probability density function can be written as

$$f(x_1, \ldots, x_p) = \sqrt{|\sigma^{ij}|} \, (2\pi)^{-p/2} e^{-0.5 Q(\underset{\sim}{x})} \tag{4.8.3}$$

where we have denoted by $\underset{\sim}{x}$ the vector $(x_1, \ldots, x_p)$ and

$$\begin{cases} Q(\underset{\sim}{x}) = \displaystyle\sum_{i,j=1}^{p} \sigma^{ij}(x_i - \mu_i)(x_j - \mu_j) \\[2mm] \mu_i \quad = E(X_i) \\[2mm] \sigma_{ij} \quad = \mathrm{Cov}(X_i, X_j) \\[2mm] (\sigma^{ij}) = (\sigma_{ij})^{-1}. \end{cases} \tag{4.8.4}$$

Note that $E((X_i - \mu_i)(X_j - \mu_j)) = E((X_j - \mu_j)(X_i - \mu_i))$, hence $\sigma_{ij} = \sigma_{ji}$ and the so-called **variance-covariance matrix** (σ_{ij}) is a $p \times p$ symmetric matrix. $\underset{\sim}{X} = (X_1, \ldots, X_p)$ is called MVN with **mean vector** $\underset{\sim}{\mu} = (\mu_1, \ldots, \mu_p)$, and covariance matrix (σ_{ij}).

The following result (proven using matrix algebra and multivariate statistics) is a key to our method of generating MVN random variables.

Theorem 4.8.5. *If $\underset{\sim}{X}^* = (X_1^*, \ldots, X_p^*)$ is a vector of independent standard normal random variables, then $\underset{\sim}{X} = (X_1, \ldots, X_p)$ defined by*

$$\underset{\sim}{X}' = L\underset{\sim}{X}^{*\prime} + \underset{\sim}{\mu}' \tag{4.8.6}$$

is MVN with mean vector $\underset{\sim}{\mu}$ and covariance matrix (σ_{ij}), where L is lower triangular such that $LL' = (\sigma_{ij})$ (and is called the lower triangular, or Cholesky, decomposition of (σ_{ij})).

Recall from Section 4.6 that if $U_1, \ldots, U_{2p}$ are independent uniform random variables on $(0, 1)$, then by the Box-Muller transformation (4.6.21) we may set

$$X_i^* = \sqrt{-2\ln(U_{2i-1})} \, \sin(2\pi U_{2i}), \qquad i = 1, 2, \ldots, p. \tag{4.8.7}$$

The following algorithm can be used to compute the Cholesky decomposition needed in Theorem 4.8.5.

Algorithm 4.8.8. *Let* (σ_{ij}) *be a* $p \times p$ *symmetric positive-definite matrix. Then the Cholesky decomposition of* (σ_{ij}), *i.e., a lower triangular matrix* L *such that* $LL' = (\sigma_{ij})$, *results from application of the following steps.*

Step 4.8.8–1. In p successive steps (for $j = 1, \ldots, p$) compute as follows at step j:

$$d_j = \sigma_{jj} - \sum_{q=1}^{j-1} l_{jq}^2 d_q$$

$$l_{ij} = \left(\sigma_{ij} - \sum_{q=1}^{j-1} l_{jq} l_{iq} d_q\right) d_j^{-1} \quad \text{for } i = j+1, \ldots, p$$

where summation terms are taken as zero when $j = 1$. From the results form the matrices $D = \text{diag}(d_i)$ and $H = (l_{ij})$.

Step 4.8.8–2. Set diagonal entries of H to 1, and set the above-diagonal entries to 0.

Step 4.8.8–3. Let $D_1 = \sqrt{D}$. (Since D is diagonal, so is D_1 and its i-th diagonal element is $\sqrt{d_i}$ where d_i is the i-th diagonal element of D.)

Step 4.8.8–4. Finally, set $L = HD_1$.

The following example makes the ease of application of Algorithm 4.8.8 clear. Suppose we wish to find the Cholesky decomposition of the $p \times p$ ($p = 3$) matrix

$$V = \begin{pmatrix} 2 & 0 & 1 \\ 0 & 1 & 0 \\ 1 & 0 & 1 \end{pmatrix}.$$

Then application of Algorithm 4.8.8 proceeds as follows.

Step 1.   With $j = 1$,

$$d_1 = \sigma_{11} - 0 = 2 - 0 = 2$$

$$l_{21} = (\sigma_{21} - 0)2^{-1} = (0 - 0)\frac{1}{2} = 0$$

$$l_{31} = (\sigma_{31} - 0)2^{-1} = (1 - 0)\frac{1}{2} = \frac{1}{2}.$$

With $j = 2$,

$$d_2 = \sigma_{22} - l_{21}^2 d_1 = 1 - 0 \cdot 1 = 1$$

$$l_{32} = (\sigma_{32} - l_{21}l_{31}d_1)d_2^{-1} = 0.$$

With $j = 3$,

$$d_3 = \sigma_{33} - (l_{31}^2 d_1 + l_{32}^2 d_2) = 1 - \left(\frac{1}{4} \cdot 2 + 0\right) = \frac{1}{2}.$$

Step 2. We now have matrices

$$D = \begin{pmatrix} 2 & 0 & 0 \\ 0 & 1 & 0 \\ 0 & 0 & 1/2 \end{pmatrix}, \quad H = \begin{pmatrix} 1 & 0 & 0 \\ 0 & 1 & 0 \\ 1/2 & 0 & 1 \end{pmatrix}.$$

Step 3. We now compute

$$D_1 = \sqrt{D} = \begin{pmatrix} \sqrt{2} & 0 & 0 \\ 0 & 1 & 0 \\ 0 & 0 & 1/\sqrt{2} \end{pmatrix}.$$

Step 4. Finally,

$$L = HD_1 = \begin{pmatrix} \sqrt{2} & 0 & 0 \\ 0 & 1 & 0 \\ 1/\sqrt{2} & 0 & 1/\sqrt{2} \end{pmatrix}.$$

Note that L is lower triangular, and that one can easily check that $LL' = V$, as was desired.

In this simple example, it is easy to verify directly that $\underset{\sim}{X}'$ of (4.8.6) has the desired mean vector $\underset{\sim}{\mu}$ and covariance matrix V as guaranteed by Theorem 4.8.5. Thus, $\underset{\sim}{X}^* = (X_1^*, X_2^*, X_3^*)$ with the X_i^* $(i = 1, 2, 3)$ independent standard normal (i.e., mean zero and variance one) random variables. Now

$$\underset{\sim}{X}' = \begin{pmatrix} X_1 \\ X_2 \\ X_3 \end{pmatrix} = L\underset{\sim}{X}^{*\prime} + \underset{\sim}{\mu}'$$

$$= \begin{pmatrix} \sqrt{2} & 0 & 0 \\ 0 & 1 & 0 \\ 1/\sqrt{2} & 0 & 1/\sqrt{2} \end{pmatrix} \begin{pmatrix} X_1^* \\ X_2^* \\ X_3^* \end{pmatrix} + \begin{pmatrix} \mu_1 \\ \mu_2 \\ \mu_3 \end{pmatrix}$$

$$= \begin{pmatrix} \sqrt{2}X_1^* + \mu_1 \\ X_2^* + \mu_2 \\ (X_1^* + X_3^*)/\sqrt{2} + \mu_3 \end{pmatrix}.$$

Therefore, the means, variances, and covariances are

$$E(X_1) = E(\sqrt{2}X_1^* + \mu_1) = 0 + \mu_1 = \mu_1,$$

$$E(X_2) = E(X_2^* + \mu_2) = 0 + \mu_2 = \mu_2,$$

$$E(X_3) = E(X_1^*/\sqrt{2} + X_3^*/\sqrt{2} + \mu_3) = 0 + 0 + \mu_3 = \mu_3;$$

$$\text{Var}(X_1) = 2\text{Var}(X_1^*) = 2 \cdot 1 = 2,$$

$$\text{Var}(X_2) = \text{Var}(X_2^*) = 1,$$

$$\text{Var}(X_3) = \frac{1}{2}\text{Var}(X_1^*) + \frac{1}{2}\text{Var}(X_3^*) = \frac{1}{2} + \frac{1}{2} = 1;$$

$$\text{Cov}(X_1, X_2) = \text{Cov}(\sqrt{2}X_1^*, X_2^*) = 0 = \text{Cov}(X_2, X_1),$$

$$\text{Cov}(X_1, X_3) = \text{Cov}(\sqrt{2}X_1^*, X_1^*/\sqrt{2} + X_3^*/\sqrt{2})$$

$$= \text{Cov}(\sqrt{2}X_1^*, X_1^*/\sqrt{2}) + \text{Cov}(\sqrt{2}X_1^*, X_3^*/\sqrt{2})$$

$$= \sqrt{2}/\sqrt{2} \cdot 1 + 0 = 1 = \text{Cov}(X_3, X_1),$$

$$\text{Cov}(X_2, X_3) = \text{Cov}(X_2^*, X_1^*/\sqrt{2} + X_3^*/\sqrt{2}) = 0 = \text{Cov}(X_3, X_2).$$

Since these are as desired, we see in this example that the result of Theorem 4.8.5 has produced an $\underset{\sim}{X}'$ with the desired mean vector $\underset{\sim}{\mu}'$ and covariance matrix V. That $\underset{\sim}{X}'$ is MVN follows from the general result that if $\underset{\sim}{X}'$ is MVN, then so is $A\underset{\sim}{X}^*$ for an arbitrary matrix A.

4.9 Confidence Regions for Multivariate Normal Random Variables

This section involves additional material on the multivariate normal distribution, and is placed here due to its close relationship to the material in Section 4.8. Suppose that the p-variate random variable $(X_1, \ldots, X_p)$ is multivariate normal in distribution with mean vector $\underset{\sim}{\mu}$ and covariance matrix (σ_{ij}). Then its probability density function is (see (4.8.3)) given by

$$f(x_1, \ldots, x_p) = \sqrt{|\sigma^{ij}|}\,(2\pi)^{-p/2}e^{-0.5Q(\underset{\sim}{x})}. \tag{4.9.1}$$

A natural question is "what does $f(x_1, \ldots, x_p)$ look like?" In the one-dimensional case $(p = 1)$, it is well known that the univariate normal probability density function is as in Figure 4.9–1.

To approach the question of **what $f(x_1, \ldots, x_p)$ looks like in the $p \geq 2$ case**, we first examine the set of points $\underset{\sim}{x}$ where $f(\underset{\sim}{x})$ is a constant, namely:

$$\{\underset{\sim}{x} : f(\underset{\sim}{x}) = c\} = \{\underset{\sim}{x} : \sqrt{|\sigma^{ij}|}\,(2\pi)^{-p/2}e^{-0.5Q(\underset{\sim}{x})} = c\}$$

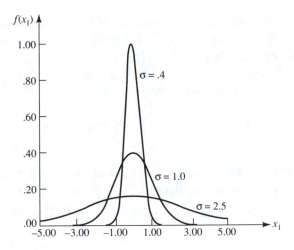

Figure 4.9–1. The univariate normal probability density function, where $\sigma^2 = 1/\sigma^{11}$ and $\mu_1 = 0$. (From Dudewicz (1976), with permission of American Sciences Press, Inc.)

$$= \{\underset{\sim}{x} : \ln\left(\sqrt{|\sigma^{ij}|}\,(2\pi)^{-p/2}\right) - \frac{1}{2}\sum_{i,j=1}^{p} \sigma^{ij}(x_i - \mu_i)(x_j - \mu_j) = \ln(c)\}$$

$$= \{\underset{\sim}{x} : \sum_{i,j=1}^{p} \sigma^{ij}(x_i - \mu_i)(x_j - \mu_j) = 2\ln\left(\sqrt{|\sigma^{ij}|}\,(2\pi)^{-p/2}\right) - 2\ln(c)\}.$$

$$(4.9.2)$$

The expression

$$\sum_{i,j=1}^{p} \sigma^{ij}(x_i - \mu_i)(x_j - \mu_j) \tag{4.9.3}$$

is a quadratic function of $x_1, \ldots, x_p$, which in this case can be shown to be an ellipsoid in p dimensions centered at $(\mu_1, \ldots, \mu_p)$. As the constant c decreases toward zero, the ellipsoid becomes larger, hence the function $f(\underset{\sim}{x})$ has the shape of a "mountain" with ellipsoidal sections. Figures illustrating this can be found in Dudewicz and Mishra (1988), pp. 166–169.

We now wish to find **the "best" confidence region R for a multivariate normal vector** $(X_1, \ldots, X_p)$ with $\underset{\sim}{\mu}$ and (σ_{ij}) known. That is, among regions R with

$$P[\underset{\sim}{X} \in R] = \gamma, \tag{4.9.4}$$

we seek the one with the smallest volume. Here γ is a preset number, such as $\gamma = 0.95$ in which case R is called a 95% confidence region. Now, by our analysis of the question "what $f(x_1, \ldots, x_p)$ looks like" we know that $f(x_1, \ldots, x_p)$ is larger at every x inside the ellipsoid $Q(x) = b^2$ than it is at any point outside that ellipsoid. Since

$$P[\underset{\sim}{X} \in R] = \int \cdots \int_{(x_1, \ldots, x_p) \in R} f(\underset{\sim}{x}) \, dx_1 \ldots dx_p \qquad (4.9.5)$$

it follows that (among all R with a fixed volume) we will make (4.9.5) maximal by taking R to be such an ellipsoid. This leads to the following theorem.

Theorem 4.9.6. *For regions R of fixed volume, $P[X \in R]$ is maximized by taking*

$$R = \{\underset{\sim}{x} : Q(\underset{\sim}{x}) \leq b^2\} \qquad (4.9.7)$$

for that value b^2 which produces the given volume.

To find the b^2 which solves (4.9.4), we need to solve

$$\gamma = P[Q(\underset{\sim}{X}) \leq b^2] = \int \cdots \int_{\underset{\sim}{x} : Q(\underset{\sim}{x}) \leq b^2} \sqrt{|\sigma^{ij}|} \, (2\pi)^{-p/2} e^{-0.5 Q(\underset{\sim}{x})} \, dx_1 \ldots dx_p$$

$$= P\left(\sum_{i,j=1}^{p} \sigma^{ij}(X_i - \mu_i)(X_j - \mu_j) \leq b^2 \right)$$

$$= P[(\underset{\sim}{X} - \underset{\sim}{\mu})'(\sigma^{ij})(\underset{\sim}{X} - \underset{\sim}{\mu}) \leq b^2].$$

$$(4.9.8)$$

In solving (4.9.8) for b^2 as a function of γ, the following theorem will be of use.

Theorem 4.9.9. *The covariance matrix (σ_{ij}) has an inverse (σ^{ij}) that is real and positive-definite. Hence the eigenvalues of (σ^{ij}), say $\lambda_1, \ldots, \lambda_p$, satisfy $\lambda_1 > 0, \ldots, \lambda_p > 0$, and there exists a real orthogonal (i.e., $L' = L^{-1}$) matrix L such that*

$$L'(\sigma^{ij})L = \operatorname{diag}(\lambda_1, \ldots, \lambda_p) \qquad (4.9.10)$$

for some ordering of $\lambda_1, \ldots, \lambda_p$.

We now solve (4.9.8) by making use of the successive changes of variables

$$\underset{\sim}{y} = \underset{\sim}{x} - \underset{\sim}{\mu}, \quad L\underset{\sim}{z} = \underset{\sim}{y}, \quad w_i = \sqrt{\lambda_i} \, z_i \quad (i = 1, \ldots, p) \qquad (4.9.11)$$

in the integral of (4.9.8). Following the first change of variables,

$$\gamma = \int \cdots \int_{\underset{\sim}{y}\,'(\sigma^{ij})\underset{\sim}{y}\,\leq b^2} \sqrt{|\sigma^{ij}|}\ (2\pi)^{-p/2}e^{-0.5\underset{\sim}{y}\,'(\sigma^{ij})\underset{\sim}{y}}\ dy_1 \ldots dy_p.$$

After the substitution $L\underset{\sim}{z} = \underset{\sim}{y}$,

$$\gamma = \int \cdots \int_{\underset{\sim}{z}\,'L'(\sigma^{ij})L\underset{\sim}{z}\,\leq b^2} \sqrt{|\sigma^{ij}|}\ (2\pi)^{-p/2}e^{-0.5\underset{\sim}{z}\,'L'(\sigma^{ij})L\underset{\sim}{z}}\ dz_1 \ldots dz_p.$$

Using $\lambda_1, \ldots, \lambda_p$, the eigenvalues of (σ^{ij}), the integral is transformed to

$$\gamma = \int \cdots \int_{\lambda_1 z_1^2 + \ldots + \lambda_p z_p^2 \leq b^2} \sqrt{\lambda_1 \ldots \lambda_p}\ (2\pi)^{-p/2}e^{-0.5(\lambda_1 z_1^2 + \ldots + \lambda_p z_p^2)}\ dz_1 \ldots dz_p.$$

Finally, using the last transformation $w_i = \sqrt{\lambda_i}z_i$ for $i = 1, \ldots p$, we find

$$\gamma = \int \cdots \int_{w_1^2 + \ldots + w_p^2 \leq b^2} (2\pi)^{-p/2}e^{-0.5(w_1^2 + \ldots + w_p^2)}\ dw_1 \ldots dw_p. \qquad (4.9.12)$$

Note that the Jacobians of the transformations in (4.9.11) are, respectively, 1, $|L| = 1$ (since $L'L = I$ implies $|L'L| = 1$ and $|L|^2 = 1$), and $\sqrt{\lambda_1 \cdots \lambda_p}$. Now the integrand of (4.9.12) is the probability density function of p independent standard (mean zero and variance one) normal random variables, say $W_1, \ldots, W_p$, hence we have

$$\gamma = P[Q(\underset{\sim}{X}) \leq b^2] = P[W_1^2 + \ldots + W_p^2 \leq b^2]. \qquad (4.9.13)$$

However, the square of a standard normal random variable is a chi-square random variable with one degree of freedom. The sum of independent chi-square random variables is chi-square, with degrees of freedom the sum of the degrees of freedom of the variables summed. Therefore $W_1^2 + \ldots + W_p^2$ is chi-square with p degrees of freedom, hence (denoting by χ_p^2 a chi-square random variable with p degrees of freedom, and recalling the probability density function of a chi-square random variable) we have

$$\gamma = P[Q(\underset{\sim}{X}) \leq b^2] = P[\chi_p^2 \leq b^2] = \int_0^{b^2} 2^{-p/2}\gamma(p/2)x^{(p/2)-1}e^{-x/2}\ dx. \qquad (4.9.14)$$

The constants b^2 needed for various p and γ are found in tables of the chi-square distribution with p degrees of freedom. For example, some values for

$p = 3$ are given in Table 4.9–2; others can be found by using the tables in Appendix E.

Table 4.9–2. Constants b^2 solving (4.9.8) for various γ when $p = 3$.

γ	.50	.75	.90	.95	.975	.99	.995
b^2	2.36597	4.10834	6.25139	7.81473	9.34840	10.3449	12.8382

The following theorem summarizes the preceding results.

Theorem 4.9.15. *If $\underset{\sim}{X}$ is p-variate normal, then*

$$P[Q(\underset{\sim}{X}) \le b^2] = P[\chi_p^2 \le b^2]$$

where χ_p^2 is a chi-square random variable with p degrees of freedom.

4.10   Fitting Distributions to Data: The GLD Family, Univariate and Bivariate

Suppose that we have a set of data $X_1, X_2, \ldots, X_n$ that are independent and identically distributed random variables. If the data is normally distributed, i.e., if X_i is $N(\mu, \sigma^2)$ for $i = 1, 2, \ldots, n$, then the mean μ and the variance σ^2 are usually estimated, respectively, by

$$\hat{\mu} \equiv \overline{X} = \frac{X_1 + \ldots + X_n}{n} \tag{4.10.1}$$

and

$$\hat{\sigma}^2 \equiv s^2 = \sum_{i=1}^{n} (X_i - \overline{X})^2 / (n-1). \tag{4.10.2}$$

Note that a random variable X that is $N(\mu, \sigma^2)$ has

$$\begin{cases} \text{(measure of center)} & = E(X) = \mu, \\ \text{(measure of variability)} & = E(X-\mu)^2 = \sigma^2, \\ \text{(measure of skewness)} & = E(X-\mu)^3/\sigma^3 = 0, \\ \text{(measure of kurtosis)} & = E(X-\mu)^4/\sigma^4 = 3. \end{cases} \tag{4.10.3}$$

If a random variable Y has a distribution other than the normal, we might attempt to approximate it by a random variable X that is $N(\mu, \sigma^2)$ for some μ and σ^2. We can do this successfully for center $E(Y)$ and variability $\text{Var}(Y) = E(Y - E(Y))^2$ by choice of μ and σ^2, in the sense that **we can make X have the same center and variability as does Y.** However, after that is done there are no free parameters in the distribution of X, and (unless the skewness and kurtosis of Y are, respectively, 0 and 3) we will not be able to match them in X. **For this reason, the normal family of distributions cannot be used to match data successfully unless the data is symmetric** (so its skewness will be 0) **and has tail weight similar to that of the normal** (so that its kurtosis will be near 3). For this reason, families of distributions with additional parameters are often used, allowing us to match more than the center and the variability of Y. Before considering the Generalized Lambda Distribution (GLD) family for this purpose, we note that **there are three common ways in which a normal distribution can be fully specified: by its probability density function** (see (4.2.15))

$$f(x) = \frac{1}{\sqrt{2\pi}\,\sigma}\, e^{-(x-\mu)^2/(2\sigma^2)},$$

by its distribution function

$$F(x) = P[X \le x] = P\left(\frac{X-\mu}{\sigma} \le \frac{x-\mu}{\sigma}\right) = \Phi\left(\frac{x-\mu}{\sigma}\right), \qquad (4.10.4)$$

and by its inverse function (or percentile function, or quantile distribution function)

$$F^{-1}(y) = (x \text{ such that } F(x) = y)$$
$$= \mu + \sigma\Phi^{-1}(y). \qquad (4.10.5)$$

While the first two specifications are more commonly used than is the third, the third is more common for certain other distributions (such as the GLD, to be discussed below). Above, note that

$$\Phi(z) = \int_{-\infty}^{z} \frac{1}{\sqrt{2\pi}}\, e^{-t^2/2}\, dt, \qquad (4.10.6)$$

and

$$\Phi^{-1}(y) = (z \text{ such that } \Phi(z) = y), \qquad (4.10.7)$$

respectively denote the standard (mean zero and variance one) normal distribution function and inverse function.

The **generalized lambda distribution family with four parameters**, **GLD**$(\lambda_1, \lambda_2, \lambda_3, \lambda_4)$, is most easily specified in terms of its percentile function

$$F^{-1}(y) = F^{-1}(y; \lambda_1, \lambda_2, \lambda_3, \lambda_4) = \lambda_1 + \frac{y^{\lambda_3} - (1-y)^{\lambda_4}}{\lambda_2}, \qquad (4.10.8)$$

where $0 \leq y \leq 1$. Here λ_1 and λ_2 are respectively location and scale parameters, while λ_3 and λ_4 determine the skewness and kurtosis of the GLD. Recall that for the normal distribution there were also restrictions on (μ, σ^2), namely $\sigma > 0$. The restrictions on $\lambda_1, \ldots \lambda_4$ that yield a valid distribution will be discussed in Section 4.10.1. It is relatively easy to find the probability density function from the percentile function of the GLD, as we now show.

Theorem 4.10.9. *For the* GLD$(\lambda_1, \lambda_2, \lambda_3, \lambda_4)$, *the probability density function is*

$$f(x) = \frac{\lambda_2}{\lambda_3 y^{\lambda_3 - 1} + \lambda_4 (1-y)^{\lambda_4 - 1}}, \qquad at \ x = F^{-1}(y). \qquad (4.10.10)$$

(Note that $F^{-1}(y)$ can be calculated from (4.10.8).)

Proof. Since $x = F^{-1}(y)$, we have $y = F(x)$. Differentiating with respect to x, we find

$$\frac{dy}{dx} = f(x)$$

or

$$f(x) = \frac{dy}{d(F^{-1}(y))} = \frac{1}{\dfrac{d(F^{-1}(y))}{dy}}. \qquad (4.10.11)$$

Since we know the form of $F^{-1}(y)$ from (4.10.8), we find directly that

$$\frac{d(F^{-1}(y))}{dy} = \frac{d}{dy}\left(\lambda_1 + \frac{y^{\lambda_3} - (1-y)^{\lambda_4}}{\lambda_2}\right)$$

$$= \frac{\lambda_3 y^{\lambda_3 - 1} + \lambda_4 (1-y)^{\lambda_4 - 1}}{\lambda_2}, \qquad (4.10.12)$$

from which the theorem follows using (4.10.12) in (4.10.11).

Note that plotting the function $f(x)$ for a density such as the normal, where $f(x)$ is given as a specific function of x in (4.2.15), proceeds by calculating $f(x)$ at a grid of x values, then plotting the pairs $(x, f(x))$ and connecting them with a smooth curve. For the GLD family, plotting $f(x)$ proceeds differently since (4.10.10) tells us the value of $f(x)$ at $x = F^{-1}(y)$. Thus, we take a grid of y values (such as .01, .02, .03, ..., .99, which will give us the 1%, 2%, 3%, ..., 99% points), find x at each of those points from (4.10.8), and find $f(x)$ at that x from (4.10.10) (which uses the corresponding y value). Then, the pairs $(x, f(x))$ are plotted and linked with a smooth curve. **GLD random variables can be generated through** $X = F^{-1}(U)$, as described in Theorem 4.2.1.

4.10.1 The Parameter Space of the GLD

We noted earlier that formula (4.10.8) does not always specify a valid distribution. The reason is that one cannot just write down any formula and be assured it will specify a distribution without checking the conditions needed for that fact to hold. In particular, a function $f(\cdot)$ is a probability density function if and only if it satisfies the conditions

$$f(x) \geq 0, \qquad \int_{-\infty}^{\infty} f(x)\, dx = 1. \tag{4.10.13}$$

From (4.10.11) we see that, for the GLD, conditions (4.10.13) are satisfied if and only if

$$\frac{\lambda_2}{\lambda_3 y^{\lambda_3-1} + \lambda_4(1-y)^{\lambda_4-1}} \geq 0, \quad \int_{-\infty}^{\infty} f(F^{-1}(y))\, dF^{-1}(y) = 1. \tag{4.10.14}$$

Since from (4.10.11) we know that

$$f(F^{-1}(y))\, d(F^{-1}(y)) = dy,$$

and y is on the range $[0, 1]$, the second condition in (4.10.14) follows. Thus, for any $\lambda_1, \lambda_2, \lambda_3, \lambda_4$ the function $f(x)$ will integrate to 1. It remains to show that the first condition in (4.10.14) holds.

Since λ_1 does not enter into the first condition in (4.10.14), this parameter will be unrestricted, leading us to the following theorem.

Theorem 4.10.15. *The* GLD$(\lambda_1, \lambda_2, \lambda_3, \lambda_4)$ *specifies a valid distribution if and only if*

$$\frac{\lambda_2}{\lambda_3 y^{\lambda_3-1} + \lambda_4(1-y)^{\lambda_4-1}} \geq 0. \tag{4.10.16}$$

The next theorem establishes the role of λ_1 as a location parameter.

Theorem 4.10.17. *If X is GLD$(0, \lambda_2, \lambda_3, \lambda_4)$, then $X + \lambda_1$ is GLD$(\lambda_1, \lambda_2, \lambda_3, \lambda_4)$.*

Proof. If X is GLD$(0, \lambda_2, \lambda_3, \lambda_4)$, by (4.10.8) we have

$$F_X^{-1}(y) = \frac{y^{\lambda_3} - (1 - y)^{\lambda_4}}{\lambda_2}.$$

Now

$$F_{X+\lambda_1}(x) = P[X + \lambda_1 \leq x] = P[X \leq x - \lambda_1] = F_X(x - \lambda_1), \quad (4.10.18)$$

hence $F_X(x - \lambda_1) = y$ also implies $F_{X+\lambda_1}(x) = y$, yielding

$$x - \lambda_1 = F_X^{-1}(y) = \frac{y^{\lambda_3} - (1 - y)^{\lambda_4}}{\lambda_2}, \quad x = F_{X+\lambda_1}^{-1}(y), \quad (4.10.19)$$

whence

$$F_{X+\lambda_1}^{-1}(y) = x = \lambda_1 + F_X^{-1}(y) = \lambda_1 + \frac{y^{\lambda_3} - (1 - y)^{\lambda_4}}{\lambda_2}. \quad (4.10.20)$$

This proves that $X + \lambda_1$ is GLD$(\lambda_1, \lambda_2, \lambda_3, \lambda_4)$.

Since $0 \leq y \leq 1$ in (4.10.16), we immediately have the following.

Corollary 4.10.21. *The GLD$(\lambda_1, \lambda_2, \lambda_3, \lambda_4)$ of (4.10.8) specifies a valid distribution if and only if*

$$g(y, \lambda_3, \lambda_4) \equiv \lambda_3 y^{\lambda_3 - 1} + \lambda_4 (1 - y)^{\lambda_4 - 1} \quad (4.10.22)$$

has the same sign (positive or negative) for all y in $[0, 1]$, as long as λ_2 takes that sign also. In particular, the GLD$(\lambda_1, \lambda_2, \lambda_3, \lambda_4)$ specifies a valid distribution if $\lambda_2, \lambda_3, \lambda_4$ all have the same sign.

To determine the (λ_3, λ_4) pairs that lead to a valid GLD, we consider the regions of (λ_3, λ_4)-space (some of these regions are depicted in Figure 4.10–1) defined by

$$\text{Region 1} = \{(\lambda_3, \lambda_4) | \lambda_3 \leq -1, \lambda_4 \geq 1\}$$
$$\text{Region 2} = \{(\lambda_3, \lambda_4) | \lambda_3 \geq 1, \lambda_4 \leq -1\}$$
$$\text{Region 3} = \{(\lambda_3, \lambda_4) | \lambda_3 \geq 0, \lambda_4 \geq 0\}$$
$$\text{Region 4} = \{(\lambda_3, \lambda_4) | \lambda_3 \leq 0, \lambda_4 \leq 0\}$$
$$V_1 = \{(\lambda_3, \lambda_4) | \lambda_3 < 0, 0 < \lambda_4 < 1\}$$
$$V_2 = \{(\lambda_3, \lambda_4) | 0 < \lambda_3 < 1, \lambda_4 < 0\}$$
$$V_3 = \{(\lambda_3, \lambda_4) | -1 < \lambda_3 < 0, \lambda_4 > 1\}$$
$$V_4 = \{(\lambda_3, \lambda_4) | \lambda_3 > 1, -1 < \lambda_4 < 0\}$$

The following lemma is a direct consequence of Corollary 4.10.21.

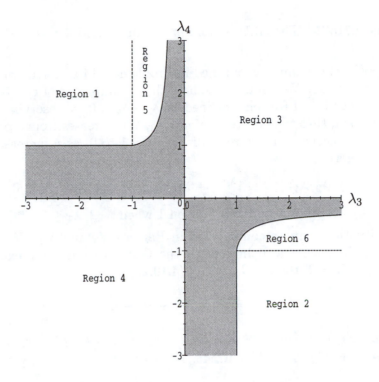

Figure 4.10–1. Regions where the GLD is valid (shaded region is where it is not valid).

Lemma 4.10.23. *The GLD is valid in Regions 3 and 4 specified in Figure 4.10–1.*

Next, we consider the other regions, starting with regions V_1 and V_2.

Lemma 4.10.24. *The GLD is not valid in regions V_1 and V_2.*

Proof. By Corollary 4.10.21, the GLD is valid at $(\lambda_1, \lambda_2, \lambda_3, \lambda_4)$ if and only if $g(y, \lambda_3, \lambda_4)$, as defined in (4.10.22), has the same sign for all y in $[0, 1]$, and λ_2 takes that same sign. In Region V_1 we have $\lambda_3 < 0$ and $0 \le \lambda_4 < 1$. It is easy to see that

$$\lim_{y \to 0^+} g(y, \lambda_3, \lambda_4) = -\infty \text{ and } \lim_{y \to 1^-} g(y, \lambda_3, \lambda_4) = +\infty,$$

so that $g(y, \lambda_3, \lambda_4)$ cannot keep the same sign over $[0, 1]$, hence the GLD is not valid for (λ_3, λ_4) in V_1. The analysis for V_2 is left as an exercise (Problem 4.13).

Lemma 4.10.25. *The GLD is valid in Regions 1 and 2 specified in Figure 4.10–1.*

The proof of this Lemma is outlined in Problems 4.14 and 4.15 with details left for the reader. We simply note here that a simple proof can be devised for the subportion of Region 1 defined by $\lambda_3 + \lambda_4 \leq 0$ by observing, through differentiation, that the first term of $g(y, \lambda_3, \lambda_4)$ increases from $-\infty$ to $\lambda_3 < -1$, and the second term decreases from $\lambda_4 > 1$ to 0, as y increases from 0 to 1. Therefore,

$$g(y, \lambda_2, \lambda_4) \leq \lambda_3 \cdot 1^{\lambda_3 - 1} + \lambda_4 (1 - 0)^{\lambda_4 - 1} = \lambda_3 + \lambda_4 \leq 0$$

and $g(y, \lambda_3, \lambda_4)$ stays negative in Region 1 when $\lambda_3 + \lambda_4 \leq 0$.

The situation is more complicated in Regions V_3 and V_4. We start by observing that at some points of V_3 the GLD is valid. For example, at $\lambda_3 = -1/2$, $\lambda_4 = 2$, the $g(y, \lambda_3, \lambda_4)$ of (4.10.22) is

$$g(y) = -\frac{1}{2} y^{-3/2} + 2(1 - y),$$

for which $g'(y) = \frac{3}{4} y^{-5/2} - 2$, which is ≤ 0 iff $(3/4) y^{-5/2} \leq 2$, i.e. iff $3/8 \leq y^{5/2}$, or $y \geq (3/8)^{2/5} = .67548$. Noting that

$$\lim_{y \to 0^+} g(y) = -\infty \text{ and } g(1) = -1/2,$$

we see that $g(y)$ increases as y increases from 0 to $(3/8)^{2/5}$, then decreases. Its maximum occurs at $y = (3/8)^{2/5}$, in which case

$$g\left(\left(\frac{3}{8}\right)^{2/5}\right) = -\frac{1}{2}((3/8)^{2/5})^{-3/2} + 2(1 - (3/8)^{2/5})$$
$$= -.90064 + .64904 = -.25160.$$

Thus, $g(y)$ is negative for all y in $[0, 1]$, and the GLD (with $\lambda_2 < 0$) is valid at the given λ_3 and λ_4.

This can be contrasted with the point $(\lambda_3, \lambda_4) = (-1/2, 1)$ where

$$g(y) = 1 - \frac{1}{2y^{3/2}}.$$

In this case

$$\lim_{y \to 0^+} g(y) = -\infty \text{ and } g(1) = \frac{1}{2},$$

establishing $(\lambda_3, \lambda_4) = (-1/2, 1)$ as a point of V_3 where the GLD is not valid.

The following result, due to Karian, Dudewicz and McDonald (1996), gives the complete characterization of the valid points of Regions V_3 and V_4.

Lemma 4.10.26. *A point in Region* V_3 *is valid if and only if*

$$\frac{(1 - \lambda_3)^{1-\lambda_3}}{(\lambda_4 - \lambda_3)^{\lambda_4-\lambda_3}}(\lambda_4 - 1)^{\lambda_4-1} < \frac{-\lambda_3}{\lambda_4}. \tag{4.10.27}$$

Proof. Let $0 < y < 1$ and $-1 < \lambda < 0$ (think of λ as λ_3), and $f(\lambda) = \alpha > 1$ (think of α as λ_4, so we are considering Region V in the second quadrant. Let

$$G(y) = \lambda y^{\lambda-1} + \alpha(1 - y)^{\alpha-1}. \tag{4.10.28}$$

Since the GLD is valid if $G(y)$ has constant sign for all y in $[0, 1]$, we examine the zeros of G. $G(y) = 0$ is equivalent to

$$\frac{-\lambda}{\alpha} = \frac{(1 - y)^{\alpha-1}}{y^{\lambda-1}}. \tag{4.10.29}$$

Through the substitutions

$$\beta = \left(\frac{-\lambda}{\alpha}\right)^{1/(\alpha-1)} \quad \text{and} \quad \gamma = \frac{1 - \lambda}{\alpha - 1},$$

(4.10.29) can be simplified to

$$\beta = y^{\gamma}(1 - y), \tag{4.10.30}$$

where β, $\gamma > 0$.

Differentiating with respect to y we obtain a relation for the critical points of $h(\lambda, y) = y^{\gamma}(1 - y)$:

$$\frac{\partial h}{\partial y} = \gamma y^{\gamma-1}(1 - y) - y^{\gamma}$$

which is zero if and only if $\gamma y^{\gamma-1} = y^{\gamma}(1+\gamma)$, i.e., if and only if y has the value $y_c = \gamma/(1 + \gamma)$. At y_c, $h(\lambda, y)$ has a maximum since $h(\lambda, 0) = h(\lambda, 1) = 0$ and $h(\lambda, y) \geq 0$. This maximum is given by

$$h(\lambda, y_c) = y_c^{\gamma}(1 - y_c) = \frac{y_c^{\gamma}}{(1 + \gamma)} = \frac{\gamma^{\gamma}}{(1 + \gamma)^{1+\gamma}}.$$

Since the difference of the two sides in (4.10.30) will go from positive at $y = 0$ to negative at $y = y_c$, G changes sign on $[0, 1]$ if and only if

$$\frac{\gamma^{\gamma}}{(1 + \gamma)^{1+\gamma}} \geq \beta. \tag{4.10.31}$$

By restating this in terms of the λ_3 and λ_4 parameters of the GLD, we see that (λ_3, λ_4) fails to yield a valid GLD if and only if

$$\frac{(1 - \lambda_3)^{1-\lambda_3}}{(\lambda_4 - \lambda_3)^{\lambda_4 - \lambda_3}}(\lambda_4 - 1)^{\lambda_4 - 1} \geq \frac{-\lambda_3}{\lambda_4}. \qquad (4.10.32)$$

The following theorem, by summarizing the results of Lemmas 4.10.23 through 4.10.26, completely characterizes the (λ_3, λ_4) pairs for which the GLD is valid.

Theorem 4.10.33. *With a suitable λ_2, the GLD is valid at (λ_3, λ_4) if and only if (λ_3, λ_4) is in one of the unshaded regions depicted in Figure 4.10–1. The curved boundaries between the valid and non-valid regions are given by.*

$$\frac{(1 - \lambda_3)^{1-\lambda_3}}{(\lambda_4 - \lambda_3)^{\lambda_4 - \lambda_3}}(\lambda_4 - 1)^{\lambda_4 - 1} = \frac{-\lambda_3}{\lambda_4} \text{ (in the second quadrant)}$$

and

$$\frac{(1 - \lambda_4)^{1-\lambda_4}}{(\lambda_3 - \lambda_4)^{\lambda_3 - \lambda_4}}(\lambda_3 - 1)^{\lambda_3 - 1} = \frac{-\lambda_4}{\lambda_3} \text{ (in the fourth quadrant)}.$$

4.10.2   Estimation of the GLD Parameters

In the literature, the use of the GLD has been restricted to Regions 3 and 4, where λ_2, λ_3, λ_4 all have the same sign. A primary reason for this is the availability of tables due to Ramberg, Dudewicz, Tadikamalla, and Mykytka (1979) and subsequently more refined tables due to Dudewicz and Karian (1996). (These tables will be discussed more extensively below, where they will play an important role in fitting a GLD to a data set.)

Earlier, we noted (see (4.10.3)) the first four moments of a normal random variable. **If X is a random variable that has the GLD$(\lambda_1, \lambda_2, \lambda_3, \lambda_4)$ distribution, we would like to know its first four moments.** These follow from the results below.

Theorem 4.10.34. *If X is a random variable with a GLD$(\lambda_1, \lambda_2, \lambda_3, \lambda_4)$ distribution, then its moments are*

$$\mu = (\text{center}) = E(X) = \lambda_1 + \frac{A}{\lambda_2},$$

$$\sigma^2 = (\text{variability}) = \text{Var}(X) = \frac{B - A^2}{\lambda_2^2},$$

$$\alpha_3 = (\text{skewness}) = E(X - E(X))^3/\sigma^3 = \frac{C - 3AB + 2A^3}{\lambda_2^3 \sigma^3},$$

$$\alpha_4 = (\text{kurtosis}) = E(X - E(X))^4/\sigma^4 = \frac{D - 4AC + 6A^2 B - 3A^4}{\lambda_2^4 \sigma 4},$$

where

$$A = \frac{1}{1 + \lambda_3} - \frac{1}{1 + \lambda_4},$$

$$B = \frac{1}{1 + 2\lambda_3} + \frac{1}{1 + 2\lambda_4} - 2\beta(1 + \lambda_3, 1 + \lambda_4),$$

$$C = \frac{1}{1 + 3\lambda_3} - 3\beta(1 + 2\lambda_3, 1 + \lambda_4) + 3\beta(1 + \lambda_3, 1 + 2\lambda_4) - \frac{1}{1 + 3\lambda_4},$$

$$D = \frac{1}{1 + 4\lambda_3} - 4\beta(1 + 3\lambda_3, 1 + \lambda_4) + 6\beta(1 + 2\lambda_3, 1 + 2\lambda_4)$$

$$- 4\beta(1 + \lambda_3, 1 + 3\lambda_4) + \frac{1}{1 + 4\lambda_4}$$

and $\beta(a, b)$ is the beta function defined in (4.6.52) by

$$\beta(a, b) = \int_0^1 x^{a-1}(1 - x)^{b-1} \, dx.$$

Proof. We first assert (see Problem 4.16) that if Z is $GLD(0, \lambda_2, \lambda_3, \lambda_4)$, then

$$E(Z^k) = \frac{1}{\lambda_2^k} \sum_{i=0}^{k} \binom{k}{i} (-1)^i \beta(\lambda_3(k - i) + 1, \lambda_4 i + 1)). \qquad (4.10.35)$$

Next, from Theorem 4.10.17 we know that X can be represented as $X = Z + \lambda_1$. Therefore,

$$E(X^k) = E((Z + \lambda_1)^k),$$

which is a simple series in $E(Z^j)$ for various js. From these terms given in (4.10.34), the moments can be computed completing the proof of the theorem.

If a random variable Y has a distribution other than the GLD, we might try to approximate it by a random variable X that is $GLD(\lambda_1, \lambda_2, \lambda_3, \lambda_4)$ for some $\lambda_1, \lambda_2, \lambda_3, \lambda_4$. Suppose that the first four moments of Y are $\mu, \sigma^2, \alpha_3, \alpha_4$. If we can choose λ_3, λ_4 so that a $GLD(0, 1, \lambda_3, \lambda_4)$ has third and fourth moments α_3 and α_4, then we can let λ_1 and λ_2 solve the equations

$$\mu = \lambda_1 + \frac{A}{\lambda_2}, \qquad \sigma^2 = \frac{B - A^2}{\lambda_2^2}. \qquad (4.10.36)$$

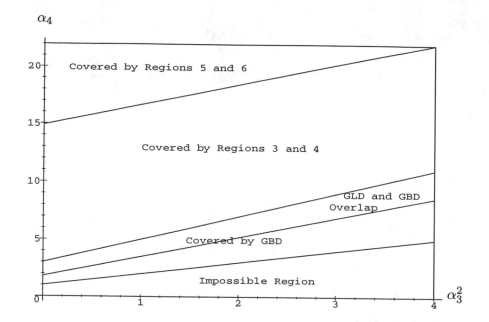

Figure 4.10–2. Regions of (α_3^2, α_4) that the GLD system can attain.

It follows from Theorem 4.10.34 that the resulting λ_1, λ_2, λ_3, λ_4 specify a GLD with the desired first four moments. Here, we note that A, B, C, D are functions only of λ_3, λ_4, and that (4.10.36) can be solved for any μ and any $\sigma^2 > 0$. We have therefore shown the following.

Corollary 4.10.37. *The* GLD($\lambda_1, \lambda_2, \lambda_3, \lambda_4$) *can match any first two moments* μ *and* σ^2, *and some third and fourth moments* α_3 *and* α_4.

Hence, if many (α_3, α_4) can be attained, the GLD family will be very useful for fitting a broad range of data sets and approximating a variety of other random variables. So **we next note how broad a set of (α_3, α_4) can be attained.** This is shown in Figure 4.10–2 on (α_3^2, α_4) axes.

The "impossible" region in Figure 4.10–2 is not valid for any distribution, since it can be shown that for any distribution the moments α_3 and α_4 obey the inequality

$$\alpha_4 > 1 + \alpha_3^2. \tag{4.10.38}$$

This is less well known than the inequality $E(X^2) \geq \mu^2$, which follows from $\mathrm{Var}(X) = E(X^2) - \mu^2 \geq 0$. Furthermore, since the second term on the right-hand side of (4.10.38) is nonnegative, we also have $\alpha_4 > 1$. Note that the GLD as well as the GBD (to be described in Section 4.10.3) both cover a portion of (α_3^2, α_4)-space. Dudewicz and Karian (1996) provide tables of solutions for all (α_3^2, α_4) shown in Figure 4.10–2, except for those in the impossible region and the uppermost region labeled "Covered by Regions 5 and 6."

In Section 4.10.3 we give an extension of the GLD that covers the part of Figure 4.10–2 not covered by the GLD but is possible (i.e., not in the impossible area). The region covered by the GLD already includes the moments of such distributions as the uniform, Student's t, normal, Weibull, gamma, lognormal, exponential, and some beta distributions. Thus, it is a rich class in terms of moment coverage. For details, see Figure 3 on p. 206 of Ramberg, Dudewicz, Tadikamalla, and Mykytka (1979).

The actual fitting of a GLD to a data set is illustrated next.

Example 4.10.39. Fitting a GLD to a data set. Suppose we have a set of data $X_1, X_2, \ldots, X_n$ that are independent and identically distributed from an unknown distribution. We may **estimate the first four moments**, respectively, by

$$\hat{\alpha}_1 = \overline{X} = \sum_{i=1}^{n} x_i/n, \quad \hat{\alpha}_2 = m_2, \quad \hat{\alpha}_3 = m_3/m_2^{3/2}, \quad \hat{\alpha}_4 = m_4/m_2^2 \quad (4.10.40)$$

where

$$m_2 = \sum_{i=1}^{n}(X_i - \overline{X})^2/n,$$

$$m_3 = \sum_{i=1}^{n}(X_i - \overline{X})^3/n, \quad (4.10.41)$$

$$m_4 = \sum_{i=1}^{n}(X_i - \overline{X})^4/n.$$

Next, we use the tables of Dudewicz and Karian (1996)—in particular Table 1—to find the $\hat{\lambda}_1(0,1)$, $\hat{\lambda}_2(0,1)$, $\hat{\lambda}_3$, $\hat{\lambda}_4$ for which the tabulated (α_3, α_4) is closest to the $(\hat{\alpha}_3, \hat{\alpha}_4)$ calculated in (4.10.40). (If $\hat{\alpha}_3 < 0$, use $|\hat{\alpha}_3|$ in the table lookup and then switch the values of $\hat{\lambda}_3$ and $\hat{\lambda}_4$, and change the sign of $\hat{\lambda}_1$.) Since the tables yield mean zero and variance one, **transform to**

$$\hat{\lambda}_1 = \hat{\lambda}_1(0,1)\sqrt{m_2} + \overline{X}_1, \quad \hat{\lambda}_2 = \hat{\lambda}_2(0,1)/\sqrt{m_2}. \quad (4.10.42)$$

This approach is called choosing a GLD by the **method of moments**. As with any fit of a distribution, **one should check the fit visually by**

plotting the probability density function given in (4.10.10) with the fitted $\hat{\lambda}_1$, $\hat{\lambda}_2$, $\hat{\lambda}_3$, $\hat{\lambda}_4$ superimposed on a histogram. One evaluates the plot visually to determine whether the fitted density appropriately represents the data set. One can also perform a chi-square test of goodness of fit; for details and an example, see Ramberg, Dudewicz, Tadikamalla, and Mykytka (1979), top of p. 208.

Example 4.10.43. Numerical example of fitting a GLD. Suppose we have a data set to which we wish to fit a GLD, and the calculations at (4.10.40) yield

$$\overline{X} = .0345, \quad \sqrt{m_2} = .0098, \quad \hat{\alpha}_3 = .87, \quad \hat{\alpha}_4 = 4.92. \qquad (4.10.44)$$

In Table 1 of Dudewicz and Karian (1996), we find that the closest pair to $(\hat{\alpha}_3, \hat{\alpha}_4) = (.87, 4.92)$ is $(.85, 4.9)$. For that pair, we read from the table the entries

$$\hat{\lambda}_1(0,1) = -.4140, \quad \hat{\lambda}_2(0,1) = .01342,$$
$$\hat{\lambda}_3 = .004581, \quad \hat{\lambda}_4 = .01022. \qquad (4.10.45)$$

Then we transform as at (4.10.42) to

$$\hat{\lambda}_1 = -.4140 \times .0098 + .0345 = .0304,$$
$$\hat{\lambda}_2 = .01342/.0098 = 1.3694. \qquad (4.10.46)$$

Thus, the fitted GLD is the one with inverse function

$$F^{-1}(y) = .0304 + \frac{y^{0.004581} - (1-y)^{0.01022}}{1.3694} \quad (0 \le y \le 1) \quad (4.10.47)$$

and the goodness of the fit to the data set can be assessed as discussed at the end of Example 4.10.39. Note that to generate random variables for a simulation from distribution (4.10.47), one simply computes $F^{-1}(U_1), F^{-1}(U_2), \ldots$ where the $U_1, U_2, \ldots$ are independent and uniform on $(0,1)$, obtaining a sequence of independent random variables with the specified GLD. Note that the range over which the GLD density is positive is not always the whole real line; for details, see Table 2 of Ramberg, Dudewicz, Tadikamalla, and Mykytka (1979).

4.10.3   The Extended GLD

The region immediately above the impossible region in Figure 4.10–2 is not covered by the GLD. Since (α_3^2, α_4) pairs from this region occur quite frequently in practice, we now give an extension of the GLD (due to Karian and

Dudewicz, and McDonald (1996)) to a system that covers all of (α_3^2, α_4)-space that can realistically occur (for a discussion of realistic (α_3^2, α_4) possibilities see Wilcox (1990), Pearson and Please (1975), and Micceri (1989)).

A random variable X has the beta distribution of Section 4.6.8 if for $\beta_3 > -1$ and $\beta_4 > -1$, the probability density function (p.d.f.) of X is given by

$$f(x) = \begin{cases} \dfrac{\Gamma(\beta_3 + \beta_4 + 2)}{\Gamma(\beta_3 + 1)\Gamma(\beta_4 + 1)} x^{\beta_3}(1 - x)^{\beta_4} & \text{for } 0 \leq x \leq 1 \\ 0 & \text{otherwise} \end{cases}$$

where $\Gamma(t)$ is the gamma function defined at (4.6.7) by

$$\Gamma(t) = \int_0^\infty x^{t-1} e^{-x} \, dx.$$

$E(X^r)$, the expectation of X^r, is given (see Dudewicz and Mishra (1988), pp. 224–225) by

$$E(X^r) = \frac{\Gamma(\beta_3 + r + 1)\Gamma(\beta_3 + \beta_4 + 2)}{\Gamma(\beta_3 + \beta_4 + r + 2)\Gamma(\beta_3 + 1)},$$

from which the following more convenient recursive expression for $E(X^r)$ can be derived:

$$E(X^r) = \left(\frac{\beta_3 + r}{\beta_3 + \beta_4 + r + 1} \right) E(X^{r-1}), \quad r = 1, 2, 3, \ldots . \quad (4.10.48)$$

If

$$Y = \beta_1 + \beta_2 X, \quad (4.10.49)$$

where β_1 is any real number and β_2 is positive, then

$$F_Y(y) = P(Y \leq y) = P(\beta_1 + \beta_2 X \leq y)$$

$$= \int_0^{(y-\beta_1)/\beta_2} C x^{\beta_3}(1 - x)^{\beta_4} \, dx, \quad (4.10.50)$$

where

$$C = \frac{\Gamma(\beta_3 + \beta_4 + 2)}{\Gamma(\beta_3 + 1)\Gamma(\beta_4 + 1)}.$$

Using the transformation $y = \beta_1 + \beta_2 x$ in the integral given in (4.10.50), we obtain

$$F_Y(y) = C \int_{\beta_1}^y \beta_2^{-(\beta_3 + \beta_4 + 1)}(x - \beta_1)^{\beta_3}(\beta_1 + \beta_2 - x)^{\beta_4} \, dx,$$

which leads us to the following definition.

Definition 4.10.51. The random variable X is said to have a **generalized beta distribution, GBD($\beta_1, \beta_2, \beta_3, \beta_4$)**, if for any β_1, $\beta_2 > 0$, $\beta_3 > -1$, and $\beta_4 > -1$, the p.d.f. of X is

$$f(x) = \begin{cases} C\beta_2^{-(\beta_3+\beta_4+1)}(x - \beta_1)^{\beta_3}(\beta_1 + \beta_2 - x)^{\beta_4} & \text{for } \beta_1 \leq x \leq \beta_1 + \beta_2 \\ 0, & \text{otherwise.} \end{cases}$$

The moments of a GBD($\beta_1, \beta_2, \beta_3, \beta_4$) random variable, $\alpha_1, \alpha_2, \alpha_3, \alpha_4$, can now be derived from (4.10.48) and the relation $Y = \beta_1 + \beta_2 X$ as

$$\alpha_1 = \mu = \beta_1 + \beta_2(\beta_3 + 1)/B_2, \tag{4.10.52}$$

$$\alpha_2 = \sigma^2 = \frac{\beta_2^2(1 + \beta_3)(1 + \beta_4)}{B_2^2 B_3}, \tag{4.10.53}$$

$$\alpha_3 = \frac{2(\beta_4 - \beta_3)\sqrt{B_3}}{B_4\sqrt{(\beta_3 + 1)(\beta_4 + 1)}}, \tag{4.10.54}$$

$$\alpha_4 = \frac{3B_3\left(\beta_3\beta_4 B_2 + 3\beta_3^2 + 5\beta_3 + 3\beta_4^2 + 5\beta_4 + 4\right)}{B_4 B_5(\beta_3 + 1)(\beta_4 + 1)}, \tag{4.10.55}$$

where

$$B_i = \beta_3 + \beta_4 + i \text{ for } i = 1, \ldots, 5.$$

It follows from (4.10.48) and (4.01.49) that the first four moments of a GBD, in fact, all of the GBD moments, exist.

We now describe the portion of the (α_3^2, α_4)-space that is covered by the GBD.

Theorem 4.10.56. *The (α_3^2, α_4) pairs of the GBD satisfy*

$$1 + \alpha_3^2 < \alpha_4 < 3 + 2\alpha_3^2. \tag{4.10.57}$$

Moreover, all (α_3^2, α_4) that satisfy (4.10.57) can be realized from the GBD.

Proof. It has already been noted that the left inequality in (4.10.57) holds for all random variables. For a GBD random variable, from (4.10.54),

$$3 + 2\alpha_3^2 = \frac{8(\beta_4 - \beta_3)^2 B_3}{B_4^2(\beta_3 + 1)(\beta_4 + 1)} + 3. \tag{4.10.58}$$

If we write (4.10.58) as N/D with $D = B_4^2(\beta_3 + 1)(\beta_4 + 1)$, we get

$$N = 8(\beta_4 - \beta_3)^2 B_3 + 3B_4^2(\beta_3 + 1)(\beta_4 + 1)$$

$$= 3[(B_3 + 1)((B_2 + 2)(\beta_3 + 1)(\beta_4 + 1) + \frac{8}{3}(\beta_4 - \beta_3)^2 B_3]$$

$$= 3B_3[(B_2 + 2)(\beta_3 + 1)(\beta_4 + 1) + \frac{8}{3}(\beta_4 - \beta_3)^2] + R$$

where

$$R = 3B_3(B_2 + 1)(\beta_3 + 1)(\beta_4 + 1) > 0.$$

Thus,

$$N > 3B_3[B_2(\beta_3 + 1)(\beta_4 + 1) + 2(\beta_3 + 1)(\beta_4 + 1) + \frac{8}{3}(\beta_4 - \beta_3)^2]$$

$$= 3B_3[B_2\beta_3\beta_4 + 3\beta_3^2 + 3\beta_4^2 + 5(\beta_3 + \beta_4) + 4 + S],$$

where

$$S = \frac{2}{3}\beta_3^2 + \frac{2}{3}\beta_4^2 - \frac{4}{3}\beta_3\beta_4 \frac{2}{3}(\beta_4 - \beta_3)^2 \geq 0.$$

Thus,

$$N > 3B_3 \left[B_2\beta_3\beta_4 + 3\beta_3^2 + 3\beta_4^2 + 5(\beta_3 + \beta_4) + 4 \right]$$

and $\alpha_4 < 3 + 2\alpha_3^2$.

Since α_3^2 and α_4 are continuous functions of β_3 and β_4 in the region described by (4.10.57), we conclude that **the (α_3^2, α_4)-space covered by the GBD is exactly the region given by (4.10.57)**.

Example 4.10.59. Fitting the GBD to a data set. As was the case with fitting a GLD, we use the method of moments described in Example 4.10.39. Thus, for data $X_1, X_2, \ldots, X_n$ that are independent and identically distributed from an unknown distribution we estimate the first four moments through equations (4.10.40) and (4.10.41). **Next, we use the tables** of Dudewicz and Karian (1996)—in particular Table 2—to find the $\hat{\beta}_3$ and $\hat{\beta}_4$ for which the tabulated (α_3, α_4) is closest to the $(|\hat{\alpha}_3|, \hat{\alpha}_4)$ calculated in (4.10.40). (If $\hat{\alpha}_3 < 0$, we interchange $\hat{\beta}_3$ and $\hat{\beta}_4$.) We now substitute $\hat{\beta}_3$ for β_3 and $\hat{\beta}_4$ for β_4 in equations (4.10.54) and (4.10.55) and solve the resulting equations for β_2 and β_1. It is worth repeating our earlier admonition: **one should check the fit visually by plotting** the probability density function given in Definition 4.10.51 with the fitted $\hat{\beta}_1, \hat{\beta}_2, \hat{\beta}_3, \hat{\beta}_4$ superimposed on a histogram.

Example 4.10.60. Numerical example of fitting an Extended GLD.
Suppose we have a data set to which we wish to fit a GLD, and the calculations at (4.10.40) and (4.10.41) yield

$$\hat{\alpha}_1 = 106.8348, \quad \hat{\alpha}_2 = 22.2988, \quad \hat{\alpha}_3 = -.1615, \quad \hat{\alpha}_4 = 2.1061.$$

In Table 2 of Dudewicz and Karian (1996), we find that the closest pair to $(|\hat{\alpha}_3|, \hat{\alpha}_4)$ is $(.150, 2.10)$. For that pair, we read the values of $\hat{\beta}_3$ and $\hat{\beta}_4$ from the table and compute $\hat{\beta}_1$ and $\hat{\beta}_2$ from (4.10.51) and (4.10.52) to get

$$\hat{\beta}_1 = 95.8461, \quad \hat{\beta}_2 = 20.0430, \quad \hat{\beta}_3 = .8980, \quad \hat{\beta}_4 = .5639.$$

Thus, the fitted GBD is the one with p.d.f.

$$f(x) = .02323(x - 95.8461)^{.8980}(115.8891 - x)^{.5639}$$

for $95.8461 \leq x \leq 115.8891$. If we locate the point $(\alpha_3^2, \alpha_4) = (.02608, 2.1061)$ on Figure 4.10–2, we see that it falls in the region where the GBD and the GLD overlap. Therefore, it is possible to obtain another fit in this system, this time using a GLD distribution. For this fit we use Table 1 of Dudewicz and Karian (1996) to obtain

$$\hat{\lambda}_1(0,1) = -1.1221, \quad \hat{\lambda}_2(0,1) = .3010, \quad \hat{\lambda}_3 = .05487, \quad \hat{\lambda}_4 = .6387,$$

which, through the transformation (4.10.42), leads us to

$$\hat{\lambda}_1 = 112.1355, \quad \hat{\lambda}_2 = .06374, \quad \hat{\lambda}_3 = .06387, \quad \hat{\lambda}_4 = .05478.$$

4.10.4 Bivariate GLD Distribution: The GLD-2

Since the GLD and EGLD are very successful in practice in fitting a great variety of data sets, and since data is often bivariate (the simplest multivariate setting), it is very desirable to have a way of fitting a bivariate GLD. This is a distribution each univariate component of which has a univariate GLD (or EGLD) distribution, but in which the components are not independent random variables. A bivariate GLD, called GLD-2, was developed by Beckwith and Dudewicz (1996), and its essentials are presented here; for full details, see Beckwith and Dudewicz (1996) and Karian and Dudewicz (1999).

The algorithm developed by Beckwith and Dudewicz (1996) is built upon the following idea. First, fit univariate GLDs to the components X and Y of the bivariate random variable (X, Y) separately. Second, use the method

R. L. Plackett (1965) proposed for generating bivariate distributions with specified marginals, to develop a bivariate distribution with these marginals. Third, optimize over the infinite set of distributions in the Plackett class, thus fitting the actual data set. Finally, produce a bivariate plot to check the quality of the fit. More specifically, the algorithm is:

Algorithm GLD–2

Step GLD-2.1. Given a bivariate data set with unknown bivariate p.d.f. $h(x,y)$, fit both marginals X and Y with GLDs, using a suitable method (e.g., the method of moments).

Step GLD-2.2. Graph the marginal p.d.f.s along with histograms of X and Y and graph the marginal d.f.'s along with the empiric d.f.s (e.d.f.s), to verify the quality of the two univariate fits.

Step GLD-2.3. Graph a scatterplot of the (x,y) data with the lines $x = \tilde{x}$ and $y = \tilde{y}$ where $\tilde{x}$ and $\tilde{y}$ are the respective medians of X and Y.

Step GLD-2.4. Count the number of points in each of the four quadrants of the scatterplot in Step GLD-2.3 and label these a, b, c, and d, for the counts in quadrants 3, 2, 4, and 1, respectively.

Step GLD-2.5. Calculate $\Psi^+ = \frac{a \cdot d}{b \cdot c}$, and use this value for Ψ in the formula for the distribution function $H(x,y)$ in (4.10.61) or in the formula for the p.d.f. $h(x,y)$ given in (4.10.62).

Step GLD-2.6. Graph the GLD-2 to visualize the quality of the fit.

The distribution function and p.d.f. formulas are, respectively:

$$H(x,y) = \begin{cases} \dfrac{S(x,y) - \sqrt{S^2(x,y) - 4\Psi(\Psi - 1)F(x)G(y)}}{2(\Psi - 1)} & (\Psi \neq 1) \\ F(x)G(y) & (\Psi = 1) \end{cases} \quad (4.10.61)$$

where Ψ is nonnegative, and

$$h(x,y) = \frac{\Psi f(x)g(y)\,[1 + (\Psi - 1)(F(x) + G(y) - 2F(x)G(y)]}{(S^2(x,y) - 4\Psi(\Psi - 1)F(x)G(y))^{3/2}} \quad (4.10.62)$$

with

$$S(x,y) = 1 + (F(x) + G(y))(\Psi - 1).$$

Note that $\Psi = 1$ is the case when X and Y are independent random variables, in which case $H(x,y) = F(x)G(y)$ and $h(x,y) = f(x)g(y)$ for all x and all y. To illustrate this method we give the results of fitting the bivariate data which arose in a study of imaging in the brain (Dudewicz, Levy, Lienhart, and Wehrli (1989)). Here the issue was to fit the measured quantity called AD, which is the ^1H hydrogen density in a specific portion of brain tissue as related to the average density in the whole brain. In this example, the AD measurements are for the Cortical White Matter Right side (CWR) (designated by X) and the Cortical White Matter Left side (CWL) (designated by Y). Thus, the data are the pairs (X,Y) and the concern of the research was to determine if there was a relationship between X and Y, or if X and Y were independent random variables. The data, from p. 337 of Dudewicz, Levy, Lienhart, and Wehrli (1989), is given in Table 4.10–3.

Table 4.10–3. Brain tissue ^1H hydrogen density data.

X	Y	X	Y	X	Y
96.8	99.2	96.8	96.8	95.6	100.8
107.6	102.9	86.5	86.1	87.3	85.7
99.6	98.1	94.5	95.5	89.8	91.2
99.5	103.9	97.0	92.9	88.2	88.8
102.9	109.2	92.2	91.0	89.4	94.1
84.8	82.8	100.3	98.2	88.6	89.9
97.9	95.9	102.5	95.6	87.7	86.4
103.2	100.2	100.6	99.4	88.1	91.0

The results of applying Algorithm GLD-2 to the data of Table 4.10–3, shown in Figures 4.10–4, 4.10–5, and 4.10–6, can be interpreted as follows:

1. As prescribed by Step GLD-2.1 of Algorithm GLD-2, Figure 4.10–4 shows the GLD p.d.f. that is fitted to X (together with the histogram for X) and the fitted c.d.f. (with the empiric distribution of X). Figure 4.10–5 does the same for Y. These figures indicate that the GLD fits the data well.

2. As required by Step GLD-2.3, a two-dimensional scatterplot, with median lines, is shown in Figure 4.10–6. The counts a, b, c, and d, specified in Step GLD-2.4, are shown on the scatterplot.

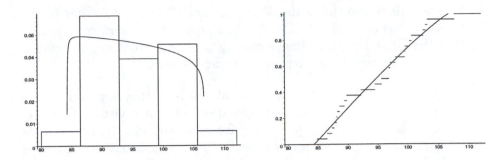

Figure 4.10–4. Histogram of X with its GLD p.d.f.; empirical distribution with its GLD d.f.

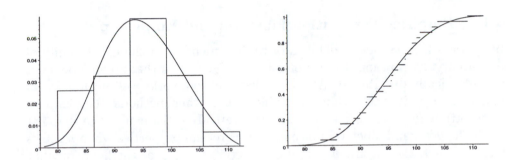

Figure 4.10–5. Histogram of Y with its GLD p.d.f.; empirical distribution with its GLD d.f.

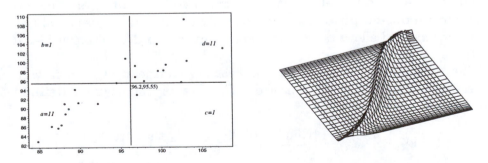

Figure 4.10–6. The (X, Y) scatterplot and the bivariate GLD–2 p.d.f. for (X, Y).

3. Figure 4.10–6 also shows the 3-dimensional GLD-2 p.d.f. that is obtained by using (4.10.62). The "snake shape" of this surface indicates that the GLD-2 fit lacks the symmetry properties that are associated with some distributions such as the bivariate normal.

We see that X and Y are not independent. In fact, they are highly correlated, and so a tumor in one side might stand out on a visual X and Y comparison. Also, the fits are quite good for each variable. Some software useful in some of these considerations, and other applications, are given in the article by Beckwith and Dudewicz (1996); for other details, also see Karian and Dudewicz (1999).

4.11   Empiric Distribution Function and Empiric p.d.f.

In Section 4.10 we saw in detail how to fit a member of the GLD family of probability distributions to a set of data $X_1, X_2, \ldots, X_n$ that are independent and identically distributed. Although that method works well in practice in many settings, it is sometimes desirable to use other methods. The empiric distribution function is one which a theoretical statistician or probabilist might suggest, and much is known of its theoretical properties when n grows large. In fact, the empiric distribution function (d.f.) is at the heart of what are called "bootstrap methods" in statistics, which have become a hot topic in the last decade with the advent of cheap and fast computations. **In this section we note details of the empiric d.f.**

We also note that the empiric d.f. is in many cases undesirable, since it is "rough" even when we know that the real underlying d.f. is "smooth." **A smooth alternative, introduced recently and called the continuous-empiric d.f.,** is more highly recommended for actual use. This new method allows one to develop **an empiric p.d.f.,** also considered in this section.

The empiric d.f. or sample d.f. based on independent random variables $X_1, \ldots, X_n$ each of which have d.f. $F(\cdot)$ (usually unknown) **is defined for all x by**

$$F_n(x|X_1, \ldots, X_n) = \frac{(\text{Number of } X_1, \ldots, X_n \text{ that are } \leq x)}{n}. \quad (4.11.1)$$

Note that the true d.f. gives us information such as $F(99.3)$, the probability that any X_i will be less than or equal to 99.3. The empiric d.f. estimates this probability by $F_n(99.3|X_1, \ldots, X_n)$, the proportion of the sample that was in fact less than or equal to 99.3. It is known that the empiric d.f. has good properties as $n \to \infty$ (e.g., see Chapters 6 and 7 of Dudewicz and Mishra (1988)).

Example 4.11.2. The seasonal snowfall at Syracuse, New York, for the three seasons 1982–83, 1983–84, 1984–85 was respectively 66, 114, and 116 (in inches, to the nearest inch). The true d.f. of the seasonal snowfall is unknown. Find the empiric d.f. $F_3(x|X_1, X_2, X_3)$, and plot it on a graph.

Since (Number of $X_1 = 66$, $X_2 = 114$, $X_3 = 116$ which are $\leq x$)

$$= \begin{cases} 0 \text{ if } x < 66 \\ 1 \text{ if } 66 \leq x < 114 \\ 2 \text{ if } 114 \leq x < 116 \\ 3 \text{ if } x \geq 116, \end{cases} \tag{4.11.3}$$

it follows (by dividing by $n = 3$) that

$$F_3(x|66, 114, 116) = \begin{cases} 0 \quad \text{if } x < 66 \\ 1/3 \text{ if } 66 \leq x < 114 \\ 2/3 \text{ if } 114 \leq x < 116 \\ 1 \quad \text{if } x \geq 116. \end{cases} \tag{4.11.4}$$

This empiric d.f. is plotted in Figure 4.11–1.

In the plot of the empiric d.f. of Example 4.11.2, in Figure 4.11–1, we see that the empiric d.f. is a step function, increasing by $1/n$ at each data point value X_i $(i = 1, 2, \ldots, n)$. Thus, it models a random variable that can take on only the values $X_1, \ldots, X_n$ of the data set, and takes them on with equal probabilities $1/n$ each, or $2/n$, $3/n$, etc. if there are equal X_i's. Thus, $F_n(x|X_1, \ldots, X_n)$ is the d.f. of a discrete random variable, and **we can generate observations from such a random variable by the methods of Section 4.3.** However, such observations will never take on any value but one of $X_1, \ldots, X_n$. In Example 4.11.2, this means that in our simulation we would never observe any snowfalls outside the set of recent historical values $\{66, 114, 116\}$. **Since we know that snowfall is a continuous variable, restricting the values generated to a finite (and small) set of possible values does not make sense.**

For example, if one is performing a simulation to decide on an amount of road deicer for winter use for a city, one would with high probability be ill prepared if one generated snowfalls from a simulation using the empiric d.f. **Even if n is large, there are still problems with this approach: one will never generate a value larger than $\max(X_1, \ldots, X_n)$, the**

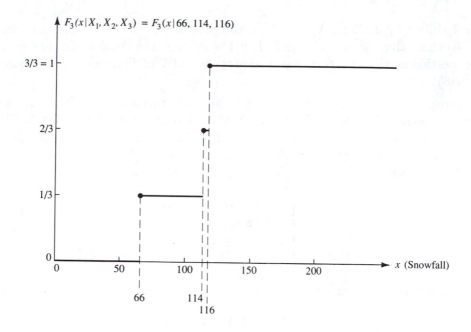

Figure 4.11–1. Empiric d.f. for $n = 3$ data points of Example 4.11.2.

historical maximum, although one knows that higher values are possible, and are in fact the most severe tests of the system under simulation in many examples, such as those dealing with storm intensities, floods, and the like. For these reasons, one should note that **there is a natural way to modify the empiric d.f. to an absolutely continuous d.f., the derivative of which is called the empiric p.d.f.**

Definition 4.11.5. *Let* $a \leq \min(X_1, \ldots, X_n) \leq \max(X_1, \ldots, X_n) \leq b$ *be given numbers. The* **continuous-empiric d.f.** *is the d.f.* $G_n(x|a, X_1, \ldots, X_n, b)$ *that equals 0 if* $x \leq a$, *equals 1 if* $x \geq b$, *and for* $a < x < b$ *has the value of the straight line segments which join the successive midpoints of the bars that constitute the empiric d.f. (The midpoint of the leftmost bar is joined to the point* $(a, 0)$, *while the rightmost bar is joined to the point* $(b, 1)$.*) The* **empiric p.d.f.** *is*

$$g_n(x|a, X_1, \ldots, X_n, b) = G'_n(x|a, X_1, \ldots, X_n, b). \qquad (4.11.6)$$

Example 4.11.7. In Figure 4.11–2 we show the continuous-empiric d.f. constructed on the same graph which has the empiric d.f.; here we use $a =$

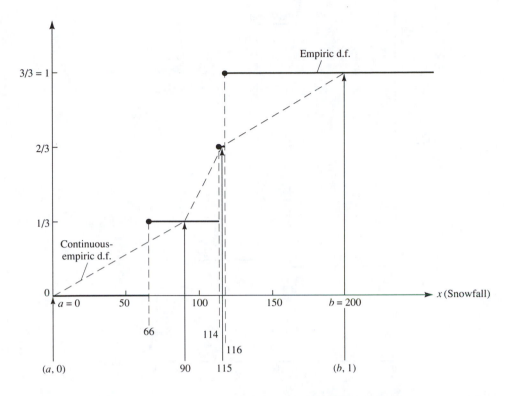

Figure 4.11–2. A continuous-empiric d.f. (dashed curve) constructed for the $n = 3$ data points of Example 4.11.2.

50, $b = 200$. The corresponding empiric p.d.f. is shown in Figure 4.11–3. **Generating random variables from the empiric p.d.f. is relatively simple**, since it is essentially a mixture of uniforms on the various intervals between a and b.

The construction of the empiric p.d.f. follows since the equations of the line segments that make up the continuous-empiric d.f. are

$$y = \frac{\frac{1}{3} - 0}{90 - 0}\, x \quad \text{(for } 0 < x < 90\text{)},$$

$$y = \frac{\frac{2}{3} - \frac{1}{3}}{115 - 90}\, x + b_1 \quad \text{(for } 90 < x < 115\text{)},$$

$$y = \frac{1 - \frac{2}{3}}{200 - 115}\, x + b_2 \quad \text{(for } 115 < x < 200\text{)},$$

and $y = 0$ (for $x < 0$), $y = 1$ (for $x > 200$). Since the empiric p.d.f. is the

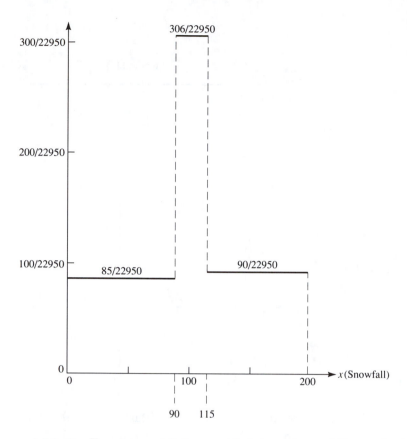

Figure 4.11–3. Empiric p.d.f. for $n = 3$ data points of Example 4.11.2.

derivative of the continuous-empiric d.f., its values are the slopes of these line segments (or, 0 when the continuous-empiric d.f. is a constant, which is for $x < 0$ and for $x > 200$). Those slopes, used in Figure 4.11–3, are

$1/270$ for $0 < x < 90$, $1/75$ for $90 < x < 115$, $1/255$ for $115 < x < 200$.

Note that the area under the curve in Figure 4.11–3 totals to

$$(90 - 0)\frac{1}{270} + (115 - 90)\frac{1}{75} + (200 - 115)\frac{1}{255} = \frac{1}{3} + \frac{1}{3} + \frac{1}{3} = 1,$$

as must be the case for any probability density function.

It can be shown that **the general formula for the empiric p.d.f.,**

$g_n(x|a, X_1, \ldots, X_n, b)$, **is** (see Dudewicz and Mishra (1988), pp. 202–203)

$$= \begin{cases} \dfrac{2}{n(Y_{j+1} - Y_{j-1})}, \text{ if } \dfrac{Y_j + Y_{j-1}}{2} \leq x < \dfrac{Y_{j+1} + Y_j}{2} \ (j = 1, \ldots, n) \\ 0, \qquad\qquad\qquad \text{otherwise,} \end{cases} \quad (4.11.8)$$

where $Y_1 \leq Y_2 \leq \ldots \leq Y_n$ are $X_1, X_2, \ldots, X_n$ in increasing numerical order (called the **order statistics** of the sample), $Y_0 = 2a - Y_1$, and $Y_{n+1} = 2b - Y_n$. Note that the continuous-empiric d.f. is continuous everywhere, but usually it is not differentiable at $(Y_{j+1}, Y_j)/2$ for $j = 1, \ldots, n$; this does not cause problems however, and we can (as in (4.11.8)) take g_n to be the left-hand value at those points.

4.12 Sampling from a Histogram

Often a characteristic for which we need to generate values for a simulation study does not have a known distribution. If we have a set of independent and identically distributed random variables from the unknown distribution, say $X_1, \ldots, X_n$, then the methods studied in previous sections may be used to estimate the distribution, and subsequently sample from it. The question we now wish to study is, **"How should we proceed if the data are presented in the form of a histogram (such as in Figure 4.12–1)?"** Two cases arise when attempting to answer this question.

Case 1. The basic data is available. The first thing one should do when facing the above question and a histogram in the form of Figure 4.12–1, is to seek the basic data used to develop the histogram. **If $X_1, \ldots, X_n$ are available**, then one is not tied to the histogram, and one can

1. Fit a GLD family member (see Section 4.10).

2. Fit the empiric probability density function (see Section 4.11).

3. Fit a member of some other family of distributions that seems appropriate (for example, for the data shown in Figure 4.12–1 a member of the exponential family may well be appropriate).

One then samples from the selected distribution by the methods discussed earlier. Although Case 1 often occurs (if one persistently seeks out the original data), sometimes the data is not available, or was only tabulated

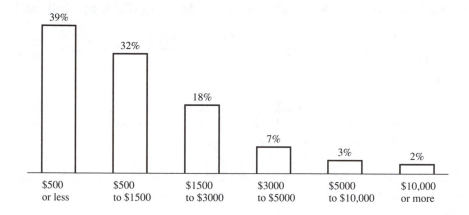

Figure 4.12–1. A histogram of the amount new homeowners spend
on landscaping. (Source: "Keeping It Green,"
Syracuse Herald-Journal, 4-4-88, p. F1.)

by category without recording the original values. In such instances, other
approaches are needed.

Case 2. The basic data are not available. In this case, one will need
to proceed from the histogram, either with direct use, or with a distribution somehow fitted using the histogram instead of the basic data. In this
instance, **one should make sure that the histogram is appropriate
(i.e., valid).** On closer examination of Figure 4.12–1, we find some flaws
with it as a histogram. First, the bars are not proportional in height to the
percentages noted, which they should be if the bases are of equal lengths.
Second, the bar heights times the base lengths are not proportional to the
percentages, which they should be in histograms with unequal base lengths.
Therefore, as is quite typical, we will need to develop a valid histogram before proceeding. This is done in Figure 4.12–2, where we now have areas
of bars equal to the percentages in the base range of expenditures. In developing this valid histogram, we face two additional problems: two of the
bases are of infinite length, and the percentages do not add to 100%. These
problems were solved by

1. Making the first bar $0 to $500 (instead of $500 or less), since there is
   a natural lower bound in this example.

2. Not attempting to draw a bar at the upper end.

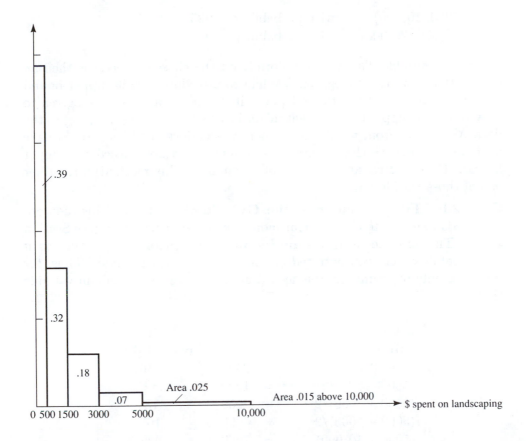

Figure 4.12–2. A histogram of landscaping expenditure.

3. Adjusting the percentages that first failed to add to 100, due to round-ing, to add to 100.

We can now take at least three approaches to the problem of sampling from the valid histogram.

Case 2.a. Sample from the histogram. Here one first selects an interval, taking

(0, 500)	with probability	.39
(500, 1500)	with probability	.32
(1500, 3000)	with probability	.18
(3000, 5000)	with probability	.07

(5000, 10, 000)   with probability   .025
(10, 000, 20, 000)   with probability   .015.

A random variable distributed uniformly on the chosen interval is then an observation from the histogram. We had to introduce a finite upper bound to use this method. Unfortunately, even if the histogram had been given to us with a finite upper bound—that might just be the $\max(X_1, \ldots, X_n)$ in the data $X_1, \ldots, X_n$ from which the histogram was developed—we would have no reason to believe that future observations could not exceed this upper bound. For these reasons, we do not recommend this method, but rather one of those that follow.

Case 2.b. Fit a member of the GLD family. To use this method, one needs to estimate the first four moments of the distribution (see Section 4.10). This is done using the **midpoint assumption**, i.e., acting as if all probability were concentrated at the center of each interval. Thus, for purposes only of estimating the needed moments, we use a random variable that

takes value			with probability
$(0 + 500)/2$	=	250	.39
$(500 + 1500)/2$	=	1000	.32
$(1500 + 3000)/2$	=	2250	.18
$(3000 + 5000)/2$	=	4000	.07
$(5000 + 10, 000)/2$	=	7500	.025
$(10, 000 + 20, 000)/2$	=	15, 000	.015.

The mean then is estimated by

$$(250)(.39) + (1000)(.32) + (2250)(.18) + (4000)(.07) + (7500)(.025)$$

$$+(15, 000)(.015) = 1515.$$

The second, third, and fourth moments are estimated similarly, then a GLD (or an Extended GLD) is fitted and sampled from as discussed in Section 4.10.

Case 2.c. Fit a member of another family. While a full discussion is beyond the scope of this text, note that some families that often arise are the normal (discussed in Section 4.13) and the exponential.

In summary, **we recommend** use of Case 1 if possible, otherwise use of Case 2.b. or Case 2.c.

4.13   Fitting a Normal (Univariate or Multivariate) Distribution

Often it is desired to fit a univariate normal distribution, specified by (4.2.15) (or, equivalently, by (4.10.4) or (4.10.5)) to a set of data $X_1, X_2, \ldots, X_n$ that is assumed to consist of independent and identically distributed random variables.

If the data is normally distributed, i.e., if X_i is $N(\mu, \sigma^2)$ for $i = 1, 2, \ldots, n$, then the mean μ and the variance σ^2 are usually estimated respectively by $\overline{X}$ and s^2 (see (4.10.1) and (4.10.2)). That is, we usually use the $N(\overline{X}, s^2)$ distribution as our fitted distribution,

$$f(x) = \frac{1}{\sqrt{2\pi} s} e^{-0.5(x-\overline{X})^2/s^2}. \tag{4.13.1}$$

These estimates arise from the "method of matching moments."

Often, in practice, it is not certain that the data is normally distributed. It may be known that the data are independent since they arise from independent simulation replications, and that they have the same distribution since the simulation replications are randomly started from the same distributions each time. **In such cases, we recommend the following.**

> **Test the hypothesis** that $X_1, X_2, \ldots, X_n$ are normal random variables; **if the hypothesis is not rejected,** then use the $N(\overline{X}, s^2)$ distribution; **if the hypothesis is rejected, transform the data** to $g(X_1), \ldots, g(X_n)$ in such a way that $g(X_i)$ is normally distributed, and then proceed using the $N(\overline{Y}, s_Y^2)$ distribution for $g(X_i)$, where
>
> $$\overline{Y} = (Y_1 + Y_2 + \ldots + Y_n)/n,$$
> $$s_Y^2 = \sum_{i=1}^{n} (Y_i - \overline{Y})^2/(n-1), \tag{4.13.2}$$
> $$Y_i = g(X_i) \quad (i = 1, \ldots, n).$$

To put this recommendation into practice, one needs to know **how to test the hypothesis that $X_1, X_2, \ldots, X_n$ are normal random variables.**

One approach that is often used is the **Lilliefors test**. In this approach, one first calculates

$$Z_1 = \frac{X_1 - \overline{X}}{s}, \quad Z_2 = \frac{X_2 - \overline{X}}{s}, \ldots, \quad Z_n = \frac{X_n - \overline{X}}{s} \qquad (4.13.3)$$

(which should be approximately standard normal if the hypothesis is true). Then the test statistic

$$D_n^* = \sup_z |F_n(z) - \Phi(z)| \qquad (4.13.4)$$

is found, where $F_n(\cdot)$ is the sample d.f. of $Z_1, Z_2, \ldots, Z_n$ (see (4.11.1)) and $\Phi(\cdot)$ is the standard normal distribution function (see (4.10.4)). One rejects normality if $D_n^* > d_{n,\alpha}^*$ (where the constant $d_{n,\alpha}^*$ gives a test of level α; see Dudewicz and Mishra (1988, p. 671) for details and references). While this test is often used, the tests recommended below may be preferable due to their better power.

A second approach which is often used is that of **probability plotting**. In this approach, one takes advantage of the key fact that if a random variable X has the $N(\mu, \sigma^2)$ distribution, then as in (4.10.4) its distribution function is

$$F(x) = P[X \le x] = \Phi(\frac{x - \mu}{\sigma}),$$

hence

$$\Phi^{-1}(F(x)) = \Phi^{-1}(\Phi(\frac{x - \mu}{\sigma})) = \frac{x - \mu}{\sigma} = \frac{1}{\sigma} x - \frac{\mu}{\sigma}, \qquad (4.13.5)$$

which is a straight line. Thus, while the plot of $(x, F(x))$ is a curve (the normal distribution function curve), **the plot of $(x, \Phi^{-1}(F(x)))$ is a straight line with slope $1/\sigma$ and intercept $-\mu/\sigma$ for any normal distribution function $F(\cdot)$**. If we don't know $F(\cdot)$, we may estimate it by the sample d.f. of $X_1, X_2, \ldots, X_n$ (see (4.11.1)). Now if we plot at the n points $x = X_{(1)}, X_{(2)}, \ldots, X_{(n)}$, where $X_{(1)} < X_{(2)} < \ldots < X_{(n)}$ are $X_1, X_2, \ldots, X_n$ in numerical order, we will be plotting the points

$$(X_{(i)}, \Phi^{-1}(F_n(X_{(i)}|X_1, X_2, \ldots, X_n))) = (X_{(i)}, \Phi^{-1}(i/n)), \qquad (4.13.6)$$

for $i = 1, \ldots, n$. Since $(x, \Phi^{-1}(F(x))) = (x, x/\sigma - \mu\sigma)$ plots as a straight line, if $F_n(\cdot)$ is a good estimate of the true $F(\cdot)$ we expect the plot of the points in (4.13.6) to look like a straight line. If one fits a straight line to the plotted points (called a probability plot), the slope will provide an estimate of $1/\sigma$ and the intercept will provide an estimate of $-\mu/\sigma$ (and these can be solved to provide estimates of μ and σ).

This test is often performed on what is called "normal probability paper;" see Nelson (1979) for further details. One problem with this probability plotting test is that assessment of the linearity of the plot is subjective. One advantage is that one "sees" from the plot such characteristics as exemplified in Figure 4.13–1, as well as many others. Such information may be very useful in selecting a transformation so that $g(X)$ is normal in distribution when X is not.

A test that has been developed to test for normality of the regression line fitted to a plot like that of (4.13.6) is **the Shapiro-Wilk test, which has been found to be very powerful in detecting nonnormality.** The Shapiro-Wilk test statistic is

$$W = \frac{\left(\sum_{i=1}^{k} a_{n-i+1}(X_{(n-i+1)} - X_{(i)})\right)^2}{\sum_{i=1}^{n} X_i^2 - \frac{1}{n}\left(\sum_{i=1}^{n} X_i\right)^2} \tag{4.13.7}$$

where $k = n/2$ if n is even (and $k = (n-1)/2$ if n is odd). While the a_js needed were originally tabulated in 1965 for n up to 50, they have now been extended up to $n = 100$, and a slightly modified procedure for which no further tables are needed was given by Weisberg and Bingham in 1975; for tables and references, see Shapiro and Brain (1982). Note that one rejects for small W (i.e., for $W < d$ where d is set to give the desired level of significance for the sample size n).

Now that we know how to test for univariate normality, we need to know **how to transform the data if normality is rejected.** Some commonly used transformations are $g(x) = \sqrt{x}$, $\sqrt{x-a}$, $\ln(x)$, etc. One can attempt

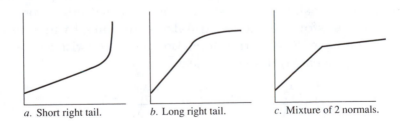

a. Short right tail.   *b.* Long right tail.   *c.* Mixture of 2 normals.

Figure 4.13–1. Information in probability plots.

to find one of these for which normality is not rejected for the transformed data. A more scientific approach, with at least as good results, is to **examine a class of transformations**. For each transformation $g(\cdot)$ in the class, compute $g(X_1), g(X_2), \ldots, g(X_n)$, and then compute W_g, the Shapiro-Wilk test statistic for normality of $g(X_1), g(X_2), \ldots, g(X_n)$. Since one rejects when $W_g < d$, **all g for which the test statistic exceeds d are acceptable, and from those one should choose one that "makes sense" when interpreted** in the terms of the basic problem. **As a class of transformations, we recommend**

$$g_\lambda(x) = \frac{x^\lambda - 1}{\lambda}, \tag{4.13.8}$$

supplemented by $\ln(x)$, and also supplemented by any transformations that seem natural in the context of the problem or that are suggested by the probability plot. The transformations given in (4.13.8) are called the Box-Cox (or power) transformations.

So far we have considered the univariate normal distribution. **If the data have the multivariate normal distribution** of (4.8.3) and (4.8.4) but the **mean vector** $\underset{\sim}{\mu} = (\mu_1, \ldots, \mu_p)'$ and the **covariance matrix** (σ_{ij}) are unknown, then (for $j = 1, 2, 3, \ldots, p$) these are usually estimated, respectively, by

$$\hat{\mu}_j = \sum_{i=1}^{n} X_{ij}/n,$$

$$\hat{\sigma}_{ij} = \sum_{r=1}^{n} (X_{ri} - \hat{\mu}_i)(X_{rj} - \hat{\mu}_j)/(n-1). \tag{4.13.9}$$

If one is not sure of the multivariate normality, one can test this hypothesis using the Malkovich-Afifi extension of the Shapiro-Wilk test (see Shapiro and Brain (1982, Section 6) for details), or the new graphical test developed by Ozturk and Romeu. However, this does not provide a way to transform if normality is rejected. An approach that often works well in practice is to **transform each coordinate to univariate normality**. If, for example, only the Box-Cox transformations are used (with easy extension if they are supplemented), these can yield

$$\left(\frac{X_1^{\lambda_1} - 1}{\lambda_1}, \frac{X_2^{\lambda_2} - 1}{\lambda_2}, \ldots, \frac{X_p^{\lambda_p} - 1}{\lambda_p} \right). \tag{4.13.10}$$

(4.13.10) is a transformation of $(X_1, X_2, \ldots, X_p)$, to a single p-variate observation that is treated as if it were multivariate normal in distribution.

This multivariate normality claim can, and often should, be tested with the Malkovich-Afifi test (or another test).

As a final note, while Box-Cox transformations may not make sense **if the data can be negative** (e.g., $\sqrt{x}$ cannot be used if x is negative), in such cases one can simply use the class of transformations

$$\frac{\text{sgn}(x)|x|^\lambda - 1}{\lambda}.$$

(4.13.11)

(Here, $\text{sgn}(x) = 1$ if $x > 0$, and $= -1$ if $x < 0$.)

Problems for Chapter 4

4.1 (Section 4.2) Suppose that we wish to generate random variables with distribution function $F(x) = x^n$ (for $0 \leq x \leq 1$, and $= 0$ for $x < 0$, and $= 1$ for $x > 1$), where $n = 5$.

   a. Use Theorem 4.2.1 to find a function of a uniform random variable U on $(0, 1)$ such that the function has the desired distribution.

   b. Using URN13 (see (3.5.9)) with seed $x_0 = 1$ to generate the needed uniform variables $U_1, U_2, \ldots$, use your results from Part a to generate 100 independent observations $X_1, X_2, \ldots, X_{100}$ from distribution function $F(x)$.

4.2 (Section 4.3) Suppose that we wish to generate random variables with the discrete distribution that takes on values $0, 1, 2, 3$ with probabilities $.01, .04, .05, .90$.

   a. Using the uniform variables generated in Part b of Problem 4.1, generate 20 random variables $X_1, X_2, \ldots, X_{20}$ that are independent and have the desired discrete distribution using the method of Theorem 4.3.6.

   b. Using the result of Lemma 4.3.10, modify your procedure in Part a so that it is most efficient.

   c. How many comparisons did you make in generation in Part a? How many in Part b?

4.3 (Section 4.4) Generate 8 independent random variables $X_1, X_2, \ldots, X_8$ that each have the Poisson distribution with mean 2.2. (If you need uniform random variables, use the random variables obtained in Part b of Problem 4.1.) How many uniform random variables did you use? What average number of uniform random variables did you need per Poisson variable generated?

4.4 (Section 4.4) Proceed as in Problem 4.3, but with mean 5.6 for the Poisson variable.

4.5 (Section 4.4) Suppose a random variable sequence that is independent with a geometric distribution with success probability .05 is desired. Using Part b of Problem 4.1 for any needed uniform random variables:

a. Use the method of Corollary 4.4.14 to produce 3 such variables. How many uniforms did you need?

b. Use the method of Theorem 4.4.20 to produce 3 such variables. How many uniforms did you need?

4.6 (Section 4.5) Suppose that the vector (X_1, X_2, X_3), where X_1 is the number of university degrees a person holds, X_2 is the number of houses they own, and X_3 is the number of spouses they have been married to, takes on values a_i with probability p_i according to the table below. Generate 20 values of the random vector (X_1, X_2, X_3) that are independent, using the most efficient version of Theorem 4.5.6. (For uniform random variables needed, use those found in Part b of Problem 4.1.)

Value	Prob.	Value	Prob.	Value	Prob.	Value	Prob.
(0,0,0)	.10	(1,0,0)	.03	(2,0,0)	.01	(3,0,0)	.003
(0,0,1)	.10	(1,0,1)	.04	(2,0,1)	.01	(3,0,1)	.002
(0,0,2)	.10	(1,0,2)	.03	(2,0,2)	.03	(3,0,2)	.005
(0,1,0)	.05	(1,1,0)	.01	(2,1,0)	.005	(3,1,0)	.001
(0,1,1)	.15	(1,1,1)	.03	(2,1,1)	.04	(3,1,1)	.01
(0,1,2)	.05	(1,1,2)	.01	(2,1,2)	.005	(3,1,2)	.009
(0,2,0)	.01	(1,2,0)	.01	(2,2,0)	.01	(3,2,0)	.001
(0,2,1)	.03	(1,2,1)	.03	(2,2,1)	.035	(3,2,1)	.018
(0,2,2)	.01	(1,2,2)	.01	(2,2,2)	.005	(3,2,2)	.001

4.7 (Section 4.6) Use the Kolmogorov-Smirnov One-Sample Test to test the hypothesis that the 100 observations generated in Problem 4.1 have the desired distribution function $F(\cdot)$. (Use level of significance .05.)

4.8 (Section 4.6) Prove Theorem 4.6.44.

4.9 (Section 4.6) Show that m and M in (4.6.48) are, respectively, the minimum and maximum. Then prove Theorem 4.6.49.

4.10 (Section 4.7) Suppose that $g(x) = 5x^4$, a p.d.f. from which we have generated 100 observations in Problem 4.1. Suppose that $f(x) = 6x^5$ (for x between 0 and 1). Find a value of the constant a such that (4.7.14) holds. Use Theorem 4.7.15, the values from Part b of Problem 4.1, and additional uniforms (as and if needed) from the sequence URN13 started in Problem 4.1, to produce 3 values from p.d.f. $f(\cdot)$.

4.11 (Section 4.8) Suppose that (X_1^*, X_2^*, X_3^*) is a vector of independent standard normal random variables. Find a vector that is MVN with mean vector $(1, 2, 3)$, and covariance matrix

$$\begin{pmatrix} 4 & .1 & .1 \\ .1 & 9 & .1 \\ .1 & .1 & 16 \end{pmatrix}.$$

(Use Theorem 4.8.5, and give the explicit final function needed.)

4.12 (Section 4.9) For the trivariate random vector constructed in Problem 4.11, specify a region of values in Euclidean 3-space that has probability .95 of containing the vector. (An explicit final solution is desired, with a graph of the region. If this is too hard, with some loss of credit work this problem for the bivariate random vector that consists of the first two components of the trivariate vector.)

4.13 (Section 4.10) Use an argument similar to the one given in the proof of Lemma 4.10.24 to show that the GLD is not valid in region V_2.

4.14 (Section 4.10) Prove Lemma 4.10.25 by verifying that for $\lambda_3, \lambda_4)$ in Region 1:

a. $g(y, \lambda_3, \lambda_4) \leq g(y, -1, \lambda_4)$ where $g(y, \lambda_3, \lambda_4)$ is defined by (4.10.22).

b. $g(y, -1, \lambda_4)$ has a minimum at $\lambda_4 = -1/\ln(1 - y)$, provided this is larger than 1. Otherwise, $g(y, -1, \lambda_4)$ has a minimum at $\lambda_4 = 1$ yielding

$$g(y, \lambda_3, \lambda_4) \leq g(y, -1, \lambda_4) \leq g(y, -1, 1) \leq 1 - \frac{1}{y^2} \leq 0.$$

c. If the minimum of $g(y, -1, \lambda_4)$ is at $\lambda_4 = -1/\ln(1-y)$, then $0 \leq y \leq 1 - 1/e$ and

$$g(y, \lambda_3, \lambda_4) \leq g(y, -1, \lambda_4) \leq g(y, 1, -1/\ln(1-y))$$
$$= -\frac{1}{y^2} - \frac{(1-y)^{-(1+\frac{1}{\ln(1-y)})}}{\ln(1-y)}.$$

Show, by substituting $u = \ln(1-y)$ in the above expression, that for $-1 \leq u \leq 0$,

$$g(y, \lambda_3, \lambda_4) \leq -\frac{u - e^{-u-1}}{u(1 - e^u)^2} \leq 0.$$

4.15 (Section 4.10) Use an argument based on symmetry to show that the GLD is valid in Region 2 of Figure 4.10–1.

4.16 (Section 4.10) Verify that if Z is $GLD(0, \lambda_2, \lambda_3, \lambda_4)$, then $E(Z^k)$ is given by formula (4.10.26).

4.17 (Section 4.10) Using the methods of Section 4.10, fit a GLD to the first 20 data points generated in Problem 4.1. Plot the true p.d.f. and the fitted GLD on the same axes, then evaluate the goodness of the fit.

4.18 (Section 4.11) Fit the empiric p.d.f. to the same data as in Problem 4.17. Plot on the same axes with the true p.d.f., and evaluate the goodness of the fit.

4.19 (Section 4.12) Using the 100 data points of Problem 4.1, construct a histogram with 5 classes. Fit a GLD to the data, using only the information in the histogram. Evaluate the fit by plotting the true p.d.f. and the fitted GLD on the same axes.

4.20 (Section 4.13) Test the data of Problem 4.1 for normality. Find a transformation of the class (4.13.8) that is most nearly normal in distribution. (Be sure to specify all powers for which normality is not rejected.)

Chapter 5

Intermediate GPSS

The discussion so far of the movement of transactions through the blocks of a GPSS model has avoided all references to the algorithm that GPSS uses to move transactions. This chapter describes how GPSS programs are executed by giving a precise formulation of the mechanism that moves transactions. The use of random number generators, functions, and a number of additional blocks is also discussed. These basic elements of GPSS are covered in sufficient detail to lay the groundwork for the development of complex simulation models. Two such studies are illustrated as examples in Sections 5.6 and 5.11, with models provided on the disk that accompanies the book.

5.1   Transaction Movement

Some general ideas about the use of linked lists in simulations were considered in Chapter 1. Here we see how GPSS uses priority lists (discussed in Section 1.5) to move transactions through the simulation. There are three types of actions that may need to be performed at any given instant during the execution of a GPSS program:

1. Activities that are scheduled to occur some time in the future.

2. Activities that should, and can, be undertaken at that instant.

3. Activities that should take place as soon as possible, but are currently unable to execute because of internal model conditions. The capturing of an already busy facility by a transaction represents this type of activity.

To execute the movement of transactions in the correct order, GPSS employs two linked lists of transactions. Each entry in these lists contains, among other things, the following information:

A Transaction Number (TN): An integer that uniquely identifies the transaction among all transaction in the system at that time

Current Block (CB): The number of the block that the transaction is currently in

Next Block (NB): The number of the block where the transaction should be sent

Depart Time (DT): A simulated time when the transaction should move to the block designated by Next Block, if it is not prevented from doing so by internal conditions

Priority (P): The priority of the transaction

Figure 5.1–1 shows the structure associated with a transaction on one of these linked lists.

| TN | CB | NB | DT | P | Other Data | Pointer |

Figure 5.1–1. The structure of a transaction on a chain.

Transactions that should move as soon as possible, whether they can move at that precise instant or not, are kept on a linked list called the **Current Events Chain** (CEC). Transactions whose movements are to occur at a scheduled time in the future are held on a linked list called the **Future Events Chain** (FEC). The CEC is operated as a priority-driven queue; that is, as a FIFO (first in, first out) system ordered by the transaction priorities. The FEC is ordered by increasing Departure Times (DTs) with FIFO discipline for transactions with identical DTs.

At any given instant during execution, every transaction is simultaneously in a block and on at least one chain. (As we will discover later, there can be chains other than the CEC and the FEC, and a transaction may be on one of these chains as well.) Once a transaction from the CEC is moved, it is moved as far as it can go at that instant. A transaction stops moving when it leaves the system through a TERMINATE block, when it enters an

ADVANCE block and hence is scheduled to move again at a later time (i.e., is transferred from the CEC to the FEC), or when it is refused entry into its next block (as when encountering a SEIZE block associated with a busy facility). In this last case, the transaction remains on the CEC, ready to be moved again when the blocking condition is changed.

When a transaction comes to rest, due to any of these conditions, the GPSS processor must pick another transaction and advance it until it can no longer move. This is done by rescanning the CEC from front to back to find the first transaction that can move. Typically, rescanning from the front rather than continuing scanning is necessary because the movement of the previous transaction may have unblocked the path of transactions that are at the front of the CEC. In certain circumstances, it may be more efficient (in terms of computation time) to continue scanning rather than to rescan from the front of the CEC, but the rescanning approach is logically somewhat simpler.

If rescanning produces a transaction that can move forward, the process continues until rescanning the CEC does not produces any transaction that can move. One of two conditions can lead to this situation: the CEC is empty, or all transactions on the CEC are blocked. In either case, the first transaction from the FEC is moved to the CEC and the simulated clock is updated to reflect the time at which this transaction was scheduled to move to its next block. Any additional transactions that are scheduled to move at the newly updated time are also transferred from the FEC to the CEC. This transfer inserts new transactions at the rear of the CEC within the proper priority grouping.

The scanning of the CEC is then restarted and all the transactions from the CEC that can move forward through the blocks of the program are again processed in the manner already described. The transaction-moving algorithm of the GPSS processor is summarized in the flowchart in Figure 5.1–2. It is, of course, essential that GPSS place at least one transaction on either the CEC or the FEC during initialization. Otherwise, both the CEC and the FEC would be empty (i.e., there would be no transactions to move) and hence there would be no simulation.

As the GPSS processor steps through a GPSS program, in addition to generating code, it assigns block numbers sequentially to every block in the program and places a transaction on the FEC whenever it encounters a GEN-ERATE block. The creation and deletion of transactions are managed through the use of a stack that contains all the transactions that potentially could enter the simulation. The stack is a linked-list data structure that is oper-ated as a LIFO (last in, first out) system. As we saw in Section 1.5, a stack

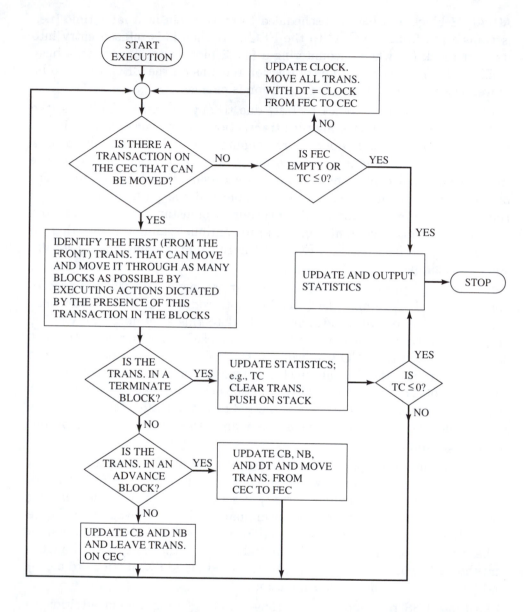

Figure 5.1–2. The GPSS algorithm for transaction movement.

is a special case of a priority queue. Each implementation of GPSS must cope with a fixed memory size by imposing a limit on the maximum number of transactions that can be made available. This limit is typically between 100 and 200. Every transaction carries with it a sequentially assigned identification number starting with 1. To create a transaction, a transaction is "popped off" the stack and placed on the FEC. There need not be a great deal of data movement associated with this process, since the movement of a transaction from the top of the stack to the FEC can be accomplished through the use of pointers. When a transaction needs to be removed from the simulation, it is "pushed down" on the stack.

During assembly (code generation and initialization before the execution of a program), whenever a GENERATE block is encountered, GPSS uses the A, B, and C operands to compute the time when the first transaction should enter the simulation through this block. It then pops a transaction from the stack; updates its CB, NB, DT, and P entries; and enters this transaction on the FEC. By the end of the code-generation phase, before the start of execution, the FEC will contain exactly one transaction corresponding to each GENERATE block of the program. This process is described by the flowchart in Figure 5.1–3.

To make sure that transactions enter the simulation at the proper times, the arrival of the next transaction through a GENERATE block is scheduled before the transaction that is in the GENERATE block is moved. This is done by using the A and B operands of the GENERATE block and the departure time of the current inhabitant of the GENERATE block to compute the departure time of the new transaction, which is entered into the FEC.

For a detailed look at the movement of transactions, suppose that in the program given in Figure 5.1–4 (with its block numbers included for convenience), the numbers 20, 22, 30, 30, 26, and 27 are sampled in successive transaction entries through the GENERATE block. This implies arrival times of 20, 42, 72, 102, 128, and 155. Suppose also that 35, 48, 25, and 39 are sampled in repeated executions of the ADVANCE block. Following the notation used by Schriber (1974, p. 62), we will represent each transaction as a 4-tuple [TN, DT, CB, NB] consisting of the transaction number, its departure time, the current block it is in, and the next block to which it will move. In the trace given in Figure 5.1–5, two actions are described in each time cycle. The first is the movement of transaction(s) from the FEC to the CEC, and the second is the forward movement of transactions through the model blocks.

Following initialization, two transactions [1,20,*,1] and [2,150,*,8] are entered into the FEC. The * indicates that these transactions do not yet have

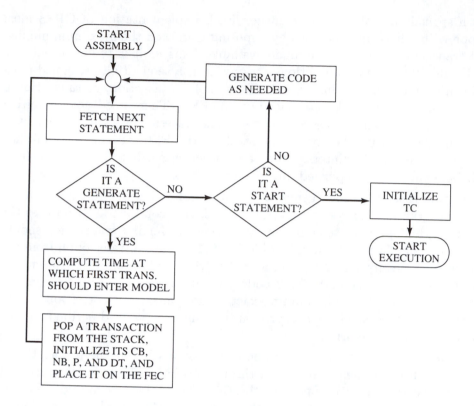

Figure 5.1–3. The GPSS assembly process.

```
              SIMULATE

     1        GENERATE      25,5
     2        QUEUE         WAIT
     3        SEIZE         FAC
     4        DEPART        WAIT
     5        ADVANCE       40,15
     6        RELEASE       FAC
     7        TERMINATE

              GENERATE      150
              TERMINATE     1

              START         1
              END
```

Figure 5.1–4. Transaction movement within a simple model.

Line	Time	CEC	FEC	Stack
1	0	EMPTY	EMPTY	1,2,3,4
2	0	EMPTY	[1,20,*,1] [2,150,*,8]	3,4,5
3	20	[1,20,*,1]	[2,150,*,8]	3,4,5
4	20	EMPTY	[3,42,*,1] [1,55,5,6] [2,150,*,8]	4,5,6
5	42	[3,42,*,1]	[1,55,5,6] [2,150,*,8]	4,5,6
6	42	[3,42,2,3]	[1,55,5,6] [4,72,*,1] [2,150,*,8]	5,6,7
7	55	[3,42,2,3] [1,55,5,6]	[4,72,*,1] [2,150,*,8]	5,6,7
8	55	EMPTY	[4,72,*,1] [3,103,5,6] [2,150,*,8]	1,5,6,7
9	72	[4,72,*,1]	[3,103,5,6] [2,150,*,8]	1,5,6,7
10	72	[4,72,2,3]	[1,102,*,1] [3,103,5,6] [2,150,*,8]	5,6,7
11	102	[4,72,2,3] [1,102,*,1]	[3,103,5,6] [2,150,*,8]	5,6,7
12	102	[4,72,2,3] [1,102,2,3]	[3,103,5,6] [5,128,*,1] [2,150,*,8]	6,7
13	103	[4,72,2,3] [1,102,2,3] [3,103,5,6]	[5,128,*,1] [2,150,*,8]	6,7
14	103	[1,102,2,3]	[5,128,*,1] [4,128,5,6] [2,150,*,8]	3,6,7
15	128	[1,102,2,3] [5,128,*,1] [4,128,5,6]	[2,150,*,8]	3,6,7
16	128	[5,128,2,3]	[2,150,*,8] [3,155,*,1] [1,167,5,6]	4,6,7
17	150	[5,128,2,3] [2,150,*,8]	[3,155,*,1] [1,167,5,6]	4,6,7
18	150	[5,128,2,3]	[3,155,*,1] [1,167,5,6] [4,300,*,8]	4,6,7

Figure 5.1–5. Status of the CEC and FEC for the model in Figure 5.1–4.

a current block. At line 3, transaction 1 is moved to the CEC and, since it still has not moved into any block, $*$ continues to be its current block designation. Transaction 1 now moves through the GENERATE block, causing the GPSS processor to create a new transaction (transaction 3) and place it on the FEC. Next, transaction 1 moves through the QUEUE, SEIZE, and DEPART blocks (blocks 2, 3, and 4), enters the ADVANCE block (block 5), and is placed on the FEC with departure time = current time + service time $(20 + 35 = 55)$.

At line 5, transaction 3 is moved from the FEC to the CEC and the simulation time is updated to 42 (the scheduled departure time of transaction 3). At line 6, the movement of transaction 3 through the GENERATE block has caused the entry of a new transaction on the FEC, and transaction 3 remains on the CEC because it is held up in block 2 waiting to capture the facility.

At line 7, transaction 1 is brought back to the CEC and the simulation time is updated to 55 (the scheduled departure time of transaction 1). When the CEC is scanned, transaction 1 is moved through the system and the memory space allocated to it is reclaimed on the stack. When the CEC is rescanned, the previously blocked transaction 3 is moved on to the ADVANCE block (and the FEC) with a departure time of 103 (current time, 55, plus service time, 48). We continue the trace at time 128 (line 15) when transactions 4 and 5 are both moved from the FEC to the CEC. Since transaction 5 was ahead of transaction 4 on the FEC, it remains ahead of it when they are transferred to the CEC. At line 16, transaction 5 moves through the GENERATE block (causing the entry of transaction 3 to the FEC) and stops in block 2. This causes a rescanning of the CEC, the movement of transaction 4 out of the system, and the reallocation of its memory space on the stack. A third scanning of the CEC moves transaction 1 to the ADVANCE block and hence to the FEC with a departure time of 167 ($128 + 39$). At line 17, transaction 2 moves to the CEC and proceeds to shut off the simulation.

5.2   The START Statement and Chain Output

With the proper use of the C and D operands of the START statement, it is possible to obtain snapshots (standard output) from a GPSS program at various stages during its execution. The intervals between the snapshots are regulated by a snap interval counter maintained by GPSS. The C operand of a START statement initializes the counter and, like the termination counter, the snap interval counter is decremented by the amount specified by the A operand of a TERMINATE block whenever a transaction enters that block. When the snap interval counter assumes a zero or negative value, GPSS produces the standard program output as of that instant and reinitializes the counter to the value of the C operand of the START statement. Thus, the program skeleton given in Figure 5.2–1 will produce the standard GPSS output as every other transaction, starting with the second one, leaves the system. If the D operand of a START statement is set to 1, output describing the status of all chains is produced (in addition to the standard GPSS output) whenever output would ordinarily occur. If the START statement in Figure 5.2–1 were replaced by

$$\text{START} \quad 10,,2,1$$

then the output associated with the departure of every other transaction would contain CEC and FEC listings. A modification of the program of Figure 5.1–4 to produce a chain listing is given in Figure 5.2–2.

```
GENERATE
    ⋮
TERMINATE      1

START          10,,2
END
```

Figure 5.2–1. Use of the snap interval counter.

```
1                 SIMULATE
2     1           GENERATE      25,5
3     2           QUEUE         WAIT
4     3           SEIZE         FAC
5     4           DEPART        WAIT
6     5           ADVANCE       40,15
7     6           RELEASE       FAC
8     7           TERMINATE     1
9
10    8           GENERATE      150
11    9           TERMINATE     5
12
13                START         5,,1,1
```

Figure 5.2–2. Modification of the program in Figure 5.1–4
to produce chain listings.

Notice that the program of Figure 5.2–2 designed to stop at the first occurrence of either of the following two conditions: the clock assumes value 150, or 5 transactions leave the system. For brevity, only the clock and chain output associated with this program are given in Figure 5.2–3. The first five columns of the chain listings give the transaction number, current block, next block, departure time, and priority of a listed transaction. The other columns in the chain printouts do not concern us at this time. A careful review of the output of Figure 5.2–3 will show that the GENERATE and ADVANCE blocks in fact produce the interarrival and service times that were assumed in Section 5.1.

```
THIS IS SNAP 1 OF 5.
RELATIVE CLOCK 55              ABSOLUTE CLOCK 55

CURRENT EVENTS CHAIN -
*TRANS  CUR BLOCK   NEXT BLOCK   DEPART TIME   PRIO    MARK TIME
3       2           3            42            0       42
        SET         SEL          DELAY         CHAIN   PREEMPT_CNT/FLAG
        3                        X             C       0

FUTURE EVENTS CHAIN -
TRANS   CUR BLOCK   NEXT BLOCK   DEPART TIME   PRIO    MARK TIME
4       1           2            72            0       -2
        SET         SEL          DELAY         CHAIN   PREEMPT_CNT/FLAG
        4                                      F       0

TRANS   CUR BLOCK   NEXT BLOCK   DEPART TIME   PRIO    MARK TIME
2       8           9            150           0       0
        SET         SEL          DELAY         CHAIN   PREEMPT_CNT/FLAG
        2                                      F       0

THIS IS SNAP 2 OF 5.
RELATIVE CLOCK 103            ABSOLUTE CLOCK 103

CURRENT EVENTS CHAIN -
*TRANS  CUR BLOCK   NEXT BLOCK   DEPART TIME   PRIO    MARK TIME
4       2           3            72            0       72
        SET         SEL          DELAY         CHAIN   PREEMPT_CNT/FLAG
        4                        X             C       0

TRANS   CUR BLOCK   NEXT BLOCK   DEPART TIME   PRIO    MARK TIME
1       2           3            102           0       102
        SET         SEL          DELAY         CHAIN   PREEMPT_CNT/FLAG
        1                        X             C       0
```

Figure 5.2–3. The chain output of the program
in Figure 5.2–2 (continues).

FUTURE EVENTS CHAIN -

TRANS	CUR BLOCK	NEXT BLOCK	DEPART TIME	PRIO	MARK TIME
5	1	2	128	0	-4
	SET	SEL	DELAY	CHAIN	PREEMPT_CNT/FLAG
	5			F	0
*TRANS	CUR BLOCK	NEXT BLOCK	DEPART TIME	PRIO	MARK TIME
2	8	9	150	0	0
	SET	SEL	DELAY	CHAIN	PREEMPT_CNT/FLAG
	2			F	0

THIS IS SNAP 3 OF 5.

RELATIVE CLOCK 128   ABSOLUTE CLOCK 128

CURRENT EVENTS CHAIN -

*TRANS	CUR BLOCK	NEXT BLOCK	DEPART TIME	PRIO	MARK TIME
1	2	3	102	0	102
	SET	SEL	DELAY	CHAIN	PREEMPT_CNT/FLAG
	1		X	C	0
*TRANS	CUR BLOCK	NEXT BLOCK	DEPART TIME	PRIO	MARK TIME
5	2	3	128	0	128
	SET	SEL	DELAY	CHAIN	PREEMPT_CNT/FLAG
	5		X	C	0

FUTURE EVENTS CHAIN -

TRANS	CUR BLOCK	NEXT BLOCK	DEPART TIME	PRIO	MARK TIME
2	8	9	150	0	0
	SET	SEL	DELAY	CHAIN	PREEMPT_CNT/FLAG
	2			F	0
TRANS	CUR BLOCK	NEXT BLOCK	DEPART TIME	PRIO	MARK TIME
3	1	2	155	0	-5
	SET	SEL	DELAY	CHAIN	PREEMPT_CNT/FLAG
	3			F	0

RELATIVE CLOCK 150   ABSOLUTE CLOCK 150

CURRENT EVENTS CHAIN -

*TRANS	CUR BLOCK	NEXT BLOCK	DEPART TIME	PRIO	MARK TIME
5	2	3	128	0	128
	SET	SEL	DELAY	CHAIN	PREEMPT_CNT/FLAG
	5		X	C	0

FUTURE EVENTS CHAIN -

TRANS	CUR BLOCK	NEXT BLOCK	DEPART TIME	PRIO	MARK TIME
3	1	2	155	0	-5
	SET	SEL	DELAY	CHAIN	PREEMPT_CNT/FLAG
	3			F	0
TRANS	CUR BLOCK	NEXT BLOCK	DEPART TIME	PRIO	MARK TIME
1	5	6	167	0	102
	SET	SEL	DELAY	CHAIN	PREEMPT_CNT/FLAG
	1			F	0

17   END

Figure 5.2–3. Chain output of the program in Figure 5.2–2 (concluded).

5.3   Random Number Generators Built into GPSS

GPSS has the built-in capacity to produce uniformly distributed "random" numbers from 0 to 0.999 .... The specific algorithms used to generate random numbers depend on the implementation. For example, GPSS/VX on VAX-11 systems uses URN22 and GPSS/H Release 1 uses URN27, both of which were discussed in Section 3.5. In most cases, the generators are congruential generators that are assigned default seed values that the users can change through the RMULT control statement.

The eight built-in random number generators are designated by RN1, RN2,..., RN8. RN1 is used by default in all GENERATE and ADVANCE blocks. To determine the time that a transaction must stay in the block

$$ADVANCE \quad A,B$$

RN1 is used to produce some value r between 0 and 0.999 .... This value is then used to compute

$$A - B + r(2B + 1).$$

Since $0 \leq r < 1$, we must have $0 \leq r(2B + 1) < 2B + 1$, and

$$A - B \leq A - B + r(2B + 1) < 2B + 1 + A - B = A + B + 1.$$

Following truncation, an integer between $A - B$ and $A + B$ inclusive is generated for use in the ADVANCE block. Random number generators were considered in some detail in Chapter 3, but it is worth restating that serious errors can result from the use of poor-quality random number generators, even if the generators are bundled with the programming language as is the case with GPSS. In Section 7.6, we give a method of incorporating high-quality generators such as URN13 of Section 3.5 into GPSS programs.

5.4   The RESET, RMULT, and CLEAR Control Statements

In many simulations, the statistics gathered at the end of a limited run are distorted because of unnatural start-up effects. In a single-line/single-server queueing model, for instance, the average waiting time, queue length, and facility utilization are all distorted because the simulation time span includes the initial period when the queue and facility were empty. If we assume that the system we are simulating reaches **steady-state** (i.e., fluctuations of model statistics diminish and eventually converge to specific values), we can think of the simulation time span as consisting of an initial **transient** period and a later steady-state period.

To counteract the distortion caused by the transient state, we may choose to run the simulation long enough that all the relevant statistics are averaged out. A far more efficient approach, however, would be to run the simulation for a brief period to get past the transient stage, zero all statistics without purging the system of its transactions, and continue the simulation to accumulate statistics over only the steady-state segment.

The RESET statement does precisely what is needed to reduce the effect of the transient state. It zeros all model statistics, leaves transactions in their current blocks and chains, does not interfere with the random number generating streams, and clears the relative clock but leaves the absolute clock intact. Suppose, through simulation, we wanted to determine the average queue length in the simple queueing situation described by the GPSS program in Figure 5.4–1. By using multiple START statements, we can study the convergence of this statistic. The program listing given in Figure 5.4–1 and the abbreviated version of this program's output given in Figure 5.4–2 show the simulated time and queue statistics associated with 5 runs produced by 5 successive

START   1

statements.

```
 1                   SIMULATE
 2
 3        1          GENERATE     250,75
 4        2          QUEUE        WAIT
 5        3          SEIZE        FAC
 6        4          DEPART       WAIT
 7        5          ADVANCE      225,50
 8        6          RELEASE      FAC
 9        7          TERMINATE
10
11        8          GENERATE     45000
12        9          TERMINATE    1
13                   START        1
14                   START        1
15                   START        1
16                   START        1
17                   START        1
```

Figure 5.4–1. Program for sampling average waiting times.

RELATIVE CLOCK 45000   ABSOLUTE CLOCK 45000

QUEUE	MAXIMUM CONTENTS	AVERAGE CONTENTS	TOTAL ENTRIES	ZERO ENTRIES	PERCENT ZEROS
WAIT	1	0.091	180	95	52.778

	AVERAGE TIME/TRANS	$AVERAGE TIME/TRANS	TABLE NUMBER	CURRENT CONTENTS
	22.667	48.000		0

RELATIVE CLOCK 90000   ABSOLUTE CLOCK 90000

QUEUE	MAXIMUM CONTENTS	AVERAGE CONTENTS	TOTAL ENTRIES	ZERO ENTRIES	PERCENT ZEROS
WAIT	1	0.103	361	182	50.416

	AVERAGE TIME/TRANS	$AVERAGE TIME/TRANS	TABLE NUMBER	CURRENT CONTENTS
	25.715	51.860		0

RELATIVE CLOCK 135000   ABSOLUTE CLOCK 135000

QUEUE	MAXIMUM CONTENTS	AVERAGE CONTENTS	TOTAL ENTRIES	ZERO ENTRIES	PERCENT ZEROS
WAIT	1	0.097	541	273	50.462

	AVERAGE TIME/TRANS	$AVERAGE TIME/TRANS	TABLE NUMBER	CURRENT CONTENTS
	24.299	49.052		1

RELATIVE CLOCK 180000   ABSOLUTE CLOCK 180000

QUEUE	MAXIMUM CONTENTS	AVERAGE ENTRIES	TOTAL ENTRIES	ZERO ZEROS	PERCENT
WAIT	1	0.117	723	354	48.963

	AVERAGE TIME/TRANS	$AVERAGE TIME/TRANS	TABLE NUMBER	CURRENT CONTENTS
	29.107	57.030		0

RELATIVE CLOCK 225000   ABSOLUTE CLOCK 225000

QUEUE	MAXIMUM CONTENTS	AVERAGE ENTRIES	TOTAL ENTRIES	ZERO ZEROS	PERCENT
WAIT	1	0.111	899	454	50.501

	AVERAGE TIME/TRANS	$AVERAGE TIME/TRANS	TABLE NUMBER	CURRENT CONTENTS
	27.733	56.027		0

21   END

Figure 5.4–2. Average queue lengths through multiple STARTs.

Average queue lengths of 0.091, 0.103, 0.097, 0.117, and 0.111 are obtained at simulated times of 45,000; 90,000; 135,000; 180,000; and 225,000; respectively. Note that the GPSS program consists of lines 1 through 20 and the program output appears following each START statement (lines 16 through 20). To show how the START and RESET combination can be used to counteract the effects of the transient state, consider the program listing and queue output given in Figure 5.4–3. The program is a slight modification of the one given in Figure 5.4–1. Note that the original model had to be simulated for 225,000 units of time before the average queue length appeared to stabilize. However, when the transient effects were reduced in the modified version, the average waiting time was obtained in only 90,000 units of simulated time.

The CLEAR statement is used when we want to restart a simulation with somewhat altered conditions. For example, the sequence given in Figure 5.4–4 will start the simulation with

$$A \quad \text{GENERATE} \quad 12,3$$

in effect. After the completion of this run, the program will rerun with

$$A \quad \text{GENERATE} \quad 12,1$$

in effect.

The CLEAR statement reinitializes all model statistics, such as those associated with queue and facility usage, to zero; removes all transactions from the model; and schedules the arrival of the first transaction to each GENERATE block. Because the CLEAR statement does not reinitialize the seeds of the random number generators, different random numbers will be used in the second run of the program skeleton given in Figure 5.4–4. In situations where it is desirable to run a simulation a number of times with different conditions but with the same arrival and/or service patterns, the RMULT statement can be used to reset the seeds of selected random number generators to fixed values immediately before each run. The form of the RMULT statement is

$$\text{RMULT} \quad \text{A,B,C,D,E,F,G,H}$$

where A, B, C, D, E, F, G, and H are integers to be used as initial seeds for the generators R1,..., R8, respectively. It is possible to default any combination of A through H and specify seeds for selected generators as in the case

$$\text{RMULT} \quad 123,,456,,,1234$$

```
1                    SIMULATE
2
3       1            GENERATE      250,75
4       2            QUEUE         WAIT
5       3            SEIZE         FAC
6       4            DEPART        WAIT
7       5            ADVANCE       225,50
8       6            RELEASE       FAC
9       7            TERMINATE
10
11      8            GENERATE      45000
12      9            TERMINATE     1
13                   START         1
```

RELATIVE CLOCK 45000   ABSOLUTE CLOCK 45000

QUEUE	MAXIMUM CONTENTS	AVERAGE CONTENTS	TOTAL ENTRIES	ZERO ENTRIES	PERCENT ZEROS
WAIT	1	0.091	180	95	52.778
		AVERAGE TIME/TRANS	$AVERAGE TIME/TRANS	TABLE NUMBER	CURRENT CONTENTS
		22.667	48.000		0

```
17                   RESET
18                   START    1
```

RELATIVE CLOCK 45000   ABSOLUTE CLOCK 90000

QUEUE	MAXIMUM CONTENTS	AVERAGE CONTENTS	TOTAL ENTRIES	ZERO ENTRIES	PERCENT ZEROS
WAIT	1	0.116	181	87	48.066
		AVERAGE TIME/TRANS	$AVERAGE TIME/TRANS	TABLE NUMBER	CURRENT CONTENTS
		28.746	55.351		0

```
19                   END
```

Figure 5.4–3. The use of RESET to counteract transient effects.

```
A            GENERATE      12,3
             ⋮
             TERMINATE

             START

             CLEAR
A            GENERATE      12,1
             START
             END
```

Figure 5.4–4. Using different GENERATE blocks in successive runs.

where 123, 456, and 1234 initialize RN1, RN3, and RN6, respectively.

Care should be taken to place the RMULT, CLEAR, and START statements in the proper order. Since initial transactions associated with GENERATE blocks are placed on the FEC by the GPSS initialization process, and subsequently by CLEAR statements, RMULT should precede all GENERATE blocks and the same RMULT statement should also precede all CLEAR statements. That way, all runs will start with the same sequence of random numbers from the generators specified by RMULT. The two-segment program skeleton given in Figure 5.4–5 shows how this can be done.

Figure 5.4–5. Use of the RMULT, CLEAR, and START statements.

5.5 GPSS Functions

There are simple mechanisms for defining and using functions in GPSS programs. One common use of functions is for sampling observations from specified probability distributions. The restriction of interarrival and service times to uniformly distributed observations on intervals from A – B to A + B (where A and B are the first two operands of GENERATE or ADVANCE blocks) is very unrealistic. It is far more likely that a simulation will require interarrival and service times that are exponentially distributed; in some cases, it may be necessary to produce these times from patterns suggested by empirical studies.

Of the five types of functions that can be used in GPSS, we will consider the two most commonly needed: the D (discrete) and the C (continuous) types. In both cases, a function definition with the following specifications is needed:

Function name: A user-defined identifier in the label field.

FUNCTION: Literally, in the operator field.

Independent variable specification: Designation of the independent variable in the operand field (the A operand position).

Function type specification: D or C to indicate discrete or continuous (the B operand position).

Number of points: The number of points to be used in the definition of the function (the B operand position immediately following the function type).

Points: The points that define the function, with coordinates separated by commas and points separated by /. The points are specified on lines immediately following the FUNCTION statement.

5.5.1   Discrete Functions

Several methods for sampling from a known distribution and from empirically determined distributions were considered in Chapter 4. Here we consider ways of dealing with such problems in the context of GPSS. For a given independent variable, a GPSS discrete function assumes one of a finite number of values. Suppose that the points (a_1, b_1), $(a_2, b_2), \ldots, (a_n, b_n)$ (with $a_1 < a_2 < \cdots < a_n$) are used in the function definition and the value of the function at x is to be determined. The independent variable x is tested against all of $a_1, a_2, \ldots, a_n$ to find the index j for which $a_{j-1} < x \leq a_j$ (assuming for convenience that $a_0 = -\infty$) and the function is assigned value

b_j. Thus,

> if $-\infty < x \le a_1$, the value of the function is b_1
>
> if $a_1 < x \le a_2$, the value of the function is b_2
>
> $\vdots$
>
> if $a_{n-1} < x$, the value of the function is b_n.

(5.5.1)

In many applications, it is desirable to use one of the random number generators as the independent variable in a function definition. The definition

> SERVICE   FUNCTION   RN2,D4
> 0.2,4/0.3,5/0.6,2/1.0,6

indicates a discrete (the D in operand B) function defined by four points (the 4 in operand B), using RN2 as its independent variable (operand A). The four points are (0.2,4), (0.3,5), (0.6,2), and (1,6). When the function SERVICE is invoked, RN2 is used to produce a random number, r. If

> $0 \le r \le 0.2$, SERVICE is assigned value 4;
>
> $0.2 < r \le 0.3$, SERVICE is assigned value 5;
>
> $0.3 < r \le 0.6$, SERVICE is assigned value 2;
>
> $0.6 < r \le 1.0$, SERVICE is assigned value 6.

The function SERVICE assumes value 4 with a probability of 0.2, value 5 with a probability of 0.1, value 2 with a probability of 0.3, and value 6 with a probability of 0.4. Figure 5.5–1 gives a graphic description of the function SERVICE. To produce the outcomes associated with the roll of an honest die, the function

> DIE   FUNCTION   RN6,D6
> 0.16667,1/0.33333,2/0.5,3/0.66667,4
> 0.83333,5/1,6

could be used to produce values of 1, 2,..., 6 with equal (or almost equal) probabilities. The graph of this function is shown in Figure 5.5–2.

For a somewhat more realistic example, suppose that the times between arrivals of transactions into a system are distributed as follows: half the transactions arrive 20 minutes after their predecessors, a fourth arrive 25 minutes after their predecessors, and the remaining fourth arrive 35 minutes after their predecessors. Furthermore, when these transactions require service at facility FAC, they need 10, 25, and 40 minutes of service time with

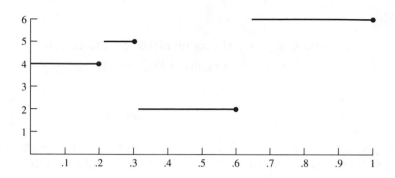

Figure 5.5–1. The discrete function SERVICE.

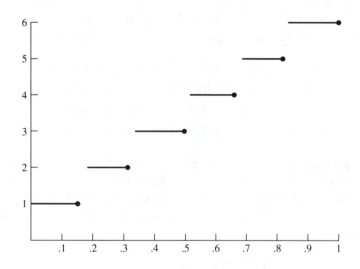

Figure 5.5–2. The discrete function DIE.

equal probability. The program skeleton given in Figure 5.5–3 simulates these conditions. The invocation of the functions IAT and ST at the GENER-ATE and ADVANCE blocks in Figure 5.5–3 indicates that FN$ must precede the function name. This is part of a more general convention of prefixing GPSS entity names with family designations (in this case, FN for function) and a $ character, which will be explored more fully in Section 7.1.

⋮

| IAT | FUNCTION | RN1,D3 |

0.5,20/0.75,25/1.0,35

⋮

| ST | FUNCTION | RN3,D3 |

0.333333,10/0.666667,25/1.0,40

⋮

| GENERATE | FN$IAT |

⋮

QUEUE	FACQ
SEIZE	FAC
DEPART	FACQ
ADVANCE	FN$ST
RELEASE	FAC

⋮

Figure 5.5–3. The definition and use of discrete functions.

5.5.2 Continuous Functions

In most applications, random variables such as interarrival times have continuous distributions. If a random variable X has a continuous distribution, the inverse distribution method (Section 4.2) can be used to generate values for X. In GPSS, a discrete function approximating the inverse distribution could be constructed for sampling puposes. It can be seen from the step function approximation of the inverse distribution function given in Figure 5.5–4 that, in general, the number and placement of points used in the definition of the discrete function dictate the quality of the approximation. A better approach to approximating continuous functions would be to choose points $(a_1,b_1),\ldots,(a_n,b_n)$ on the function to be approximated and then use the polygonal function defined by joining these points as the approximating function. It can be seen from Figure 5.5–5 that the number and placement of points dictate the quality of approximation. Continuous functions use this linear interpolation method to compute function values for specific values of the independent variable. The graph of the continuous function defined by

ARRIVE FUNCTION RN3, C3

0,2/0.5,8/1.0,11

is given in Figure 5.5–6. To compute the value of ARRIVE for $r = 0.3$ (generated by RN3), the equation of the straight line joining (0,2) and (0.5,8)

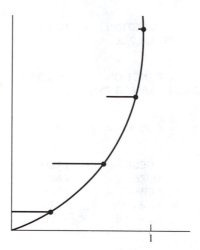

Figure 5.5–4. A discrete function's approximation
to a continuous function.

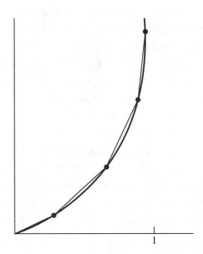

Figure 5.5–5. A linear interpolation of a continuous function.

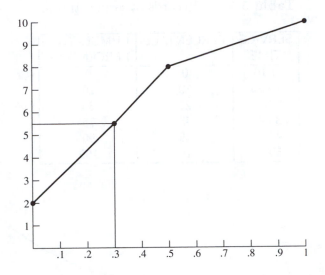

Figure 5.5–6. The continuous function ARRIVE.

is computed (this turns out to be ARRIVE $= 12r + 2$), which in turn yields ARRIVE $= 5.6$ for $r = 0.3$. If the function ARRIVE were being used in a GENERATE block, as in

GENERATE   FN$ARRIVE

RN3 would be used to provide a random number r, which would in turn be used to evaluate ARRIVE, which then would be truncated to an integer value. Thus, if RN3 produced 0.3, the final interarrival time used by the GENERATE block would be 5. As in the case of discrete functions (see (5.5.1)), if the independent variable x is outside the range specified by the function definition, the value assigned to the function is the value the function attains at the point closest to x.

As a more realistic example, suppose that the records of service times at a facility indicate the distribution of service times given in Table 5.5–7. In GPSS, the function ST (Figure 5.5–8) defined by

ST   FUNCTION   RN5,C6
0.0,10/0.1,20/0.3,30/0.75,40/0.9,50/1.0,60

can be used in conjunction with

ADVANCE   FN$ST

Table 5.5–7. Records of service times.

SERVICE TIME	PERCENTAGE	CUMULATIVE PERCENTAGE
0-10	0	0
10-20	10	10
20-30	20	30
30-40	45	75
40-50	15	90
50-60	10	100

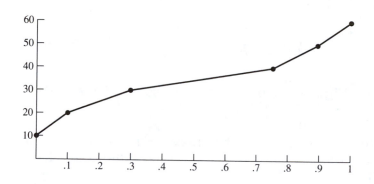

Figure 5.5–8. The function ST defined by Table 5.5–7.

to simulate the pattern of service times that were observed empirically.

There are situations where either a discrete or a continuous function may be used. Suppose the ADVANCE block in a simulation is to use delay times of 1, 2,..., 6, each with a probability of 1/6. The discrete function

```
DDIE       FUNCTION          RN4,D6
0.16667,1/0.33333,2/0.5,3/0.66667,4
0.83333,5/1,6
```

may be used to provide the ADVANCE block with appropriate time delays. It is also possible, however, to use the continuous function

```
CDIE      FUNCTION      RN4,C2
0,1/1,7
```

In this case, if RN4 were to yield value r with $0 \leq r < 1/6$, then $1 \leq$ CDIE $(r) < 2$ and, following truncation, the ADVANCE block would use the value 1. Next, considering values of r restricted to the intervals $1/6 \leq r < 2/6, \ldots, 5/6 \leq r < 1$, we see that 2, 3,... and 6 will be used by the ADVANCE block. Furthermore, the numbers 1, 2,..., 6 will all occur with equal probability. The CDIE function and the step function describing the actual values that would be used by the ADVANCE block are shown in Figure 5.5–9.

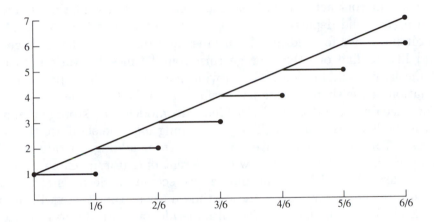

Figure 5.5–9. The CDIE and DDIE functions yielding the same values.

5.6   Example: A Manufacturing Shop

A manufacturing shop has seven mechanics and three identical machines, each of which needs two mechanics to operate. When a job arrives, a clerk does some recordkeeping paperwork on it and passes the job request on to a mechanic, who spends some time setting up the appropriate tools to be used in connection with this job. When the actual machine work is to start, a second mechanic helps with the machine operation. Before the job is considered finished, the clerk must complete the recordkeeping and billing paperwork. The shop accepts jobs during the first 7 hours of its 8-hour-per-day operation. During the last hour, work is done only on jobs that have already been accepted. In the following information obtained from shop logs, all service and interarrival time distributions are uniform on the indicated intervals:

1. The clerk uses 4 ± 2 minutes to do the initial paperwork and 9 ± 4 minutes to do the follow-up work to complete the job.

2. Job interarrival times are 15 ± 3 minutes during the 7-hour period.

3. The time that a mechanic takes setting up tools before using a machine is 10 ± 3 minutes and the time spent using a machine is 55 ± 30 minutes.

Because of contract stipulations, the average turnaround time (time between arrival and departure) of jobs should not exceed 1.5 hours. Before committing funds for additional machines and operators, the management would like to find out the average turnaround times for various machine and mechanic combinations. To provide insight into these problems, the simulation of the shop operations shown in Figure 5.6–1 is designed.

The machine operators and machines are modeled as storages, and the clerk is modeled as a facility. The jobs coming into the shop are the transactions. To obtain perspective on the effect of adding operators and/or machines, the simulation is run with different operator/machine configurations under a fixed job arrival and service scenario. Seconds rather than minutes are chosen for the simulation time units so that there will be sufficient detail in the simulation. The initial tendency might be to produce the job (transaction) arrivals by

$$\text{GENERATE} \quad 900,180,,32$$

and to schedule the initial service time by the clerk by

$$\text{ADVANCE} \quad 240,120$$

This approach, however, would make it impossible to replicate the same interarrival and service times in successive simulations that use different machine and operator configurations. For this reason, user-defined functions, each with a different random number generating stream, are used for the interarrival and service times. The function IAT produces interarrival times, the functions CLERK1 and CLERK2 produce service times for the initial and final work done by the facility CLERK, and the functions STME1 and STME2 simulate the setup and machine use times, respectively. The queue LINE captures statistics related to the time that jobs spend in the shop.

The GENERATE statement

$$\text{L} \quad \text{GENERATE} \quad \text{FN\$IAT}$$

is operative for seven counts of the termination counter (i.e., for 7 hours of simulated time). Following this period, the control statements

```
1                        SIMULATE
2                        RMULT        1111,2222,3333,4444,5555,6666
3
4           OPER         STORAGE      7              Machine operators
5           MACH         STORAGE      3              Machines
6
7           IAT          FUNCTION     RN2,C2         Interarrival times
8           0,720/1,1081
9
10          CLERK1       FUNCTION     RN3,C2         Initial work time by clerk
11          0,120/1,361
12
13          STME1        FUNCTION     RN4,C2         Setting-up time
14          0,420/1,781
15
16          STME2        FUNCTION     RN5,C2         Machine work time
17          0,1500/1,5101
18
19          CLERK2       FUNCTION     RN6,C2         Final work time by clerk
20          0,300/1,781
21
22    1     L            GENERATE     FN$IAT         Jobs arrive,
23    2                  QUEUE        LINE               get in line,
24    3                  SEIZE        CLERK              get service from clerk.
25    4                  ADVANCE      FN$CLERK1
26    5                  RELEASE      CLERK
27
28    6                  ENTER        OPER,1         Get set up by an operator
29    7                  ADVANCE      FN$STME1
30    8                  ENTER        OPER,1         Get another operator,
31    9                  ENTER        MACH,1             a machine,
32    10                 ADVANCE      FN$STME2           and receive work time.
33    11                 LEAVE        MACH,1
34    12                 LEAVE        OPER,2
35
36    13                 SEIZE        CLERK          Return to the clerk,
37    14                 ADVANCE      FN$CLERK2          for final paperwork.
38    15                 RELEASE      CLERK
39    16                 DEPART       LINE
40    17                 TERMINATE
41
42    18                 GENERATE     3600           Hourly timer
43    19                 TERMINATE    1
44
45                       START        7,NP           Simulate for 7 hours,
46    1     L            GENERATE     3601               stop job arrivals and
47                       START        1                  simulate for another hour.
48
```

Figure 5.6–1. A simulation of the manufacturing shop.

```
L             GENERATE           3601
              START              1
```

make the simulation progress for another hour with no additional arrivals.

The following control statements, repeated with operator and machine storages of 9 and 4, 10 and 5, and 11 and 5, reexecute the simulation with identical arrival and service patterns for these combinations:

```
OPER    STORAGE     8
MACH    STORAGE     4
L       GENERATE    FN$IAT
        RMULT       1111,2222,3333,4444,5555,6666
        CLEAR
        START       7,NP
L       GENERATE    3601
        START       1
```

The NP designation for operand B of the START statement suppresses the output at the end of the 7-hour period. The output at the end of 8 hours of operation, shown in Figure 5.6–2, indicates that the mix of 9 mechanics and 4 machines, giving an average turnaround time of 5402 seconds (1.50 hours), is the best combination for the desired objectives.

Before the problem is declared solved, two issues must be considered. First, it should be realized that 1.50 hours is the average time spent in the queue by all jobs—not just those that were completed by the end of the 8-hour period. Since in the case that is of interest (the 9-operator and 4-machine combination) only one job was still in the shop (block 10) at the end of the simulation, 1.50 hours should be very close to the average turnaround time for that day. A method for computing average turnaround times that do not include partially processed transactions is given in Section 7.7. Second, there is a problem with accepting the result of a single simulated day as a reasonably faithful representation of the shop's operations. At the very least, the 9-mechanic and 4-machine configuration must be simulated for several days to see if the preliminary results obtained in Figure 5.6–2 are reasonable. This can be done by using a sequence of the following control statements:

```
CLEAR
START    1

CLEAR
START    1
```
⋮

RELATIVE CLOCK 28800   ABSOLUTE CLOCK 28800

BLOCK	CURRENT	TOTAL		BLOCK	CURRENT	TOTAL
1	1	31		11	0	15
2	0	29		12	0	15
3	0	29		13	0	15
4	0	29		14	0	15
5	7	29		15	0	15
6	0	22		16	0	15
7	7	22		17	0	15
8	0	15		18	0	8
9	0	15		19	0	8
10	0	15				

QUEUE	MAXIMUM CONTENTS	AVERAGE CONTENTS	TOTAL ENTRIES	ZERO ENTRIES	PERCENT ZEROS
LINE	14	8.790	29	0	0.000

	AVERAGE TIME/TRANS	$AVERAGE TIME/TRANS	TABLE NUMBER	CURRENT CONTENTS
	8729.586	8729.586		14

FACILITY	AVERAGE UTILIZATION	NUMBER ENTRIES	AVERAGE TIME/TRANS	SEIZING TRANS. NO.	PREEMPTING TRANS. NO.
CLERK	0.511	44	334.295		

STORAGE	CAPACITY	AVERAGE CONTENTS	TOTAL ENTRIES	AVERAGE TIME/TRANS	AVERAGE UTILIZ.
MACH	3	1.622	15	3114.333	0.541
OPER	7	6.282	37	4889.622	0.897

	CURRENT CONTENTS	MAXIMUM CONTENTS
	0	3
	7	7

48			
49		OPER	STORAGE	8
50		MACH	STORAGE	4
51	1	L	GENERATE	FN$IAT
52			RMULT	1111,2222,3333,4444,5555,6666
53			CLEAR	
54			START	7,NP
55	1	L	GENERATE	3601
56			START	1

RELATIVE CLOCK 28800   ABSOLUTE CLOCK 28800

BLOCK	CURRENT	TOTAL		BLOCK	CURRENT	TOTAL
1	1	31		11	0	23
2	0	29		12	0	23
3	0	29		13	0	23
4	0	29		14	0	23
5	1	29		15	0	23
6	0	28		16	0	23

Figure 5.6–2. The output of the manufacturing shop simulation of Figure 5.6–1 (continues).

7	2	28	17	0	23
8	0	26	18	0	8
9	0	26	19	0	8
10	3	26			

QUEUE	MAXIMUM CONTENTS	AVERAGE CONTENTS	TOTAL ENTRIES	ZERO ENTRIES	PERCENT ZEROS
LINE	9	6.373	29	0	0.000

		AVERAGE TIME/TRANS	$AVERAGE TIME/TRANS	TABLE NUMBER	CURRENT CONTENTS
		6328.931	6328.931	6	

FACILITY	AVERAGE UTILIZATION	NUMBER ENTRIES	AVERAGE TIME/TRANS	SEIZING TRANS. NO.	PREEMPTING TRANS. NO.
CLERK	0.650	52	360.115		

STORAGE	CAPACITY	AVERAGE CONTENTS	TOTAL ENTRIES	AVERAGE TIME/TRANS	AVERAGE UTILIZ.
MACH	4	2.716	26	3008.539	0.679
OPER	8	7.030	54	3749.185	0.879
		CURRENT CONTENTS	MAXIMUM CONTENTS		
		3	4		
		8	8		

57				
58		OPER	STORAGE	9
59		MACH	STORAGE	4
60	1	L	GENERATE	FN$IAT
61			RMULT	1111,2222,3333,4444,5555,6666
62			CLEAR	
63			START	7,NP
64	1	L	GENERATE	3601
65			START	1

RELATIVE CLOCK 28800   ABSOLUTE CLOCK 28800

BLOCK	CURRENT	TOTAL	BLOCK	CURRENT	TOTAL
1	1	31	11	0	28
2	0	29	12	0	28
3	0	29	13	0	28
4	0	29	14	0	28
5	0	29	15	0	28
6	0	29	16	0	28
7	0	29	17	0	28
8	0	29	18	0	8
9	0	29	19	0	8
10	1	29			

QUEUE	MAXIMUM CONTENTS	AVERAGE CONTENTS	TOTAL ENTRIES	ZERO ENTRIES	PERCENT ZEROS
LINE	8	5.440	29	0	0.000

Figure 5.6–2. The output of the manufacturing shop
simulation of Figure 5.6–1 (continued).

		AVERAGE TIME/TRANS 5402.414	$AVERAGE TIME/TRANS 5402.414	TABLE NUMBER 1	CURRENT CONTENTS
FACILITY	AVERAGE UTILIZATION	NUMBER ENTRIES	AVERAGE TIME/TRANS	SEIZING TRANS. NO.	PREEMPTING TRANS. NO.
CLERK	0.740	57	373.737		

STORAGE	CAPACITY	AVERAGE CONTENTS	TOTAL ENTRIES	AVERAGE TIME/TRANS	AVERAGE UTILIZ.
MACH	4	3.068	29	3046.862	0.767
OPER	9	6.937	58	3444.604	0.771
		CURRENT CONTENTS 1 2	MAXIMUM CONTENTS 4 9		

66					
67			OPER	STORAGE	10
68			MACH	STORAGE	5
69	1	L	GENERATE	FN$IAT	
70			RMULT	1111,2222,3333,4444,5555,6666	
71			CLEAR		
72			START	7,NP	
73	1	L	GENERATE	3601	
74			START	1	

RELATIVE CLOCK 28800   ABSOLUTE CLOCK 28800

BLOCK	CURRENT	TOTAL	BLOCK	CURRENT	TOTAL
1	1	31	11	0	29
2	0	29	12	0	29
3	0	29	13	0	29
4	0	29	14	1	29
5	0	29	15	0	28
6	0	29	16	0	28
7	0	29	17	0	28
8	0	29	18	0	8
9	0	29	19	0	8
10	0	29			

QUEUE	MAXIMUM CONTENTS	AVERAGE CONTENTS	TOTAL ENTRIES	ZERO ENTRIES	PERCENT ZEROS
LINE	8	5.328	29	0	0.000

		AVERAGE TIME/TRANS 5291.621	$AVERAGE TIME/TRANS 5291.621	TABLE NUMBER 1	CURRENT CONTENTS
FACILITY	AVERAGE UTILIZATION	NUMBER ENTRIES	AVERAGE TIME/TRANS	SEIZING TRANS. NO.	PREEMPTING TRANS. NO.
CLERK	0.745	58	369.862	5	

Figure 5.6–2. The output of the manufacturing shop
simulation of Figure 5.6–1 (continued).

STORAGE	CAPACITY	AVERAGE CONTENTS	TOTAL ENTRIES	AVERAGE TIME/TRANS	AVERAGE UTILIZ.
MACH	5	3.083	29	3061.483	0.617
OPER	10	6.800	58	3376.552	0.680
		CURRENT CONTENTS	MAXIMUM CONTENTS		
		0	5		
		0	10		

75		OPER	STORAGE	11
76		MACH	STORAGE	5
77	1	L	GENERATE	FN$IAT
78			RMULT	1111,2222,3333,4444,5555,6666
79			CLEAR	
80			START	7,NP
81	1	L	GENERATE	3601
82			START	1

RELATIVE CLOCK 28800   ABSOLUTE CLOCK 28800

BLOCK	CURRENT	TOTAL	BLOCK	CURRENT	TOTAL
1	1	31	11	0	29
2	0	29	12	0	29
3	0	29	13	0	29
4	0	29	14	1	29
5	0	29	15	0	28
6	0	29	16	0	28
7	0	29	17	0	28
8	0	29	18	0	8
9	0	29	19	0	8
10	0	29			

QUEUE	MAXIMUM CONTENTS	AVERAGE CONTENTS	TOTAL ENTRIES	ZERO ENTRIES	PERCENT ZEROS
LINE	8	5.266	29	0	0.000
		AVERAGE TIME/TRANS	$AVERAGE TIME/TRANS	TABLE NUMBER	CURRENT CONTENTS
		5229.207	5229.207	1	

FACILITY	AVERAGE UTILIZATION	NUMBER ENTRIES	AVERAGE TIME/TRANS	SEIZING TRANS. NO.	PREEMPTING TRANS. NO.
CLERK	0.755	58	374.759	10	

STORAGE	CAPACITY	AVERAGE CONTENTS	TOTAL ENTRIES	AVERAGE TIME/TRANS	AVERAGE UTILIZ.
MACH	5	3.083	29	3061.483	0.617
OPER	11	6.770	58	3361.810	0.615
		CURRENT CONTENTS	MAXIMUM CONTENTS		
		0	5		
		0	11		
83		END			

Figure 5.6–2. The output of the manufacturing shop simulation of Figure 5.6–1 (concluded).

These statements simulate the same conditions on different days. The simulation for 9 additional different days (not shown here) gives 1.33, 1.50, 1.40, 1.42, 1.58, 1.44, 1.58, 1.50, and 1.49 hours for the daily turnaround time averages. The problem of how many days of simulation are needed before sufficient confidence can be placed in the average work time per day will be taken up in Section 6.1.

5.7 Sampling: The Exponential Distribution

It was shown in Section 1.6 that, under reasonable assumptions, when there are random arrivals into a system, the number of arrivals per unit of time has a Poisson distribution and the interarrival times have an exponential distribution. Thus, it becomes particularly important to be able to sample from exponential distributions and to have the GENERATE block use such sampled values in scheduling arrivals.

Recall that $f(x)$, the probability density function of the exponential distribution with parameter λ, is given by

$$f(x) = \begin{cases} \lambda e^{-\lambda x}, & \text{if } x \geq 0 \\ 0, & \text{otherwise} \end{cases}$$

and the cumulative distribution function is

$$F(x) = \begin{cases} \int_0^x \lambda e^{-\lambda t} dt = 1 - e^{-\lambda x}, & \text{if } x > 0 \\ 0, & \text{otherwise.} \end{cases}$$

To sample from an exponential distribution in GPSS, a variation on the inverse distribution function method (Section 4.2) is used to obtain a function approximating $F^{-1}(x)$. This function is defined by connecting appropriately chosen points from $F^{-1}(x)$. In case of the exponential distribution, this is not a serious problem, since $F^{-1}(x)$ can be conveniently represented by

$$F^{-1}(x) = -\frac{1}{\lambda} \ln(1 - x), \qquad 0 \leq x < 1$$

as was shown in Section 4.2.1 (equations (4.2.5) and (4.2.6)). The points chosen in Figure 5.7–1 have been used extensively in the literature to construct an approximation to the inverse of the exponential distribution when

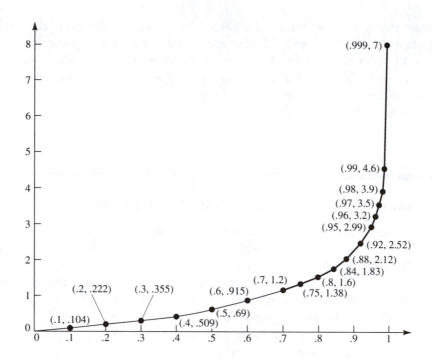

Figure 5.7–1. Points used in the approximation of $F^{-1}(x)$ for the exponential distribution with mean 1.

$\lambda = 1$. This suggests the use of the following GPSS function for sampling from the exponential distribution with mean 1:

```
EXPON    FUNCTION    RN3,C24
0,0/.1,.104/.2,.222/.3,.355/.4,.509/.5,.69
.6,.915/.7,1.2/.75,1.38/.8,1.6/.84,1.83/.88,2.12
.9,2.3/.92,2.52/.94,2.81/.95,2.99/.96,3.2/.97,3.5
.98,3.9/.99,4.6/.995,5.3/.998,6.2/.999,7/.9998,8
```

To sample from an arbitrary exponential distribution with mean λ, a value x can be sampled from the exponential distribution with mean 1, making λx a random observation from the desired distribution. Special provision is made in the GENERATE and ADVANCE blocks of GPSS to allow easy sampling from arbitrary exponential distributions. When operand B in a GENERATE or ADVANCE block designates a function, the interarrival or service time is computed by multiplying the value of the function with the value of the A operand. For example,

GENERATE   250,FN$EXPON

will make interarrival times exponential with mean 250 and

ADVANCE   232,FN$EXPON

will attempt to move transactions from the ADVANCE block after a delay sampled from the exponential distribution with mean 232. In both cases, the product of the function value and the A operand value is truncated to an integer before it is used.

5.8   Arithmetic Variables in GPSS

Variables are used in GPSS to express relationships among various model entities. Evaluations of arithmetic expressions are done through the VARIABLE construct. As with functions, a variable definition only establishes the expression that is to be evaluated. Specific references to an already defined variable, later in the program, cause the actual computation of values of the variable. The variable definition consists of

1. The variable's name in the label field.

2. VARIABLE literally in the operator field.

3. The arithmetic expression in the operand field.

The operators used in the expression portion of the variable definition are +, −, *, /, and @. The +, −, and * are the usual addition, subtraction, and multiplication operators. The / operator designates **integer** division and computes the truncated form of the quotient of two integers. The @ operator is the modular (mod) operator introduced in Section 2.1 that gives the remainder when one integer is divided by another.

The use of parentheses follows the usual algebraic conventions. Portions of expressions may be enclosed in parentheses so that parenthetical expressions can be computed before their values are used in subsequent computations. The *, /, and @ operators have a higher precedence than the + and − operators. Therefore, during expression evaluation, all multiplications, integer divisions, and remainder computations are done before any additions or subtractions, unless otherwise dictated by parenthetical subexpressions. Within any parenthetical grouping, the results of *, /, and @, as well as the results of + and −, are evaluated in left-to-right order.

All initial, intermediate, and final results in the evaluation of a variable are integers. Suppose that the variable AVG is defined by

$$\text{AVG} \quad \text{VARIABLE} \quad (\text{FN\$A+FN\$B+FN\$C})/3.$$

If reference to this variable is made in a program block such as

$$\text{ADVANCE} \quad \text{V\$AVG}$$

then the three functions A, B, and C are evaluated and added, then the sum is divided by 3 and the truncated result used in the ADVANCE block. Note that the letter V and $ preceded the name, AVG, to indicate that the value of the variable called AVG was to be used as the operand in the ADVANCE block. The definition of

$$\text{AVG1} \quad \text{VARIABLE} \quad \text{FN\$A}/3+\text{FN\$B}/3+\text{FN\$C}/3$$

is likely to produce quite different results from the function AVG, due to a different pattern of truncation.

There are occasions when too much accuracy is lost by having all intermediate results in expression evaluation be integers. In such situations, FVARIABLE, for floating-point variable, may be substituted for VARIABLE. As a result of the FVARIABLE specification, all arithmetic computations are performed in a floating-point format. This implies that the results of the / operation are left in fractional form and are not truncated. If functions are used in the expression portion of the definition, the values of these functions are also left in fractional form for subsequent computations. In short, all computations are done in floating-point format and the final value is truncated. Assuming that EXPON is the exponential distribution function defined in Section 5.7, the combination

$$\text{EXP} \quad \quad \text{FVARIABLE} \quad \quad 250*\text{FN\$EXPON}$$
$$\vdots$$
$$\text{GENERATE} \quad \quad \text{V\$EXP}$$

is equivalent to

$$\text{GENERATE} \quad \quad 250,\text{FN\$EXPON}.$$

However, different results would have been obtained if VARIABLE had been used instead of FVARIABLE, since then the FN\$EXPON value would have been truncated **before** the multiplication by 250.

5.9 Sampling: The Normal Distribution

As in the exponential distribution, an approximation of the inverse distribution function method described in Section 4.2 is used to sample from normal distributions. Unlike the exponential distribution, cumulative normal distribution functions cannot be expressed in closed form. Hence, numerical methods must be used to obtain points on the inverse distribution that then define a continuous function approximating the inverse distribution. To sample from the standard normal distribution with mean 0 and variance 1, commonly available tables such as those in Appendix C can be used to obtain points for the construction of an approximating function. The function SNORM, given below, is frequently used for sampling from the standard normal distribution.

```
SNORM    FUNCTION    RN2,C23
0,-4/.00138,-3/.00621,-2.5/.02275,-2/.06681,-1.5
.11507,-1.2/.15866,-1/.21186,-.8/.27425,-.6
.34458,-.4/.42074,-.2/.5,0/.57926,.2/.65542,.4
.72575,.6/.78814,.8/.84134,1/.88493,1.2/.93319,1.5
.97725,2/.99379,2.5/.99862,3/1,4
```

To sample from an arbitrary normal distribution with mean μ and variance σ^2, the results developed in Section 4.2.3 can be used. It was noted there that if Z is a random variable with standard normal distribution, then $X = \mu + \sigma Z$ is a normally distributed random variable with mean μ and variance σ^2. For the particular instance of $\mu = 100$ and $\sigma^2 = 225$, the variable

NORM   FVARIABLE   15*FN$SNORM+100

can be defined to sample values from the normal distribution with mean 100 and variance 225. If needed, the variable NORM can be used in

ADVANCE V$NORM

to produce delay times from this particular distribution.

5.10 Transaction Parameters and the ASSIGN Block

The F and G operands of the GENERATE block (Section 2.3) can be used to provide each transaction entering the simulation with parameters attached to it. Transaction parameters are integer-valued attributes associated with a transaction. The F and G operands of the GENERATE block determine the number of parameters to be allocated to a transaction and the type (fullword or halfword) of those parameters. On most 32-bit computers, fullword parameters use 32 bits and are restricted to integer values between $-2,147,483,648$ and $2,147,483,647$; halfword parameters use 16 bits and are constrained to the range $-32,768$ to $32,767$. If the F and G operands are defaulted, the transaction is allocated 12 halfword parameters. The initial value of all parameters is always 0.

When a GPSS block refers to a k-th parameter, the reference is to the parameter of the transaction that activated that block. In successive uses of this same block, different values will be used, all of which are values of the k-th parameter of successive transactions entering that block. Depending on the context, Pk or k will be used to refer to the k-th parameter. If a block operand is supposed to be a parameter number, then the integer k is used in this context to refer to the k-th parameter. If, on the other hand, the block operand could assume a variety of forms (e.g., the value of a function), then Pk is used to refer to the k-th parameter. For example,

ADVANCE   P5,5

uses the 5th parameter of the transaction entering the ADVANCE block as the A operand and the number 5 as the B operand to compute the delay time.

The ASSIGN block is used to modify values of transaction parameters. Although the ASSIGN block has 3 operands, we will be concerned only with its A and B operands here.

Operand A: This operand, which cannot be defaulted, is the parameter number that is to be modified.

Operand B: This operand specifies the value that will be assigned to the parameter indicated by the A operand. This operand cannot be defaulted either.

For example,

ASSIGN   4,72

changes the value of the 4th parameter of the transaction entering the assign block to 72. The previous value of the 4th parameter is lost. Likewise,

ASSIGN   11,FN$EXPON

causes an evaluation of the function EXPON to take place. The evaluated value is then assigned to the 11th parameter.

It is also possible to increment or decrement a parameter by using the ASSIGN block in **increment mode** or **decrement mode**. A plus sign + is placed immediately following the A operand for incrementation and a minus sign − is similarly placed for decrementation. For example, the block

$$\text{ASSIGN} \quad 4+,72$$

will increment the value of parameter 4 by 72, whereas the block

$$\text{ASSIGN} \quad 11-,\text{FN\$EXPON}$$

will decrement the value of parameter 11 by an amount equal to an evaluated value of the function EXPON.

Indirect addressing can also be used in connection with the ASSIGN statement by using the form

$$\text{ASSIGN} \quad \text{P2},200$$

which assigns 200 not to parameter 2 but to the parameter whose number is in P2. Thus, if parameter 2 had a value of 4 when a transaction entered this block, the value of the 4th parameter of this transaction would be changed to 200. Indirect addressing and incrementation/decrementation may both be used in a statement, as in

$$\text{ASSIGN} \quad \text{P4+},\text{FN\$EXPON}$$

which will cause the incrementation of the parameter whose number is in P4 by a value obtained by evaluating the function EXPON.

5.11 Example: A Batch Computer System Operation

A batch computer system operation receives jobs from a variety of customers via dedicated communication lines. A major component of the charge to customers for the use of computer services is determined by the amount of use that is made of the central processing unit (CPU). The practice has been to provide service on a first-come, first-served basis to all customer jobs. Some customers have indicated to the manager of operations that some of their work requires faster response time and that they are willing to pay a higher rate for CPU time if specific jobs could receive preferential treatment. Other customers are satisfied with things as they are, and still others do not mind if certain response times deteriorate, provided that the associated computation costs are correspondingly reduced.

The manager of operations decides to establish four different CPU usage rates for all customers, who may submit jobs at one of four priority levels. The commitment to the users is that an incoming job at a given priority level will execute ahead of all other jobs with a lower priority that are waiting to access the CPU, but not ahead of a job that might already have captured it. To determine the average waiting times of jobs at each priority level, the manager requests that customers indicate the priority they would have chosen for jobs that had been submitted to the system on a typical day. Analysis of this information reveals that

1. The choice of priority level is independent of the CPU time required by the jobs.

2. The number of jobs at each priority level follows a Poisson pattern with averages of 100.43, 352.94, 230.77, and 148.148 jobs per hour for jobs at priority levels 4, 3, 2, and 1, respectively, priority 4 being the highest.

A study of jobs across all priorities shows that CPU times are clustered around 800, 2675, and 11,950 ms (milliseconds) with a frequency graph as shown in Figure 5.11–1.

Careful analysis shows that normal distributions with standard deviations of 125, 270, and 1623 ms are good fits for the clusters centered at 800, 2675, and 11,950 ms, respectively. Also, the three clusters represent 15%, 65%, and 20% of all incoming jobs. CPU times are summarized in Table 5.11–2.

To determine the expected waiting times of jobs at each of the four priority levels, the simulation given in Figure 5.11–3 is executed. The first segment of the program simulates the priority 1 jobs. Recall that operand E of the GENERATE block (the first four segments) designates the priority levels of the transactions entering the simulation through these blocks.

Since the next three segments perform essentially the same function for jobs with priorities 2, 3, and 4, the discussion here will center on the first segment. The Poisson arrival pattern with an average of 148.148 arrivals per hour implies an average of 148.148/3600 arrivals per second and an average interarrival time of 24.3000 seconds. Since the time unit chosen for the simulation is the millisecond, exponential interarrival times with mean 24,300 are used in the GENERATE block at line 35.

The JOBTPE function at line 36 determines the type of this job and saves that type as parameter 1. Parameters 2 and 3 are used to store the mean

Figure 5.11–1. The relative frequencies of jobs by
their CPU time requirements.

Table 5.11–2. The CPU time distribution (in milliseconds)
of incoming jobs.

JOB TYPE	MEAN	STANDARD	PERCENT OF TOTAL
1	800	125	15
2	2675	270	65
3	11,950	1623	20

and the standard deviation of the CPU time that will be required by the transaction that is entering blocks 3 and 4. To obtain waiting-time information, the **PRIOR1Q** queue is used before the CPU is captured. To obtain overall waiting-time information, the **OVERALLQ** queue is used in the first four segments.

The results of the simulation given in Figure 5.11–4 show average waiting times of 382491.063, 26741.369, 9529.258, and 4413.125 ms for jobs of priority levels 1 through 4, respectively. Priority 1 jobs have a significant average waiting time of over 6 minutes. The average waiting time for priority 2 jobs is less than half a minute; the waiting times for priority 3 and 4 jobs are about 10 and 4 seconds, respectively. It may well be worth knowing about the pattern of waiting times for priority 1 jobs in addition to the average waiting time, since it is possible that the actual waiting times may be widely dispersed. A mechanism for obtaining this type of information will be described in Section 7.8.

```
1
2                            SIMULATE
3
4
5                  *
6                  *  FUNCTION DEFINITIONS
7                  *
8                     SNORM  FUNCTION    RN2,C23           Standard normal Dist.
9                  0,-4/.00138,-3/.00621,-2.5/.02275,-2/.06681,-1.5
10                 .11507,-1.2/.15866,-1/.21186,-.8/.27425,-.6
11                 .34458,-.4/.42074,-.2/.5,0/.57926,.2/.65542,.4
12                 .72575,.6/.78814,.8/.84134,1/.88493,1.2/.93319,1.5
13                 .97725,2/.99379,2.5/.99862,3/1,4
14
15                    EXPON  FUNCTION    RN3,C24           Exp. dist. with mean 1
16                 0,0/.1,.104/.2,.222/.3,.355/.4,.509/.5,.69
17                 .6,.915/.7,1.2/.75,1.38/.8,1.6/.84,1.83/.88,2.12
18                 .9,2.3/.92,2.52/.94,2.81/.95,2.99/.96,3.2/.97,3.5
19                 .98,3.9/.99,4.6/.995,5.3/.998,6.2/.999,7/.9998,8
20
21                    JOBTPE FUNCTION    RN4,D3            Job type function
22                 .15,1/.80,2/1,3
23
24                    MST    FUNCTION    P1,D3             Mean CPU time. P1 is job type
25                 1,800/2,2675/3,11950
26
27                    SDST   FUNCTION    P1,D3             St. dev. CPU time P1 is job type
28                 1,125/2,270/3,1623
29
30                 *
31                 *      VARIABLE DEFINITIONS
32                 *
33                    STME    FVARIABLE   P3*FN$SNORM+P2  CPU time. P2=mean, P3=st.dev.
34
35       1          GENERATE     24300,FN$EXPON,,,1        Priority 1 jobs.
36       2          ASSIGN       1,FN$JOBTPE              P1 ⟵ job type.
37       3          ASSIGN       2,FN$MST                P2 ⟵ mean CPU time.
38       4          ASSIGN       3,FN$SDST               P3 ⟵ st. dev. CPU time.
39       5          QUEUE        OVERALLQ                Place job in overall queue.
40       6          QUEUE        PRIOR1Q                 Place job in priority 1 queue.
41       7          SEIZE        CPU                     Capture CPU.
42       8          DEPART       PRIOR1Q
43       9          DEPART       OVERALLQ
44       10         ADVANCE      V$STME                  Obtain CPU service.
45       11         RELEASE      CPU
46       12         TERMINATE
47
48       13         GENERATE     15600,FN$EXPON,,,2       Priority 2 jobs
49       14         ASSIGN       1,FN$JOBTPE
```

Figure 5.11–3. Program for the batch computing system (continues).

50	15	ASSIGN	2,FN$MST	
51	16	ASSIGN	3,FN$SDST	
52	17	QUEUE	OVERALLQ	
53	18	QUEUE	PRIOR2Q	
54	19	SEIZE	CPU	
55	20	DEPART	PRIOR2Q	
56	21	DEPART	OVERALLQ	
57	22	ADVANCE	V$STME	
58	23	RELEASE	CPU	
59	24	TERMINATE		
60				
61	25	GENERATE	10200,FN$EXPON,,,3	Priority 3 jobs
62	26	ASSIGN	1,FN$JOBTPE	
63	27	ASSIGN	2,FN$MST	
64	28	ASSIGN	3,FN$SDST	
65	29	QUEUE	OVERALLQ	
66	30	QUEUE	PRIOR3Q	
67	31	SEIZE	CPU	
68	32	DEPART	PRIOR3Q	
69	33	DEPART	OVERALLQ	
70	34	ADVANCE	V$STME	
71	35	RELEASE	CPU	
72	36	TERMINATE		
73				
74	37	GENERATE	32600,FN$EXPON,,,4	Priority 4 jobs
75	38	ASSIGN	1,FN$JOBTPE	
76	39	ASSIGN	2,FN$MST	
77	40	ASSIGN	3,FN$SDST	
78	41	QUEUE	OVERALLQ	
79	42	QUEUE	PRIOR4Q	
80	43	SEIZE	CPU	
81	44	DEPART	PRIOR4Q	
82	45	DEPART	OVERALLQ	
83	46	ADVANCE	V$STME	
84	47	RELEASE	CPU	
85	48	TERMINATE		
86				
87	49	GENERATE	10800000	
88	50	TERMINATE	1	
89				
90		START	1	

Figure 5.11–3. Program for the batch computing system (concluded).

RELATIVE CLOCK 10800000   ABSOLUTE CLOCK 10800000

BLOCK	CURRENT	TOTAL	BLOCK	CURRENT	TOTAL
1	1	444	11	0	398
2	0	443	12	0	398
3	0	443	13	1	655
4	0	443	14	0	654
5	0	443	15	0	654
6	45	443	16	0	654
7	0	398	17	0	654
8	0	398	18	1	654
9	0	398	19	0	653
10	0	398	20	0	653

BLOCK	CURRENT	TOTAL	BLOCK	CURRENT	TOTAL
21	0	653	31	0	1095
22	0	653	32	0	1095
23	0	653	33	0	1095
24	0	653	34	0	1095
25	1	1099	35	0	1095
26	0	1098	36	0	1095
27	0	1098	37	1	321
28	0	1098	38	0	320
29	0	1098	39	0	320
30	3	1098	40	0	320

BLOCK	CURRENT	TOTAL	BLOCK	CURRENT	TOTAL
41	0	320			
42	0	320			
43	0	320			
44	0	320			
45	0	320			
46	1	320			
47	0	319			
48	0	319			
49	0	1			
50	0	1			

QUEUE	MAXIMUM CONTENTS	AVERAGE CONTENTS	TOTAL ENTRIES	ZERO ENTRIES	PERCENT ZEROS
OVERALLQ	56	18.408	2515	21	0.835
PRIOR1Q	49	15.689	443	6	1.354
PRIOR2Q	8	1.619	654	6	0.917
PRIOR3Q	9	0.969	1098	7	0.638
PRIOR4Q	3	0.131	320	2	0.625

		AVERAGE TIME/TRANS	$AVERAGE TIME/TRANS	TABLE NUMBER	CURRENT CONTENTS
		79049.570	79715.180		49
		382491.063	387742.656		45
		26741.369	26988.975		1

Figure 5.11–4. The output of the batch computing
system simulation (continues).

		9529.258	9590.398		3
		4413.125	4440.880		0
FACILITY	AVERAGE UTILIZATION	NUMBER ENTRIES	AVERAGE TIME/TRANS	SEIZING TRANS. NO.	PREEMPTING TRANS. NO.
CPU	0.993	2466	4348.351	16	
91		END			

Figure 5.11–4. The output of the batch computing system simulation (concluded).

Problems for Chapter 5

5.1 (Section 5.1) Suppose that in the following program segment, the first five random values produced by the GENERATE block are 8, 9, 11, 9, and 12 and the first five values produced by the ADVANCE block are 13, 8, 12, 10, and 7. Trace the GPSS assembly process (Figure 5.1–3) and the movement of transactions (Figure 5.1–2) for this program until the third transaction leaves the system.

```
GENERATE      10,3
SEIZE         F
ADVANCE       9,4
RELEASE       F
TERMINATE
```

5.2 (Section 5.1) For the situation described in Problem 5.1, give the contents of the CEC and FEC at the end of the first 7 clock cycles, in a manner similar to that in Figure 5.1–5.

5.3 (Section 5.2) At what simulated times will the following program produce output?

```
GENERATE      5
TERMINATE     2
START         30,,3
```

5.4 (Section 5.2) From the output of the program

```
SIMULATE
GENERATE        10,3
QUEUE           Q
SEIZE           F
DEPART          Q
ADVANCE         10.5
RELEASE         F
TERMINATE       1
START           5,,1,1
```

determine the successive arrival and service times generated by the program. Use these times to trace the movement of transactions through the CEC and the FEC in a manner similar to that in Figure 5.1–5.

5.5 (Section 5.2) The GPSS PRINT statement will be discussed in some detail in Section 7.4. For this problem, it suffices to know that

```
PRINT           ,,MOV,1
PRINT           ,,FUT,1
```

will produce a listing of the CEC and FEC, respectively, whenever a transaction enters these blocks. Modify the program in Problem 5.1 by inserting these PRINT statements immediately following the GENERATE and RELEASE blocks and by changing the START statement to

```
START           5.
```

Use the output to trace the movement of transactions through the CEC and FEC in a manner similar to that given in Figure 5.1–5.

5.6 (Section 5.4) What will be the effect of each of the following control sequences on the ouput of a GPSS program?

a.	b.	c.
START 100	START 100	START 100
START 100	RESET	CLEAR
⋮		
START 100	START 100	START 100
	RESET	CLEAR
	⋮	⋮
	START 100	START 100

5.7 (Section 5.5) We need to produce values of $1, 2, 3$, or 4 with equal probability in a GPSS program. Show how this can be done by using a discrete function. Also show how a continuous function could be used for this purpose.

5.8 (Section 5.5) Suppose that interarrival times at a queueing system have been observed to follow the pattern indicated in the following table. Devise a continuous GPSS function and use it in a GENERATE block to produce the interarrival times described in the table.

INTERARRIVAL TIME	PERCENTAGE	CUMULATIVE PERCENTAGE
0-15	0	0
15-25	9	9
25-35	16	25
35-45	24	49
45-60	20	69
60-75	18	87
75-95	13	100

5.9 (Section 5.7)

a. Write a GPSS program to simulate the $M/M/1/\infty$ queueing system for $\rho = 0.1, 0.2, \ldots, 0.9, 0.95$, and 0.99 (you can, for example, choose an average interarrival time of 100 and average service times of $10, 20, \ldots, 90$, 95, and 99).

b. Compare in tabular form the empirical values of average facility utilization, average number in the system, average number waiting, average time in the system, and average time waiting to their theoretical counterparts developed in Section 1.6.

c. Plot the average utilization and the average number waiting as functions of ρ.

5.10 (Section 5.7) Suppose that a queueing system has exponential interarrival times.

a. Assuming exponential service times, write a GPSS program to obtain values for the average waiting times when $\rho = 0.1, 0.2, \ldots, 0.9, 0.95$, and 0.99, and plot the average waiting time as a function of ρ. (If you have done Problem 5.9, you already have this result.)

b. Repeat Part a assuming uniformly distributed service times (take $\mu = \sigma$) and again assuming fixed service times.

c. Is there anything you can conclude by superimposing all three graphs of average waiting times?

5.11 (Section 5.7) A tax consulting office uses two part-time employees to cover its two 5-hour shifts (7:00 a.m. to noon and noon to 5:00 p.m.). The consultant on the morning shift is more experienced and her service time is exponential with mean 23 minutes. The service time for the less experienced afternoon consultant is exponential with mean 30 minutes. Arrivals throughout the 10-hour day are Poisson with a mean of 2 arrivals per hour. Simulate the operation of this office for one day.

5.12 (Section 5.7) Suppose that in the model described in Problem 5.11, clients prefer to seek advice in the afternoon and the arrival times after 1:00 p.m. are Poisson with mean 3.5 arrivals per hour. To accommodate the higher load during the afternoon, the manager hires a second inexperienced consultant for the second shift. Simulate the operation of the office under these conditions.

5.13 (Section 5.7) Obtain the average of the average waiting times by simulating the model of Problem 5.12 for 10 consecutive days and calculating the average.

5.14 (Section 5.7) A theater box office is open between 6:00 and 8:00 p.m. each day. Customers begin to arrive at 5:45 p.m. and must wait until the box office opens, and only customers who arrive by 8:00 p.m. are served. If arrivals are Poisson with an average of 110 arrivals per hour and service is exponential with a mean of 30 seconds, will the box office be able to close before the 8:15 curtain time?

5.15 (Section 5.8) What values will be computed by each of the following variables?

V1	VARIABLE	10/3*5
V2	VARIABLE	10*5/3
V3	FVARIABLE	10/3*5
V4	FVARIABLE	10*5/3
V5	VARIABLE	10/3@3
V6	VARIABLE	10@3/3

5.16 (Section 5.8) Define a variable that assumes a value of 1 if FN$X (a function defined elsewhere in the program) is odd and 0 if FN$X is even.

5.17 (Section 5.8) Determine the first 10 random numbers (to 5 significant digits) produced by RN1 when the seed is defaulted. Hint: You can define a function to be $10^5 \times$ RN1 and generate interarrivals according to this variable; then with successive START and RESET combinations, you can obtain output giving $10^5 \times$ RN1 as clock times.

5.18 (Section 5.9) Define a variable, IAT, that samples from a normal distribution with mean 10 and standard deviation 3. Are there any problems associated with using V$IAT in

$$\text{GENERATE} \qquad \text{V\$IAT}$$

later in the program? Explain.

5.19 (Section 5.10) From probability theory, we know that observations X_i from a chi-square distribution with ν degrees of freedom can be generated through $X = Z_1^2 + Z_2^2 + \cdots + Z_\nu^2$ where $Z_1, Z_2, \ldots, Z_\nu$ are ν independent standard normal random variables. Define a GPSS variable CHISQ3 that samples from a chi-square distribution with 3 degrees of freedom. If you use transaction parameters in the definition of CHISQ3, make clear what these parameters represent and how the variable CHISQ3 is invoked.

5.20 (Section 5.10) Suppose that at a point during a simulation, P3 has value 2 and variables V1 and V2 have values 17 and -5, respectively. As a consequence of the execution of each of the following ASSIGN blocks, state which transaction parameter is altered and what its new value will be.

$$
\begin{array}{ll}
\text{ASSIGN} & \text{3, V\$V1} \\
\text{ASSIGN} & \text{3+, V\$V1} \\
\text{ASSIGN} & \text{P3, V\$V1} \\
\text{ASSIGN} & \text{P3+, V\$V2} \\
\text{ASSIGN} & \text{P3-, V\$V2}
\end{array}
$$

Chapter 6

Statistical Design and Analysis of Simulations

In various parts of Chapters 1, 2, and 5 we considered in detail **how** a simulation model is constructed, and in Chapters 3, 4, and 5 **how** random variables can be generated by a computer, using a theoretical or a fitted distribution. In this chapter, as a capstone, we discuss **how** the great control the analyst has over the model can be used **to optimize the system** (Sections 6.1, 6.2, 6.3, 6.4), and **how** the simulation process can be used **to solve complex mathematical and statistical problems that defy analytical solution** (Section 6.5). Although, in the past, simulation was naive in its statistical component, the best simulations today are being run in ways that make proper analysis of the resulting data possible; for example, see Chapter 16 of Schriber (1990).

6.1 How Long to Simulate: Goal of Estimating the Mean

The answer to the question "How long to simulate?" should always be "Long enough to satisfy the knowledge goal." In this section, as well as in Sections 6.2 and 6.3, we examine this question for some of the most important and commonly occurring goals. Note that the answer sometimes given is "One run will show me how the system operates." However, in reality, one run will only show 1 period (e.g., one day or year) of operation, and this will not allow achievement of **a quantitative goal such as "with 95% confidence, we want to specify the true average queue size to within .01 after 1 day of operation."**

To achieve such a quantitative goal, suppose that we

> Simulate for an 8-hour day and find X_1, the queue size at day's end
> Simulate for a second 8-hour day to obtain X_2
>
> $\vdots$
>
> Simulate for an nth 8-hour day to obtain X_n.

If we start each day's simulation independently (using independent random numbers), then $X_1, X_2, \ldots, X_n$ **will be independent random variables, each with the same distribution with mean μ (unknown) and variance σ^2.** We may estimate μ by

$$\overline{X} = \left(\frac{1}{n}\right) \sum_{i=1}^{n} X_i, \qquad (6.1.1)$$

which by the Central Limit Theorem has an approximate $N(\mu, \sigma^2/n)$ distribution. This is the exact distribution if $X_1, X_2, \ldots, X_n$ are normal random variables. Now **our goal is to find n (which tells how long to simulate) such that**

$$P(|\overline{X} - \mu| \leq d) = P^* \qquad (6.1.2)$$

with $d = .01$ and $P^* = .95$. Since $(\overline{X} - \mu)/(\sigma/\sqrt{n})$ is approximately $N(0, 1)$, the left-hand side of (6.1.2) equals

$$P\left(\frac{-d}{\sigma/\sqrt{n}} \leq \frac{\overline{X} - \mu}{\sigma/\sqrt{n}} \leq \frac{d}{\sigma/\sqrt{n}}\right) = \Phi\left(\frac{d}{\sigma/\sqrt{n}}\right) - \Phi\left(-\frac{d}{\sigma/\sqrt{n}}\right) \qquad (6.1.3)$$

$$= 2\Phi(d\sqrt{n}/\sigma) - 1,$$

which equals P^* if and only if

$$2\Phi(d\sqrt{n}/\sigma) - 1 = P^*$$

or

$$\Phi(d\sqrt{n}/\sigma) = (1 + P^*)/2$$

or

$$\frac{d\sqrt{n}}{\sigma} = \Phi^{-1}\left(\frac{1 + P^*}{2}\right). \qquad (6.1.4)$$

So, to satisfy (6.1.2) we need to simulate for n periods where

$$n = \left\lceil \left(\frac{\sigma}{d} \Phi^{-1}\left(\frac{1 + P^*}{2}\right)\right)^2 \right\rceil, \qquad (6.1.5)$$

where $\lceil \cdot \rceil$ denotes rounding up to an integer (e.g., $\lceil 16.23 \rceil = 17$). If, for example, $d = .01$ and $P^* = .95$, and **if we know** that $\sigma = 0.5$, then

$$n = \left\lceil \frac{\sigma^2 \times (1.96)^2}{(.01)^2} \right\rceil = \lceil (196\sigma)^2 \rceil = (98)^2 \doteq 10,000 \, .$$

If we can afford only the computer time for a smaller n, such as $n = 400$, then with 95% confidence we pin down the mean queue size (still with known σ such as $\sigma = 0.5$) within width

$$d = \frac{\sigma}{\sqrt{n}} \, \Phi^{-1}((1 + P^*)/2) = 1.96\sigma/20 = .05 \, .$$

Also note that **if $X_1, X_2, \ldots, X_n$ are not normal,** then instead of relying on the Central Limit Theorem (which assures that $\overline{X}$ will be approximately normal) one could use a transformation such as X_i^λ (perhaps finding the power λ by the methods of Section 4.13) for which the transformed variable is approximately normal ($i = 1, 2, \ldots$). This possibility is discussed further later in this section at (6.1.14).

Above, we determined that the length of simulation given by (6.1.5) satisfied goal (6.1.2) when the variance σ^2 was known. Now we discuss how to proceed if σ^2 is not known. One might at first reason as follows. Once n is set, we may estimate σ^2 by the sample variance $\sum_{i=1}^{n}(X_i - \overline{X})^2/(n-1)$, usually denoted by s^2. But we cannot set n by (6.1.5) with σ^2 replaced by s^2, since we cannot find s^2 until n is set, and thus this method involves circular reasoning. One may dream up more intricate ways to proceed, but they are all doomed to failure in the sense of the following theorem.

Theorem 6.1.6. *If $X_1, X_2, \ldots, X_n$ are independent $N(\mu, \sigma^2)$ random variables with μ and σ^2 both unknown, then there is no single-stage statistic $T(X_1, \ldots, X_n)$ such that*

$$P(T(X_1, \ldots, X_n) - d \leq \mu \leq T(X_1, \ldots, X_n) + d) = P^* \, . \tag{6.1.7}$$

Here a "single-stage" statistic is one for which n is specified in advance (one cannot look at any portion of the data to set n or any other part of the statistical procedure). Simply stating that "μ is between $\overline{X} - d$ and $\overline{X} + d$" with some arbitrarily chosen n will not work, since this statement is correct only with probability

$$P(\overline{X} - d \leq \mu \leq \overline{X} + d) = P(-d \leq \overline{X} - \mu \leq d)$$
$$= 2\Phi(d\sqrt{n}/\sigma) - 1,$$

which (for large σ) can be much smaller than the desired P^*.

One might next try to solve the problem with σ^2 unknown by using the Student's t confidence interval "μ is between $\overline{X} \pm t_{n-1}^{-1}(.975)s/\sqrt{n}$ with probability .95," but the length of this interval is $2t_{n-1}^{-1}s/\sqrt{n}$. Since s is a random variable, this length is random (and in particular not $2d$, as is desired).

A solution when σ^2 is unknown can be obtained in two stages of simulation, as follows.

Theorem 6.1.8. *For goal (6.1.2), when σ^2 is unknown, steps HM–1 through HM–7 lead to an estimate of μ within d, with confidence $100P^*\%$.*

Step HM–1. Set n_0 as a positive integer ≥ 2. Set $w = t_{n_0-1}^{-1}((1+P^*)/2)/d$.

Step HM–2. Observe $X_1, \ldots, X_{n_0}$.

Step HM–3. Calculate

$$\overline{X}(n_0) = \frac{X_1 + \cdots + X_{n_0}}{n_0}, \quad s^2 = \sum_{i=1}^{n_0}(X_i - \overline{X}(n_0))^2/(n_0 - 1).$$

Step HM–4. Set

$$n = \max\{n_0 + 1, \; \lceil (t_{n_0-1}^{-1}((1+P^*)/2))^2 s^2/d^2 \rceil\}$$

where $\lceil x \rceil$ denotes the smallest integer which is $\geq x$.

Step HM–5. Observe $X_{n_0+1}, \ldots, X_n$.

Step HM–6. Calculate

$$\overline{\overline{X}} = \frac{X_1 + \cdots + X_{n_0} + X_{n_0+1} + \cdots + X_n}{n}.$$

Step HM–7. Claim, with confidence $100P^*\%$, that

$$\overline{\overline{X}} - d \leq \mu \leq \overline{\overline{X}} + d.$$

The procedure specified in Theorem 6.1.8 is called **Procedure HM for estimating μ within d with confidence $100P^*\%$.** The proof follows easily from the more general result in Theorem 6.1.9, which we use later in this chapter. This result was first proven in 1945 by C. Stein; see Dudewicz and Mishra (1988, pp. 505–507) for details of the proof.

Theorem 6.1.9. *Suppose that $X_1, X_2, \ldots$ are independent $N(\mu, \sigma^2)$ random variables. If we*

1. *Set $n_0 \ (\geq 2)$ and $w = t_{n_0-1}^{-1}((1+P^*)/2)/d$;*

2. *Take a first sample $X_1, X_2, \ldots, X_{n_0}$, and calculate its sample mean $\overline{X}(n_0)$ and sample variance s^2;*

3. *Let $n = \max\{n_0 + 1, \lceil (ws)^2 \rceil\}$ and take $n - n_0$ additional observations $X_{n_0+1}, \ldots, X_n$;*

4. *Calculate the sample mean $\overline{Y}(n - n_0) = (X_{n_0+1} + \cdots + X_n)/(n - n_0)$ of the second set of observations;*

5. *And finally, also calculate*

$$\widetilde{X} = b\overline{X}(n_0) + (1-b)\overline{Y}(n - n_0) \tag{6.1.10}$$

   where

$$b = \frac{n_0}{n}\left(1 + \sqrt{1 - \frac{n}{n_0}\left(1 - \frac{n - n_0}{(ws)^2}\right)}\right), \tag{6.1.11}$$

then

$$\frac{\widetilde{X} - \mu}{1/w} \tag{6.1.12}$$

has Student's t-distribution with $n_0 - 1$ degrees of freedom.

It follows from Theorem 6.1.9 that

$$P(\widetilde{X} - d \leq \mu \leq \widetilde{X} + d) = P\left(\frac{-d}{1/w} \leq \frac{\widetilde{X} - \mu}{1/w} \leq \frac{d}{1/w}\right)$$

$$= 2t_{n_0-1}(wd) - 1 = 2t_{n_0-1}(t_{n_0-1}^{-1}((1+P^*)/2)) - 1$$

$$= 2(1 + P^*)/2 - 1 = P^*.$$

Also,

$$P(\overline{\widetilde{X}} - d \leq \mu \leq \overline{\widetilde{X}} + d) \geq P(\widetilde{X} - d \leq \mu \leq \widetilde{X} + d) = P^*. \tag{6.1.13}$$

One practical question regarding the use of Procedure HM of Theorem 6.1.8 is **what n_0 to use**. Table 6.1–1 gives values of w (see Step HM–1) that are used in setting the total sample size n (in Step HM–4) for $P^* = .95$.

Table 6.1–1. Values of w for $P^* = .95$.

n_0	2	5	15	60	∞
$t_{n_0-1}^{-1}(.975)$	12.706	2.776	2.145	2.00	1.96
w	161.44	7.71	4.60	4.00	3.84

(Appendix D contains a Student's t-distribution table that can be used to find w for other cases.) **Thus, an n_0 of about 15 (or greater) will suffice.** Such a value of n_0 will guarantee that the value n is not large due to the s^2 in the first stage being a poor estimate of σ^2, which is why for small n_0 one finds a large w.

Note that this gives us a solution for the goal of estimating the queue size after 8 hours of operation, with $X_1, X_2, \ldots$ queue sizes after 8 hours on successive occasions. Now **suppose we used a transformation**, such as

$$Y_i = \frac{X_i^\lambda - 1}{\lambda}, \tag{6.1.14}$$

to achieve normality for some λ. Then the 95% confidence interval on $E(Y_i)$ would be $\overline{\overline{Y}} \pm d$. But, **what is the 95% interval on $E(X_i)$?** Such intervals, which are the best possible in a specific sense, were developed by Neyman, Scott, and Hoyle (for details, see Dudewicz (1983)).

Remark. Procedure HM of Theorem 6.1.8 (for estimating μ within d with confidence $100P^*\%$), is called a Heteroscedastic Method (hence the HM designation) since, when there are two or more sources of variation (as we will have in later sections of this chapter), it does not assume the variances of different sources are equal. An overview of such procedures for many different goals is given in Dudewicz (1995).

6.1.1   The Proportional Closeness Goal

We have shown how to give a confidence interval for μ with width $\pm d$. Instead, sometimes one finds that experimenters ask for a 95% confidence interval for μ of width $\pm 0.10\mu$. Since the width is proportional to μ, this is called a **proportional closeness** goal. The following are some comments on this goal.

1. The proportional closeness goal does not make sense if $\mu < 0$; one might use $\pm 0.10|\mu|$ when $\mu < 0$.

2. To our knowledge, no statistical procedures have been derived to solve the proportional closeness goal, not even if the observations are exactly $N(\mu, \sigma^2)$.

3. There are a number of procedures proposed in the simulation literature for the proportional closeness goal (e.g., see Law and Kelton (1982, pp. 291–293) and the references cited there). These procedures are adhocery, i.e., no theory backs them up. Only a small simulation study was done to assess the percent confidence of the procedures proposed by Law and Kelton.

4. Chow and Robbins have given sequential procedures, taking one observation at a time, for which

$$\lim_{n \to \infty} P(T(X_1, \ldots, X_n) - 0.10\mu \leq \mu \leq T(X_1, \ldots, X_n) + 0.10\mu) = .95 \,.$$

However, when to stop increasing n does not seem to be known.

5. The proportional closeness goal is, in our opinion, **not** one that most researchers, on reflection, would desire. For example, if a researcher says that a $0.10|\mu|$ proportional closeness is desired, this means having

$$\begin{array}{llll}
\text{an interval of width} & \pm 1 & \text{if} & \mu = 10 \\
\text{an interval of width} & \pm.1 & \text{if} & \mu = 1 \\
\text{an interval of width} & \pm 10^{-10} & \text{if} & \mu = 10^{-9} \\
\text{an interval of width} & \pm 0 & \text{if} & \mu = 0.
\end{array}$$

We believe that this realization will dissuade experimenters who, at first glance, may think they desire a proportional closeness goal.

6. A possible meaningful reformulation of the proportional closeness goal is for the experimenter to specify a function $g(\mu)$ that gives the $\pm$ width desired as a function of μ. Three possibilities are shown in Figure 6.1–2. Possibility (a) was solved earlier in this section; (b) is unsuitable (see item 5 above); and (c) is a reasonable $g(\mu)$ for which no statistical theory is yet available. We conjecture that one can develop a statistical procedure for this problem whenever $g(\cdot)$ is such that

$$\inf_{\mu} g(\mu) > 0.$$

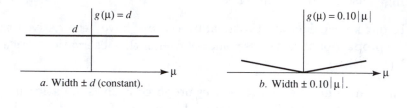

a. Width $\pm d$ (constant).   *b.* Width $\pm 0.10|\mu|$.

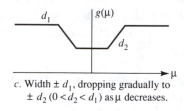

c. Width $\pm d_1$, dropping gradually to
$\pm d_2\ (0 < d_2 < d_1)$ as μ decreases.

Figure 6.1–2. Possible width functions $g(\cdot)$.

7. If Y is $N(\mu, \mu^2 b^2)$, then Y/μ is $N(1, b^2)$ and we can find a 95% confidence interval of half-width d for $E(Y/\mu)$, namely

$$\frac{\widetilde{\overline{Y}}}{\mu} - d \le E\left(\frac{Y}{\mu}\right) \le \frac{\widetilde{\overline{Y}}}{\mu} + d,$$

$$\widetilde{\overline{Y}} - d\mu \le E(Y) \le \widetilde{\overline{Y}} + d\mu.$$

Thus, $\widetilde{\overline{Y}}$ solves the proportional closeness goal when the variance is proportional to the square of the mean, since by step HM–4 of Theorem 6.1.8, $\widetilde{\overline{Y}}$ will have probability .95 of being within $\pm d\mu$ of $E(Y) = \mu$.

8. If, as often happens, the experimenter knows that

$$0 < A \le \mu \le B,$$

then using $d = 0.10A$ in Theorem 6.1.8 will in effect guarantee a proportional closeness goal for all **possible** μ.

6.1.2   Case Study of Estimating Demand for Emergency Services

A common situation in practice is that a new facility (hospital, fire station, emergency medical unit, police helicopter team, or the like) is to be built.

The demand for the facility, while not known exactly, can be estimated to have parameters similar to those found for similar operations in the past. The question then arises whether the staff, equipment, and other features planned for the new facility will be adequate.

As an example of a simple situation of this type, suppose that a fire station is to be built serving an area of a city of such composition (commercial vs. residential, etc.) that past experience suggests we will find an average of four fires per 24-hour day with the number on any given day following a Poisson distribution. Also from past experience, we know that 3/4 of the fires will require only one response unit, while the other 1/4 of the fires will require two response units. At a fire that requires only one response unit, the time needed from the unit to service the fire is normally distributed with a mean of 3 hours and a standard deviation of 0.5 hour; while at a fire that requires two response units, the time required from each unit to service the fire is normally distributed with a mean of 4 hours and a standard deviation of 1 hour.

The problem is to determine the total minutes of unit time used per day (out of the $24 \times 2 \times 60 = 2880$ minutes available theoretically) to within 60 minutes with 90% confidence.

To solve this problem, we use procedure HM of Theorem 6.1.8. We wish to estimate μ (the mean minutes per day used) within $d = 60$ with confidence $100P^*\%$ with $P^* = .90$. Suppose we choose $n_0 = 10$ and simulate 10 days of operation. The GPSS program for this simulation, developed in Section 7.5, gives the following minutes of unit time required:

$$X_1 = 1505, \quad X_2 = 804, \quad X_3 = 864, \quad X_4 = 869,$$
$$X_5 = 1314, \quad X_6 = 1599, \quad X_7 = 215, \quad X_8 = 1232,$$
$$X_9 = 885, \quad X_{10} = 603.$$

From these, we then calculate (see Step HM–3)

$$\overline{X}(n_0) = 989, \quad s = 425.07.$$

Since $t_9^{-1}(.950) = 1.83$, we have (see Step HM–4)

$$n = \max\{11, \; \lceil (1.83)^2 (425.07)^2 / (60)^2 \rceil\}$$
$$= \max\{11, \; \lceil 168.08 \rceil\} = \max\{11, 169\} = 169.$$

We therefore simulate an additional $169 - 10 = 159$ days, and find the sample mean of all $n = 169$ days of operation, which is $\overline{\overline{X}} = 1015$. We state that we are at least 90% sure that μ is between 1015.68 ± 60, i.e.,

$$955 \leq \mu \leq 1075.$$

Several observations are worth noting.

1. The 10 days of simulation suggested by Shannon (1975, p. 94, Problem 9) for this problem is woefully inadequate for our goal.

2. A person who believes that **"one run will show me how the system operates"** would estimate μ as $X_1 = 1505$, which is quite far from the true utilization per day.

3. If we want greater confidence, we will need more sampling; for 95% confidence we have $t_9^{-1}(.975) = 2.262$, and so would need a total of

$$n = \max\{11, \; \lceil (2.262)^2 (425.07)^2/(60)^2 \rceil\}$$
$$= \max\{11, \; 257\} = 257$$

days of simulated operation.

4. If we can simulate only for 100 days total but still want 90% confidence in our statement, we will need

$$100 = \max\{11, \; \lceil (1.83)^2 (425.07)^2/d^2 \rceil\} = \frac{(1.83)^2 (425.07)^2}{d^2}$$

or

$$d = 77.79\,.$$

Thus, based on the sample mean of 100 days of operation, we are able to claim 90% confidence of being within 78 minutes of the true mean.

5. Finally, in this simple example, we can find a theoretical solution to the problem (and so have simulated for illustration, as now we will be in a position to compare the simulated results with the exact result). Namely, it is clear that the exact daily mean demand is

(3 fires) × (180 min.) + (1 fire) × (2 units) × (240 min.) = 1020 min.

which is in our 90% confidence interval on μ.

6. If we always use 90% intervals, our statements will be correct in 9 out of 10 cases on the average. If 1 out of 10 is too many errors, we can increase to 95% confidence (at a cost of the additional simulation thus needed) and then have only 1 out of 20 incorrect statements on the average. With 99% confidence, we will err only 1 time out of 100 on the average; in this case we need $t_9^{-1}(.995) = 3.25$, which results in

a need for $n = \max\{11, 531\} = 531$ simulated days. In some cases the extra confidence is "worth" the cost in simulation time, in other cases it is not. Assessment of the needed confidence involves statistical decision theory, which takes account of the losses that will be produced by erroneous statements, and is beyond the scope of this text.

6.2   How Long to Simulate: Estimating the Difference $\mu_1 - \mu_2$

When a real-world system under simulation may be operated in two ways, a goal often of interest to the experimenter is to quantify the difference in a key output measure's mean value under the two modes of operation. Thus, if

runs under mode 1 yield $X_1, X_2, \ldots$ with mean μ_1 and variance σ_1^2,

while

runs under mode 2 yield $Y_1, Y_2, \ldots$ with mean μ_2 and variance σ_2^2,

the goal will be a $100P^*\%$ confidence interval on $\mu_1 - \mu_2$ with width $\pm d$. Of course, here $\mu_1, \mu_2, \sigma_1^2, \sigma_2^2$ are unknown.

A solution can be obtained as follows. Sample from $X_1, X_2, \ldots$ as in Steps HM–1 through HM–4 of Theorem 6.1.8 for some fixed w, and calculate $\widetilde{\overline{X}}$ as in (6.1.10). Then (see (6.1.12))

$$T_1 = \frac{\widetilde{\overline{X}} - \mu_1}{1/w} \tag{6.2.1}$$

will have Student's t-distribution with $n_0 - 1$ degrees of freedom. Independently, sample from $Y_1, Y_2, \ldots$ as in Steps HM–1 through HM–4 of Theorem 6.1.8 for the same fixed w, and calculate $\widetilde{\overline{Y}}$ as in (6.1.10) (but from the Ys). Then (see (6.1.12))

$$T_2 = \frac{\widetilde{\overline{Y}} - \mu_2}{1/w} \tag{6.2.2}$$

will have Student's t-distribution with $n_0 - 1$ degrees of freedom, and will be independent of T_1. Thus, if we take as our interval

$$\widetilde{\overline{X}} - \widetilde{\overline{Y}} - d \leq \mu_1 - \mu_2 \leq \widetilde{\overline{X}} - \widetilde{\overline{Y}} + d, \tag{6.2.3}$$

this interval has probability of coverage

$$P\left(\widetilde{X} - \widetilde{Y} - d \leq \mu_1 - \mu_2 \leq \widetilde{X} - \widetilde{Y} + d\right)$$

$$= P\left(\frac{-d}{1/w} \leq \frac{\widetilde{X} - \mu_1}{1/w} - \frac{\widetilde{Y} - \mu_2}{1/w} \leq \frac{d}{1/w}\right) \qquad (6.2.4)$$

$$= P(-wd \leq T_1 + T_2 \leq wd) = F_{T_1+T_2}(wd) - F_{T_1+T_2}(-wd)$$

$$= 2F_{T_1+T_2}(wd) - 1.$$

Note that $T_1 - T_2$ has the same distribution as does $T_1 + T_2$, allowing us to replace the former by the latter in the third line of (6.2.4). The distribution of $T_1 + T_2$ is the distribution of the sum of two independent Student's t random variables each with $n_0 - 1$ degrees of freedom. That distribution has been tabulated—for example, see the table of c such that $P(T_1 + T_2 \leq c) = 1 - \gamma$ on p. 511 of Dudewicz and Mishra (1988)—so we are in a position to set the confidence to P^* by taking

$$2F_{T_1+T_2}(wd) - 1 = P^*, \quad \text{i.e. } w = \frac{F_{T_1+T_2}^{-1}(\frac{1+P^*}{2})}{d}. \qquad (6.2.5)$$

For 95% confidence ($P^* = .95$), we need $F_{T_1+T_2}^{-1}(.975)$, which depends on n_0. Some typical values are given below (for additional values, see the table with $k = 2$ in Appendix H).

n_0	2	5	20
$F_{T_1+T_2}^{-1}(.975)$	25.42	3.94	2.95

The problem we have just solved, phrased in terms of a statistical hypothesis test, is called the Behrens-Fisher problem. Thus, we have just provided an exact confidence interval (which implies an exact test) for the Behrens-Fisher problem, when the Xs and Ys have an exact normal distribution. In other cases, the procedure is justified approximately by the Central Limit Theorem.

6.3   How Long to Simulate: Goal of Selection of the Best

When a system may be operated in some number k of ways (with k being at least 2), often the experimenter wants to choose the "best" way. "Best" is usually defined in terms of some output measure of the goodness of the results. **We will first look at the problem when the output is univariate (one number), and goodness is taken as meaning a large mean output.** Now, if:

runs under mode 1 yield $X_{11}, X_{12}, \ldots$ with mean μ_1 and variance σ_1^2,
runs under mode 2 yield $X_{21}, X_{22}, \ldots$ with mean μ_2 and variance σ_2^2,

$\vdots$

runs under mode k yield $X_{k1}, X_{k2}, \ldots$ with mean μ_k and variance σ_k^2,

we might think of sampling some number n of observations from each mode (i.e., taking $X_{i1}, X_{i2}, \ldots, X_{in}$ from mode i for $i = 1, 2, \ldots, k$), forming the sample means (i.e., $\overline{X}_i = (X_{i1} + X_{i2} + \ldots + X_{in})/n$ from mode i for $i = 1, 2, \ldots, k$), and selecting the mode with the largest sample mean (i.e., the mode with the largest of $\overline{X}_1, \overline{X}_2, \ldots, \overline{X}_k$). The key question with regard to sample size is whether we can set n so that we are highly certain that the mode selected is either the best one, or close to the best one, in the sense that

$$P[\text{The mode selected is within } \delta^* \text{ of the best}] \geq P^*. \tag{6.3.1}$$

Here $\delta^* > 0$ is set by the experimenter. The answer is that if the observations have normal distributions, then such an n cannot be found with a single-stage statistical procedure (which does not look at any of the data until all of the data has been obtained). However, a two-stage procedure is obtained as follows.

Sample from $X_{i1}, X_{i2}, \ldots$ as in Steps HM–1 through HM–4 of Theorem 6.1.8 from some fixed w, and proceed to calculate $\widetilde{\overline{X}}_i$ as in (6.1.10). Then (see (6.1.12))

$$T_i = \frac{\widetilde{\overline{X}}_i - \mu_1}{1/w} \tag{6.3.2}$$

will have Student's t-distribution with $n_0 - 1$ degrees of freedom. This sampling is done independently for each mode $i = 1, 2, \ldots, k$. Then select the mode that yields $\max(\widetilde{\overline{X}}_1, \widetilde{\overline{X}}_2, \ldots, \widetilde{\overline{X}}_k)$. The probability the best mode is chosen will equal

$P[\text{The } \widetilde{\overline{X}}_j \text{ from the best method is the largest}]$

$$= P[\widetilde{\overline{X}}_{(i)} < \widetilde{\overline{X}}_{(k)} \text{ for } i = 1, 2, \ldots, k - 1] \tag{6.3.3}$$

$$= P\left(\frac{\widetilde{\overline{X}}_{(i)} - \mu_{[i]}}{1/w} \leq \frac{\widetilde{\overline{X}}_{(k)} - \mu_{[k]}}{1/w} + \frac{\mu_{[k]} - \mu_{[i]}}{1/w} \text{ for } i = 1, 2, \ldots, k - 1\right)$$

$$= P[T_{(i)} \leq T_{(k)} + (\mu_{[k]} - \mu_{[i]})w \text{ for } i = 1, 2, \ldots, k - 1]$$

where the notation used is:

$$\widetilde{\overline{X}}_{(i)} \text{ is the } \widetilde{\overline{X}}_j \text{ from the mode with mean } \mu_{[i]} \qquad (6.3.4)$$

$$\mu_{[1]} \leq \mu_{[2]} \leq \cdots \leq \mu_{[k]} \text{ are } \mu_1, \ldots, \mu_k \text{ in increasing order.} \qquad (6.3.5)$$

Since $T_{(1)}, T_{(2)}, \ldots, T_{(k)}$ are independent Student's t random variables with $n_0 - 1$ degrees of freedom each, it can easily be shown that

$$P[\text{The } \widetilde{\overline{X}}_j \text{ from the best method is the largest}]$$

$$= \int_{-\infty}^{\infty} \left(\prod_{i=1}^{k-1} F_{n_0}\left(z + (\mu_{[k]} - \mu_{[i]})w\right)\right) f_{n_0}(z) \, dz \qquad (6.3.6)$$

where $F_{n_0}(\cdot)$ and $f_{n_0}(\cdot)$ are, respectively, the distribution function and probability density function of the Student's t-distribution with $n_0 - 1$ degrees of freedom. It is easily seen that (6.3.6) is smallest, for all $\mu_1, \ldots, \mu_k$ such that

$$\mu_{[k]} - \mu_{[i]} \geq \delta^* \quad (i = 1, 2, \ldots, k - 1), \qquad (6.3.7)$$

when $\mu_{[1]} = \mu_{[2]} = \cdots = \mu_{[k-1]} = \mu_{[k]} - \delta^*$ (which is therefore called the **least-favorable configuration** of the means, or the **LFC**). This follows because the integrand in (6.3.6) is positive, and becomes uniformly smaller when the differences on the left-hand side of (6.3.7) decrease. Thus, we have

$$\inf_{\mu_{[k]} - \mu_{[k-1]} \geq \delta^*} P[\text{The } \widetilde{\overline{X}}_j \text{ from the best method is the largest}]$$

$$= \int_{-\infty}^{\infty} (F_{n_0}(z + \delta^* w))^{k-1} f_{n_0}(z) \, dz, \qquad (6.3.8)$$

which will be equal to P^* if we set

$$w = h_{n_0}(k, P^*)/\delta^* \qquad (6.3.9)$$

where $h_{n_0}(k, P^*)$ is the solution h of the equation

$$\int_{-\infty}^{\infty} (F_{n_0}(z + h))^{k-1} f_{n_0}(z) \, dz = P^* . \qquad (6.3.10)$$

The values of $h_{n_0}(k, P^*)$ are tabled in Dudewicz and Mishra (1988, pp. 588–593). Some typical values for $P^* = .95$ and $n_0 = 15$ are given below (for additional values, see the table with $k = 2$ in Appendix H).

k	3	4	5	10	15
$h_{15}(k, .95)$	2.94	3.17	3.34	3.79	4.02

It can be shown that (6.3.1) is also satisfied, but that proof is beyond the scope of this book.

In summary, we have the following result.

Theorem 6.3.11. *Suppose that* $X_{i1}, X_{i2}, \ldots$ *are independent* $N(\mu_i, \sigma_i^2)$ *random variables. If we*

1. *Set* $n_0(\geq 2)$ *and* $w = h_{n_0}(k, P^*)/\delta^*$, *where* h *is the solution of equation (6.3.10);*

2. *Take a first sample* $X_{i1}, X_{i2}, \ldots, X_{in_0}$, *and calculate its sample mean* $\overline{X}_i(n_0)$ *and sample variance* s_i^2.;

3. *Let* $n_i = \max\{n_0 + 1, \lceil (ws_i)^2 \rceil\}$ *and take* $n_i - n_0$ *more observations* $X_{i,n_0+1}, \ldots, X_{in_i}$;

4. *Calculate the sample mean* $\overline{Y}_i(n_i - n_0) = (X_{i,n_0+1} + \cdots + X_{in_i})/(n_i - n_0)$ *of the second set of observations;*

5. *Calculate*
$$\widetilde{\overline{X}}_i = b_i \overline{X}_i(n_0) + (1 - b_i)\overline{Y}_i(n_i - n_0) \qquad (6.3.12)$$
   where
$$b_i = \frac{n_0}{n_i}\left(1 + \sqrt{1 - \frac{n_i}{n_0}\left(1 - \frac{n_i - n_0}{(ws_i)^2}\right)}\right); \qquad (6.3.13)$$

6. *Repeat the above for each mode; i.e., for* $i = 1, 2, \ldots, k$;

7. *Select as best the mode that produced the largest of* $\widetilde{\overline{X}}_1, \widetilde{\overline{X}}_2, \ldots, \widetilde{\overline{X}}_k$,

then

$$P[\text{The mode selected is either best or within } \delta^* \text{ of the best}] \geq P^*.$$
$$(6.3.14)$$

Above, we have considered selection of the best of k ways of operating a system, when the output of the system is univariate (one number), and "goodness" is associated with a large mean output. However, in many cases the output of a system consists of several numbers; if there are p numbers, we call the output p-variate. For example, if the outputs of interest are profit and sales, then $p = 2$. **Suppose the output from** the system under mode i (which for simplicity we will call **system** i) **has the multivariate normal distribution** of (4.8.3) and (4.8.4) with mean vector $\mu_i = (\mu_{i1}, \ldots, \mu_{ip})'$ and covariance matrix Σ_i, which we will briefly denote by $N_p(\mu_i, \Sigma_i)$, for $i = 1, 2, \ldots, k$. Some possible goals are:

Select the system for which $\max(\mu_{i1}, \mu_{i2}, \ldots, \mu_{ip})$ is largest;
Select the system with the largest value of

$$\sum_{j=1}^{p} a_j \mu_{ij} = a_1 \mu_{i1} + a_2 \mu_{i2} + \cdots + a_p \mu_{ip};$$

Select the system with the largest value of $\sum_{j=1}^{p} \mu_{ij}^2$;
Select the system with the largest value of $\mu_i' \, \Sigma_i^{-1} \mu_i$.

We will give a very general solution of this problem which includes most of the above goals and, in fact, many more. **Suppose that the experimenter specifies a function (called the preference function)** $g(\mu_1, \mu_2, \ldots, \mu_k)$ **with range** $\{1, 2, \ldots, k\}$. If $g(\mu) = j$ where $\mu = (\mu_1, \mu_2, \ldots, \mu_k)$, then among the systems with means $\mu_1, \mu_2, \ldots, \mu_k$, the experimenter would prefer the system with mean vector μ_j.

Let $D_j = \{\mu : \ g(\mu) = j\}$ for $j = 1, 2, \ldots, k$. Thus, D_j consists of all those mean vectors (one from each system) where, faced with the choice, the experimenter would prefer system j. Let

$$d_B(\mu) = \inf_{b} \ \{d(\mu, b) : \ b \notin D_{g(\mu)}\}, \tag{6.3.15}$$

where

$$d(\mu, b) = \sqrt{\sum_{i=1}^{p} \sum_{j=1}^{k} (\mu_{ij} - b_{ij})^2} . \tag{6.3.16}$$

Thus, $d(\mu, b)$ is the distance from μ to b, and $d_B(\mu)$ is the distance from the set of mean vectors μ to the closest set of mean vectors b where some decision other than $d(\mu)$ is preferred by the experimenter.

Suppose that the experimenter wishes to have the probability of correct selection be at least P^* whenever the true μ is at least d^* from the boundary of its preference set; i.e.,

$$P(CS) \geq P^* \text{ whenever } d_B(\mu) \geq d^*, \tag{6.3.17}$$

where $P^* > 1/k$ and $d^* > 0$. It is reasonable for the experimenter to sample the systems, estimate the mean vectors, and select the one that seems best for these estimates. The question is how much to sample each system; an answer is to sample as follows. Take n_0 $(\geq p+1)$ vectors from each system, and compute the usual estimates of the mean vector and covariance matrix (given by (4.13.9)), i.e., calculate the estimates

$$\overline{X}^{(i)} = \frac{1}{n_0} \sum_{r=1}^{n_0} X_r^{(i)}, \quad S_i = \frac{1}{n_0-1} \sum_{r=1}^{n_0} (X_r^{(i)} - \overline{X}^{(i)})(X_r^{(i)} - \overline{X}^{(i)})' \quad (6.3.18)$$

where $X_1^{(i)}, X_2^{(i)}, \ldots, X_{n_0}^{(i)}$ denote the observation vectors from system i, $(i = 1, 2, \ldots, k$. Then set

$$N_i = \max\{n_0, \ [cl_i]^* + 1\}, \quad i = 1, 2, \ldots, k \quad (6.3.19)$$

where

l_i is the largest characteristic root of S_i,
$c > 0$ is fixed,
$[q]^*$ is the greatest integer not greater than q.

Take $N_i - n_0$ additional observations from system i, and compute $\overline{X}_i$ based on all N_i observations; do this for $i = 1, 2, \ldots, k$. Then select the system with index

$$g(\overline{X}_1, \ldots, \overline{X}_k). \quad (6.3.20)$$

Theorem 6.3.21. *The requirement (6.3.17) is guaranteed, i.e., $P(CS) \geq P^*$ for $d_B(\mu) \geq d^*$, as long as c is chosen so that*

$$P[U_1 + \cdots + U_k \leq c(d^*)^2] = P^* \quad (6.3.22)$$

where $U_1, \ldots, U_k$ are independent random variables such that

$$\frac{n_0 - p}{n_0 - 1} \frac{U_i}{p}$$

has the $F(p, n_0 - p)$ distribution.

In practice, an easy way to find c is given by the following lemma, which should be satisfactory as long as n_0 is not very small.

Lemma 6.3.23. *As $n_0 \to \infty$, c tends in the limit to*

$$\chi^2_{kp}(P^*)/(d^*)^2. \tag{6.3.24}$$

Remarks.

1. The multivariate formulation, and its solution, were first given by Dudewicz and Taneja (1981). The form presented above is a modification of that solution given recently by Hyakutake (1988), which preserves the Dudewicz–Taneja formulation, but makes the solution more efficient and easier to use in practice.

2. Simulation for selection of the best is now well established. In one of the earlier studies in this area, Mamrak and Amer (1979) dealt with examples where one wishes to select that one of four computer services which has the largest proportion of response times that are less than 30 seconds for the execution of a FORTRAN synthetic program.

6.4   System Optimization via Statistical Design and Regression

Often, the interest in a simulation is an output measure Y that depends on input characteristics $x_1, x_2, \ldots, x_k$ for some $k \geq 1$. In such a setting, we set numerical values of the **factors** $x_1, x_2, \ldots, x_k$, then run a simulation and observe a value of Y. For example, suppose that in a medical emergency system there are x_1 drivers and x_2 vehicles, and that the time Y from logging of a call for service until the vehicle arrives on the scene is of interest. Then we may wish to study the mean time $E(Y)$ as a function $f(x_1, x_2)$ of x_1 and x_2, perhaps with a view to selecting the least cost (i.e., smallest x_1 and x_2) that can assure some goal such as $E(Y) \leq c$ where c is a service goal in minutes. Since vehicles may be out of service due to breakdown or maintenance and drivers may be ill, it is not necessarily the case that $x_1 = x_2$ at the optimum. Generally, the mean output $E(Y)$ is a smooth function of the experimental conditions $x_1, x_2, \ldots$ and can, therefore, be accurately represented by a polynomial equation of sufficiently high order; in the case of $k = 4$, by a polynomial of type

$$\begin{aligned} E(Y) = {}& \beta_0 + \beta_1 x_1 + \beta_2 x_2 + \beta_3 x_3 + \beta_4 x_4 \\ & + \beta_{12} x_1 x_2 + \beta_{13} x_1 x_3 + \beta_{14} x_1 x_4 + \beta_{23} x_2 x_3 + \beta_{24} x_2 x_4 + \beta_{34} x_3 x_4 \end{aligned}$$

$$+ \beta_{11}x_1^2 + \beta_{22}x_2^2 + \beta_{33}x_3^2 + \beta_{44}x_4^2 \qquad (6.4.1)$$
$$+ \beta_{123}x_1x_2x_3 + \beta_{124}x_1x_2x_4 + \beta_{134}x_1x_3x_4 + \beta_{234}x_2x_3x_4$$
$$+ \beta_{1234}x_1x_2x_3x_4$$
$$+ \cdots .$$

By the **experimental design**, we mean the sets of values $(x_1, x_2, \ldots, x_k)$ at which the simulation is run. The experimental design used determines which of the various β's are able to be estimated. Often one assumes that terms higher than the second order are negligible (i.e., $0 = \beta_{123} = \beta_{124} = \beta_{134} = \beta_{234} = \beta_{1234} = \cdots$), uses a **screening design** to allow reduction of k—by discarding factors that have little effect on $E(Y)$—and then uses a **central composite design** on the remaining factors. For details, see Dudewicz and Karian (1985, pp. 205–240). We illustrate this below for the simplest case, $k = 1$ factor. Further illustration, in complete detail, is then given for the case $k = 2$, which is used in Chapter 8 in an extensive case study.

6.4.1   Example: A Time-Shared Computer System

Consider a computer system with one CPU and n terminals that it services. Suppose that each terminal is idle (corresponding to some transaction time, such as thinking by a student, teller interaction with a customer in a bank, or the like) for a length of time that has an exponential distribution with a mean of 25 seconds, after which the terminal asks for CPU service that also follows an exponential distribution, but with a mean of 0.8 second. Assume that the CPU has a service queue for jobs, and serves the first job in line for up to 0.1 second (plus overhead of 0.015 second to maintain the queue, etc.). If that job is then finished, it returns to its terminal; otherwise, it goes to the end of the service queue, with its service needs diminished by 0.1 second. Such a system has been studied in various aspects by Adiri and Avi-Itzhak (1969) and by Law and Kelton (1982). A typical question in practice is: If Y denotes the response time (the time from a job's being issued by the terminal to its being completed by the CPU) and x_1 is the number of terminals, how large can x_1 be (**how many terminals can be connected to the CPU**) **if we must keep** $E(Y) \leq 6$ **seconds** to provide true interactive computing?

A GPSS program for this time-shared computer system problem is given in Section 7.12. That program, when run at $x_1 = 25$, produces 3.247 seconds as the estimate of $E(Y$ at $x_1 = 25)$. With this information, we are not even able to fit the model $E(Y) = \beta_0 + \beta_1 x_1$, since we need two points to determine a line and we have only one, the data at $x_1 = 25$. So suppose that we have run the simulation at $x_1 = 15,\ 25,\ 35,\ 45$, with the following results:

No. of Terminals x_1	15	25	35	45
Observed $E(Y)$ at x_1	1.637	3.125	7.613	16.518

From the theory of regression—for example, see Dudewicz and Mishra (1988, p. 696)—we know that **the usual (least squares) estimates are given by**

$$\hat{\underset{\sim}{\beta}} = (\underset{\sim}{X}\,\underset{\sim}{X}')^{-1}\underset{\sim}{X}\,\underset{\sim}{Y} \tag{6.4.2}$$

where

$$\hat{\underset{\sim}{\beta}} = \begin{pmatrix} \hat{\beta}_0 \\ \hat{\beta}_1 \\ \hat{\beta}_2 \end{pmatrix}, \quad \underset{\sim}{X} = \begin{pmatrix} 1 & 1 & 1 & 1 \\ 15 & 25 & 35 & 45 \\ 225 & 625 & 1225 & 2025 \end{pmatrix}, \quad \underset{\sim}{Y} = \begin{pmatrix} 1.637 \\ 3.125 \\ 7.613 \\ 16.518 \end{pmatrix}. \tag{6.4.3}$$

Using direct calculation (or, preferably, high-quality statistical software), we then find that

$$\hat{\underset{\sim}{\beta}} = \begin{pmatrix} \hat{\beta}_0 \\ \hat{\beta}_1 \\ \hat{\beta}_2 \end{pmatrix} = \begin{pmatrix} 4 & 120 & 4100 \\ 120 & 4100 & 153{,}000 \\ 4100 & 153{,}000 & 6{,}042{,}500 \end{pmatrix}^{-1} \begin{pmatrix} 28.983 \\ 1114.69 \\ 45{,}152.6 \end{pmatrix}$$

$$= \frac{1}{8 \times 10^4} \begin{pmatrix} 1{,}365{,}250 & -97{,}800 & 1550 \\ -97{,}800 & 7360 & -120 \\ 1550 & -120 & 2 \end{pmatrix} \begin{pmatrix} 28.983 \\ 1114.69 \\ 45{,}152.6 \end{pmatrix}$$

$$= \begin{pmatrix} 6.72955 \\ -.60865 \\ .0183177 \end{pmatrix}.$$

Thus, we have obtained the model

$$E(Y) = 6.72955 - 0.60865x_1 + 0.0183177x_1^2. \tag{6.4.4}$$

Using this model to predict, and then comparing the prediction with the data we obtained, we find:

No. of Terminals x_1	15	25	35	45
Observed Mean	1.637	3.125	7.613	16.518
Prediction (6.4.4)	1.721	2.962	7.866	16.433

Since the predictions are "close" to the data, we would generally regard this as a good model over the range of x_1s where it was developed. To answer the question of how many terminals we may connect and still keep $E(Y) \leq 6$, we solve

$$6.72955 - 0.60865x_1 + 0.0183177x_1^2 = 6$$

to find

$$x_1 = \frac{0.60865 \pm 0.3170}{0.036635} = 31.98.$$

Thus, **we conclude that up to about 31 terminals may be in operation without degrading the true interactive nature desired.**

Note that the solution above does not consider such aspects as:

1. How long one must observe the system before starting to take data (to assure that transient behavior, assumed not to be of interest, is not observed)

2. The desirability and effects of taking observations on multiple jobs at each x_1 in a single run (which is desirable, since once long-run behavior is being observed, it is wasteful to take only one observation on it—the simulation that produced the data in our example was run until 5000 CPU requests were completed)

3. How to test for what order the model really is (above, we assumed it to be quadratic and saw that is a good fit to the data; we did not consider whether a lower order model, in this case a line, might be as good a fit)

4. How to take into account the variability in the fitted model (6.4.4), using confidence intervals.

Many of these aspects will be treated in the $k = 2$ factor model case study in Chapter 8.

It is important to note that **optimization using statistical design and regression is much more efficient than naive optimization.** In the above example, we needed 4 units of simulation time, one at each of $x_1 = 15, 25, 35, 45$. A naive approach might be to simulate at each of $x_1 = 15, 16, 17, 18, \ldots, 44, 45$ and select the smallest for which $E(Y)$ is estimated to not exceed 6, but that would require $45 - 14 = 31$ units of simulation time, or 675% more than our study did. In addition, each of the estimates in the naive approach is variable and the variability is not smoothed by using the other data points. Thus, one might find time 5.837 at $x_1 = 27$, 6.001 at $x_1 = 28$, and yet 5.999 at $x_1 = 29$. One then has a problem of what action to take: one knows from the nature of the system that $E(Y)$ must increase as x_1 increases, but has not done the smoothing needed to make the fitted model have this increasing nature. In our opinion, the naive approach should never be used in simulation studies.

6.4.2   Central Composite Design for $k = 2$ Variables

As we have seen, the output Y of a simulation often has a mean $E(Y)$ that is a function of experimental conditions $x_1, x_2, \ldots$, as at (6.4.1). The "experimental design" used means the sets of values $(x_1, x_2, \ldots)$ at which simulations are run. In deciding which sets of values to use, we need to balance two needs:

1. With more sets of values, we will be able to fit more of the coefficients in the true model (6.4.1).

2. With fewer sets of values, we will more easily be able to afford the computer time to run the simulation.

The **Central Composite Design** is one that allows us (in the case of $k = 2$ variables) to fit the full quadratic model

$$E(Y) = \beta_0 + \beta_1 x_1 + \beta_2 x_2 + \beta_{11} x_1^2 + \beta_{22} x_2^2 + \beta_{12} x_1 x_2. \qquad (6.4.5)$$

In the case $k = 2$, this design needs 9 simulation runs; namely, those with "coded" values of (x_1, x_2) as specified in Table 6.4–1. In general, this design with k factors needs $2^k + 2k + 1$ points; hence, $2^2 + 2 \times 2 + 1 = 9$ when $k = 2$. The points of the design are shown on (x_1, x_2) axes in Figure 6.4–2, with the "factorial points" marked with xs, the "star points" with *s, and the "center point(s)" with a •. The "coded" levels represent a high (1) and a low (-1) level for each variable, their average being the center (0) level of that variable. The factor α specifies how far from the center level we go on the star points; e.g., with $\alpha = 2$, we go twice as far from the center as we did for the -1 and 1 levels. The coding is specific to each variable (so that the high and low levels of x_1 and x_2 are not necessarily the same except when coded).

For example, suppose the (uncoded) levels we desire for x_1 are 3 for low (-1) and 13 for high (1). Then the center level for x_1 will be $(3 + 13)/2 = 8$. If we choose $\alpha = 1.4$, then the star points on x_1 will be 1.4 as far from the center as the high or low points are from the center. Since $8 - 3 = 5 = 13 - 8$ and $(1.4)(5) = 7$, the star points are at $8 - 7 = 1(-\alpha)$ and at $8 + 7 = 15(\alpha)$.

Table 6.4–1. Simulation runs needed for a central composite design with $k = 2$ factors x_1, x_2 (coded levels).

Name of Runs	$(x_1, \; x_2)$
Factorial Points	$(-1, \; -1)$
	$(-1, \;\; 1)$
	$(\;\; 1, \; -1)$
	$(\;\; 1, \;\; 1)$
Star Points	$(-\alpha, \;\; 0)$
	$(\;\; \alpha, \;\; 0)$
	$(\;\; 0, \; -\alpha)$
	$(\;\; 0, \;\; \alpha)$
Center Point(s)	$(\;\; 0, \;\; 0)$

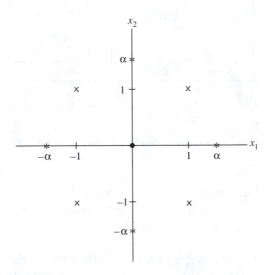

Figure 6.4–2. A central composite design with $k = 2$ factors, showing 4 factorial points (x), 4 star points (∗), and one center point (•), coded levels.

Continuing this same example, if the (uncoded) levels we desire for x_2 are 5 for low (-1) and 20 for high (1), then the center level for x_2 will be $(5+20)/2 = 12.5$. With the same $\alpha = 1.4$ (traditionally, the CCD design uses the same star point multiplier on all factors), since $12.5-5 = 7.5 = 20-12.5$, the star points will be at $(1.4)(7.5) = 10.5$ distance. Thus, the star points are at $12.5 - 10.5 = 2$ and at $12.5 + 10.5 = 23$.

The points in the design for this example are shown in Table 6.4–3.

Table 6.4–3. Simulation runs (uncoded levels) needed for the example central composite design with $\alpha = 1.4$.

Name of Runs	$(x_1,\ x_2)$
Factorial Points	(3, 5)
	(3, 20)
	(13, 5)
	(13, 20)
Star Points	(1, 12.5)
	(15, 12.5)
	(8, 2)
	(8, 23)
Center Point(s)	(8, 12.5)

Some modifications often made in this basic central composite design are the following:

1. If some of the points would be in a region where the performance is known to be undesirable, those points are altered to be in regions where performance does not have that property.

2. If the variable has integer values (or other restrictions on values), such as with variables like the number of workers, then the design is altered to meet that restriction.

3. Sometimes the factor α is allowed to be different for the various factors, in order to "cover" the design space well, so that experiments are run in a configuration that more nearly covers points thought to be of great interest.

In making the types of changes described above, one must be careful lest the design become one that is simply ad hoc (without any statistical theory behind it to support its use, which might in the worst cases not even allow estimation of all parameters in the model). For this reason, we urge minimal use of such modifications.

6.5 Statistical Simulation and the "Bootstrap" Method

Simulation is a powerful method not only for studying and optimizing real-world systems (as studied extensively in the previous chapters of this text), but also **for solving complex mathematical and statistical problems that defy analytical solution.** In recent years, what is called the "bootstrap method" has gained popularity in both statistical and nonstatistical circles for the solution of such problems. In this section, we consider this method, some of its drawbacks, and some ways to avoid them. In Section 6.6, we summarize a recent improvement, called the Generalized Bootstrap, which can avoid the drawbacks of the naive bootstrap.

Suppose that we are interested in $\theta = \theta(F)$, some function of the distribution function $F(\cdot)$, but $F(\cdot)$ is unknown. (If it were known, we could simply calculate $\theta(F)$.) **However, we have a random sample $X_1, X_2, \ldots, X_n$ from $F(\cdot)$. How can we use this sample to estimate $\theta(F)$?**

The **Bootstrap Method (BM)** approaches the above problem as follows. Suppose that one would typically estimate θ by $\hat{\theta} = \hat{\theta}(X_1, X_2, \ldots, X_n)$. Then, instead, proceed as follows:

Step BM–1. Take a random sample of size n from $\{X_1, X_2, \ldots, X_n\}$, (with replacement) and calculate the estimate $\hat{\theta}_1$ using this first sample.

Step BM–2. Take a second random sample of size n with replacement from $\{X_1, X_2, \ldots, X_n\}$ and calculate the estimate $\hat{\theta}_2$ using this second sample.

$\vdots$

Step BM–N. Take an N-th random sample of size n with replacement from $\{X_1, X_2, \ldots, X_n\}$ and calculate the estimate $\hat{\theta}_N$ using this N-th sample.

Next, calculate

$$\bar{\theta} = \left(\frac{1}{N}\right) \sum_{i=1}^{N} \hat{\theta}_i \tag{6.5.1}$$

and

$$\hat{\sigma}^2 = \left(\frac{1}{N-1}\right) \sum_{i=1}^{N} (\hat{\theta}_i - \bar{\theta})^2. \tag{6.5.2}$$

Equations (6.5.1) and (6.5.2) give the sample mean and sample variance, respectively, of the bootstrap sample estimators found in steps BM–1 through BM–N. Then

$$\mathrm{Var}(\bar{\theta}) = \mathrm{Var}(\hat{\theta}_i)/N \approx \hat{\sigma}^2/N, \tag{6.5.3}$$

and **an approximate** $100(1-\alpha)\%$ **confidence interval for** θ **is**

$$\hat{\theta} \pm \Phi^{-1}(1 - \alpha/2)\hat{\sigma}. \tag{6.5.4}$$

Thus, the bootstrap method uses the original estimator based on the original sample, but resamples the sample to arrive at an estimate of the variability of the estimate.

In **statistical simulation methods**, we might use a procedure similar to the one above. However, there are problems with sampling from $\{X_1, X_2, \ldots, X_n\}$ each time. For example, we usually know that other values between these are possible, and such sampling will never yield a value larger than $\max(X_1, X_2, \ldots, X_n)$. Since we know that larger values are possible, we can fit a density to $X_1, X_2, \ldots, X_n$, such as the empiric p.d.f. discussed in Section 4.11, and then sample from the fitted p.d.f. at each step. For further discussion, see Section 6.6.

6.6   The Generalized Bootstrap

In Section 6.5, we discussed the following **problem:** We are interested in

$$\theta = \theta(F), \tag{6.6.1}$$

some function of the distribution $F(\cdot)$, but $F(\cdot)$ is unknown. However, we have a random sample $X_1, X_2, \ldots, X_n$ from $F(\cdot)$. How can we estimate $\theta(F)$?

The Bootstrap Method, specified by Steps BM–1 through BM–N followed by (6.5.1) through (6.5.4) of Section 6.5, caught the popular statistical imagination of theoreticians, practitioners, and laymen (such as readers of *Scientific American*). This may be due, as some have suggested, to its connection to the advent of inexpensive computing power. Or, it may relate to the name "Bootstrap," which could be taken to imply

- A good deal, or

- Something for nothing, or

- Powerful statistical results from one's data sets *without* a need for any complicated theory.

It is left for others such as historians to sort out the true reasons. Nevertheless serious questions about the wisdom, and even the correctness, of things done using the BM in the decade of the 1980s have played a role in the organization of conferences devoted to the Bootstrap Method in the 1990s. Out of this effort has come a generalization that avoids problems with the BM (such as that mentioned in the end of Section 6.5: we know that larger values than $\max(X_1, X_2, \ldots, X_n)$ are possible, but the BM never yields a larger value). **The Generalized Bootstrap (GB)**, developed by Dudewicz (1992) and given below, is proposed as

- Still a good deal

- Perhaps something for nothing

- Bringing powerful statistical results from distribution fitting and testing to bear on the problem to be solved.

The Generalized Bootstrap (GB) approaches the problem as follows. Suppose that one would typically estimate θ by $\hat{\theta} = \hat{\theta}(X_1, X_2, \ldots, X_n)$. Then, instead, proceed as follows:

Step GB–1. Estimate F by $\hat{F}$. (The estimating $\hat{F}$ should always be one that has the properties known for the true F. For example, $\hat{F}$ should be continuous if it is known that F is continuous. Fitting with the Generalized Lambda Distribution or with the empiric p.d.f., both discussed in Chapter 4, will often be appropriate.)

Step GB–2. Independently generate N random samples of size n from $\hat{F}$.

From
$$Y_1, Y_2, \ldots, Y_n \text{ estimate } \theta(F), \text{ calling the estimate } \hat{\theta}_1.$$

From
$$Y_{n+1}, Y_{n+2}, \ldots, Y_{2n} \text{ estimate } \theta(F), \text{ calling the estimate } \hat{\theta}_2.$$
$\vdots$

From
$$Y_{(N-1)n+1}, \ldots, Y_{Nn} \text{ estimate } \theta(F), \text{ calling the estimate } \hat{\theta}_N.$$

Step GB–3. Use the sample $\hat{\theta}_1, \hat{\theta}_2, \ldots, \hat{\theta}_N$ to estimate $\hat{\theta}$. (For example, one may calculate (6.5.1) and (6.5.2), then take (6.5.4) as an approximate confidence interval for θ. Or, the upper and lower $\alpha/2$ percentiles might be used as the confidence interval.)

The following theorem, discussed in Dudewicz (1992), implies that by proper use of the GB one will obtain better results than with the BM.

Theorem 6.6.2. *The Bootstrap Method is a special case of the Generalized Bootstrap (namely, the BM takes the empiric d.f. in Step GB–1).*

Sun and Müller-Schwarze (1995) compared the performances of the Bootstrap Method of Section 6.5 and the Generalized Bootstrap of this section (as well as another method—the jackknife). Focusing on a particular applied problem and data set, they provided the necessary computer code in the C language and drew general implications. They reported two important conclusions.

1. The Generalized Bootstrap is more consistent in parameter estimation than is the Bootstrap Method.

2. The number of bootstrap samples N should be at least 500.

Asymptotic properties of GB have been shown by Lin (1997), who notes (p. 302) that "... some advantages of using [GB] ... over the ordinary bootstrap [BM] have been found ... ." For further discussion and examples, see Dudewicz (1992).

6.7 Other Statistical Design and Analysis Aspects of Simulation

In Sections 6.1 through 6.6 we have investigated, in some detail, most of the important aspects of statistical design and analysis of simulations. In this section, we briefly consider several additional issues that are of import in a significant number of simulations. Whenever appropriate, we give sources where readers can obtain a fuller discussion of these topics.

6.7.1 Estimation of a Percentile Point

If data produced by a simulation $X_1, X_2, \ldots, X_n$ have an underlying common distribution function $F(x) = P(X_i \leq x)$, a common goal is to table the values x_p such that

$$F(x_p) = p, \tag{6.7.1}$$

for a number of values p (such as $p = .90, .95, .975, .99$). x_p is called the pth **percentile** point (or p^{th} **quantile**) of the distribution function $F(x)$.

An efficient nonparametric approach to the computation of x_p uses the order statistics, $Y_1 \leq Y_2 \leq \ldots \leq Y_n$, (which are $X_1, \ldots, X_n$ in non-decreasing numerical order) **to estimate x_p by $Y_{[np]}$** where $[a]$ is the largest integer which does not exceed a (e.g., $[5.4] = 5$, and $[6] = 6$). For example, suppose that $n = 10,000$ samples have been obtained and that $p = .05$; one then estimates $x_{.05} = Y_{500}$.

In order to tabulate x_p properly we need to know, **how many decimals of the estimate of x_p should be used** (i.e., how many of the decimals can we "trust")? A 95% confidence interval for the true value of x_p is given by (Y_r, Y_s) where

$$r = \left[-1.96(np(1-p))^{0.5} + np + 0.5 \right],$$
$$s = \left[1.96(np(1-p))^{0.5} + np + 2.5 \right].$$

If, in addition to our previous assumptions ($n = 10,000$ and $p = .05$), we find $Y_{500} = -.773908$, we then obtain the 95% confidence interval

$$(Y_{457}, Y_{544}) = (-.786353, -.761449).$$

Since this interval has half-width .01, we should report our estimate as $x_p = -.77$ (i.e., our results are good to two decimal places, and when so stated are accurate within $\pm.01$, the usual criterion for tabling two decimal places, with 95% confidence). If we wanted to table to more decimal places, we would know our accuracy was not sufficient. In that case, n could be increased until we had the needed accuracy.

The above method replaces ad hoc procedures which were in use until this method was developed (see Dudewicz and van der Meulen (1984), and Dudewicz and Mishra (1988), pp. 660–661).

6.7.2   Estimation of Var($\bar{X}$)

Suppose that a simulation has produced the values $X_1, X_2, \ldots, X_n$, that have the same distribution but may not be independent. For purposes such as the construction of confidence intervals we may need to know **how to estimate Var($\bar{X}$)**. Since the variables may not be independent, one needs to bear in mind that Var($\bar{X}$) will not be simply Var(X_1)/n, but a value that accounts for the correlation structure of the X_is:

$$
\mathrm{Var}(\bar{X}) = \frac{1}{n^2}\mathrm{Var}(X_1 + X_2 + \cdots + X_n)
$$

$$
= \frac{1}{n^2}\left\{\sum_{i=1}^{n}\mathrm{Var}(X_i) + \sum_{i=1}^{n}\sum_{j=1}^{n}\mathrm{Cov}(X_i, X_j)\right\} \qquad (6.7.2)
$$

(see, e.g., Section 5.3 of Dudewicz and Mishra (1988)). One widely used approach is to "group" into each X_i enough of the simulation that the covariances are near zero (this is the **batch means** approach). Another approach is to estimate the covariances which arise in (6.7.2) from the X_is. For additional information, see Schriber and Andrews (1984) and its references. For the continuous-time process problem, see the results in Gafarian and Danesh-Ashtiani (1981).

6.7.3   Variance Reduction Techniques

Suppose one wants to estimate $\mu_1 - \mu_2$ and observes $X_1, X_2, \ldots, X_n$ and $Y_1, Y_2, \ldots, Y_m$, which are independent variables where the X_is have mean μ_1 and variance σ_1^2, while the Y_js have mean μ_2 and variance σ_2^2. One often would estimate $\mu_1 - \mu_2$ with

$$
\bar{X} - \bar{Y}, \qquad (6.7.3)
$$

which is an unbiased estimator of $\mu_1 - \mu_2$, and has variance

$$
\mathrm{Var}(\bar{X} - \bar{Y}) = \frac{\sigma_1^2}{n} + \frac{\sigma_2^2}{m}. \qquad (6.7.4)
$$

For example, one might proceed in this way in order to estimate the difference in system cost of two operating procedures.

   The question arises: if we did not have independence between the X_is and Y_js, could we have a smaller variance than that obtained by using (6.7.4) for our estimator? Similar to the development of (6.7.2), one has (when the

X_is are not independent of the Y_js, but $X_1, X_2, \ldots, X_n$ are independent of each other and $Y_1, Y_2, \ldots, Y_m$ are independent of each other)

$$\mathrm{Var}(\bar{X} - \bar{Y}) = \frac{\sigma_1^2}{n} + \frac{\sigma_2^2}{m} - 2\mathrm{Cov}(\bar{X}, \bar{Y}). \qquad (6.7.5)$$

Therefore, if we could induce **positive correlation** between X_is and Y_js, we could attain **variance reduction**.

Often one attempts to induce the needed positive correlation by using the same random numbers when one simulates the Y_js as when one simulates the X_is. For this to work, the random numbers must be used "in the same way" in the two simulations. This is not a trivial task, since the procedures yielding the two sets of observations often use the random numbers in different ways (and the random number streams get "out of synchronization"). In the worst case, one may actually induce a negative correlation, in which case the $\mathrm{Var}(\bar{X} - \bar{Y})$ will be larger than if one used independent random number streams. For a good discussion of these points, and some more advanced methods (and excellent references), see Wilson (1981). For a clever and successful use of variance reduction in an applied problem, see Mazumdar, Coit, and Shih (1998).

Problems for Chapter 6

6.1 (Section 6.1) In Section 6.1.2, we studied how long to simulate a system to estimate its mean within 60 with 90% confidence. Suppose that the data there is replaced by taking the square root of each number. For this new data set $Y_1, \ldots, Y_{10}$ (we are still using $n_0 = 10$), how many additional days of simulation are needed for this goal?

6.2 (Section 6.2) Suppose that the data in Section 6.1.2 comes from "mode 1" of operation of a system, and that in Problem 6.1 (after the square root is taken) comes from "mode 2" of operation of that system. Taking $n_0 = 5$, show how to find a 95% confidence interval for the difference in the means of the output of the two systems.

6.3 (Section 6.3) Consider the two modes discussed in Problem 6.2. Suppose that our goal is to select the better of the two modes, and we wish to be at least 95% sure that the mode selected is either the best or within 100 units of the best. Take $n_0 = 10$ and use the data of Problem 6.2 to show how to make the selection.

6.4 (Section 6.4) In constructing the design in Table 6.4–3, the factor $\alpha = 1.4$ was used. Construct a similar design (with the same high/low levels for each factor), but which uses $\alpha = 2$.

6.5 (Section 6.4) Proceed as in Problem 6.4, but let $\alpha = 1$. This is called a face-centered design; on a diagram like Figure 6.4–2, explain how the name arises.

6.6 (Section 6.4) Instead of the design in Table 6.4–3, suppose that a person uses the design with points

$(x_1, \ x_2)$
(3, 5)
(3, 20)
(13, 5)
(13, 20)
(1, 15)
(15, 15)
(8, 1)
(8, 20)
(8, 15)

and claims that this was done due to the facts that (a) their variable x_2 cannot take noninteger values, (b) it is known that values of x_2 above 20 yield poor performance, and (c) the value $x_2 = 1$ is the smallest one possible and therefore should be included in the simulation runs. For each of the 9 points, discuss which of these (a), (b), (c) might justify changing it (if it was changed). On a diagram like Figure 6.4–2 show the original (uncoded) design, and the new one just given.

6.7 (Section 6.5) Take a random sample of size 10 from a uniform density on $(0, \theta)$, with $\theta = 8$. Then generate 50 bootstrap samples of size 10 each to estimate the variance of $\hat{\theta}$, the maximum likelihood estimator (MLE) of θ. Construct a 95% confidence interval for θ. (Be sure to present your computer code, and the samples, in detail. Compare to the known variance.)

6.8 (Sections 6.5 and 6.6) Proceed as in Problem 6.7, but instead of the bootstrap samples, fit the empiric p.d.f. to the data, and then proceed as in Problem 6.7. (Use $a = 0$ and $b = 10$ in fitting the empiric p.d.f.) Also plot the empiric p.d.f. and the true p.d.f. on a graph.

6.9 (Sections 6.5 and 6.6) Compare the results of Problems 6.7 and 6.8, and make inferences about the relative merits of naive (bootstrap) methods versus smoothing (simulation or generalized bootstrap) methods.

6.10 (Sections 6.5 and 6.6) Proceed as in Problem 6.7, but instead of using bootstrap samples, fit a GLD p.d.f. to the data, and then proceed as in Problem 6.7. Also plot the GLD p.d.f. and the true p.d.f. on a graph. Compare the results with those of the methods in Problems 6.7 and 6.8.

Chapter 7

Advanced GPSS Features

It is possible to develop GPSS models of considerable complexity with only the concepts developed in Chapters 2 and 5. The advanced GPSS features covered in this chapter, however, provide additional flexibility so that more sophisticated models can be designed. So far, the use of system-supplied parameters has been limited to the built-in random number generators and transaction parameters. Access to many internal GPSS attributes, such as the current simulation time, and to attributes of model entities, such as the length of a queue, can be used to great advantage in a variety of modeling situations. These attributes and their use in model development are major topics of this chapter. The chapter also covers a number of additional GPSS blocks that are needed to develop complex models.

GPSS/PC models of the simulations of Sections 7.1, 7.3, 7.5, 7.10, 7.12, and 7.16 are included on the disk accompanying this book. Due to the limited memory addressing capability of the Educational Version of GPSS/PC, the large (in terms of memory requirement) models developed in Sections 7.1, 7.12, and 7.16 will not execute in the environment provided by this version. For this reason, in addition to the complete models, the disk contains "scaled down" renditions of these models.

7.1 Standard Numerical Attributes

During the execution of a program, the GPSS processor must manage a variety of data structures associated with the model execution. For example, data concerning the busy or idle status of a facility at a given instant during the simulation is essential for the proper execution of the program. Other data items, such as block counts and average queue residence times, are also continually updated so that statistics relevant to the model can be produced following execution. Model attributes in GPSS are called **Standard Numerical Attributes** (SNAs). SNAs include system-defined numerical entities, as well as user-defined items such as functions, variables, and constants used as operands in various blocks.

The use and utility of certain SNAs (random number generators, function values, variable values, transaction parameters, and constants) have been illustrated in previous chapters. Before additional SNAs are discussed, the limited GPSS entity-naming convention that was adopted in Chapter 2 must be modified. Internally, GPSS references model entities by numeric values rather than by user-defined names. Thus, if a program contains three function definitions for functions named A, B, and C, GPSS renames them as functions 1, 2, and 3 in the order in which their definitions appear in the program. This allows the modeler to refer to function B by either its user-supplied name (FN$B) or its GPSS-supplied numeric designation (FN2), assuming that B was the second function defined in the program. The general convention for referencing model entities is

<family name> $ <symbolic name of family member>

if the user-defined name is to be used and

<family name> <integer specifying family member>

if the numeric identification is used. Specific names or integers must be substituted for the portions enclosed in angle brackets. From previous chapters, we know that FN, V, and RN are the family names for function entities, variable entities, and random number generators, respectively.

Throughout the remainder of this text, "**<family name>j" will refer to the *j*-th member of the entity of the designated family**. Thus, FN5 and FNj refer to the fifth and *j*-th functions of the GPSS model, respectively. Analogously, "**<family name>$sn" will refer to the member of the designated family that has been given the user-defined symbolic name "sn"**. Thus, V$CHARGE refers to the variable with the user-defined name CHARGE. There are, of course, instances where only one of these modes of reference is reasonable, as in the case of the random number generating SNAs RN1,..., RN8. Since SNAs can be used in a variety of forms, GPSS specifies the forms that are allowed within each GPSS block. For example, Appendix B lists

k, SNAj, SNA$sn

as possibilities for operand A of a GENERATE block. This means that any constant (k), SNA with numeric reference (SNAj), or SNA with symbolic reference (SNA$sn) may be used as the A operand in a GENERATE block.

As an example of the use of numeric references, consider the simulation of the batch computing system in Section 5.11. The program, given in Figure 5.11–2, has four similar segments, each dealing with computer jobs with a given priority. The first segment of this program is reproduced here for convenience as Figure 7.1–1. If PRIOR1Q, PRIOR2Q, PRIOR3Q, PRIOR4Q, and OVERALLQ are referenced by 1, 2, 3, 4, and 5, respectively, the program can be significantly shortened. Transaction parameter 4 can then be used to hold 1, 2, 3, or 4, the queue number to which the transaction should be directed, and the QUEUE blocks of the first segment can be changed to

$$\text{QUEUE} \quad 5$$
$$\text{QUEUE} \quad \text{P4.}$$

Since the same thing can be done in all segments, all but the GENERATE blocks of the four segments can be made identical and the entire program of Figure 5.11–2 can now be rewritten in a more compact form (Figure 7.1–2). Since SNAs can be used in more significant ways than in program compaction, we now describe many of the SNAs involving specific entities.

35	1	GENERATE	24300,FN$EXPON,,,1	Priority 1 jobs.
36	2	ASSIGN	1,FN$JOBTPE	P1 ⟵ job type.
37	3	ASSIGN	2,FN$MST	P2 ⟵ mean CPU time.
38	4	ASSIGN	3,FN$SDST	P3 ⟵ st. dev. CPU time.
39	5	QUEUE	OVERALLQ	Place job in overall queue.
40	6	QUEUE	PRIOR1Q	Place job in priority 1 queue.
41	7	SEIZE	CPU	Capture CPU.
42	8	DEPART	PRIOR1Q	
43	9	DEPART	OVERALLQ	
44	10	ADVANCE	V$STME	Obtain CPU service.
45	11	RELEASE	CPU	
46	12	TERMINATE		

Figure 7.1–1. A segment of the simulation of a batch computing system.

7.1.1 Facilities

There are four SNAs, with family names F, FC, FR, and FT describing facility entities:

Fj, F$sn: The facility status SNA; it has value 0 if the facility is idle and 1 if it is busy.

FCj, FC$sn: The number of times that a transaction has entered the facility.

```
 1                    SIMULATE
 2
 3        SNORM      FUNCTION      RN2,C23    Standard normal distribution
 4        0,-4/.00138,-3/.00621,-2.5/.02275,-2/.06681,-1.5/.11507,-1.2 /.15866,-1
 5        .21186,-.8/.27425,-.6/.34458,-.4/.42074,-.2/.5,0/.57926,.2 /.65542,.4/.72575,.6
 6        .78814,.8/.84134,1/.88493,1.2/.93319,1.5/.97725,2 /.99379,2.5/.99862,3/1,4
 7
 8        EXPON      FUNCTION      RN3,C24    Exp. dist. with mean 1
 9        0,0/.1,.104/.2,.222/.3,.355/.4,.509/.5,.69/.6,.915/ .7,1.2/.75,1.38
10        .8,1.6/.84,1.83/.88,2.12/.9,2.3/.92,2.52/.94,2.81/ .95,2.99/.96,3.2
11        .97,3.5/.98,3.9/.99,4.6/.995,5.3/.998,6.2/.999,7/ .9998,8
12
13        JOBTPE     FUNCTION      RN4,D3     Job types
14        .15,1/.80,2/1,3
15        MST        FUNCTION      P1,D3      Mean of CPU time. P1 is job type
16        1,800/2,2675/3,11950
17        SDST       FUNCTION      P1,D3      St. dev. of CPU time, P1 is job type
18        1,125/2,270/3,1623
19
20   STME     FVARIABLE      P3*FN$SNORM+P2     CPU time: P2=mean, P3=st. dev.
21
22    1       GENERATE       24300,FN$EXPON,,,1   Priority 1 jobs
23    2       ASSIGN         4,1                  P4 is the job priority
24    3       TRANSFER       ,NEXT
25    4       GENERATE       15600,FN$EXPON,,,2   Priority 2 jobs
26    5       ASSIGN         4,2                  P4 is the job priority
27    6       TRANSFER       ,NEXT
28    7       GENERATE       10200,FN$EXPON,,,3   Priority 3 jobs
29    8       ASSIGN         4,3                  P4 is the job priority
30    9       TRANSFER       ,NEXT
31   10       GENERATE       32600,FN$EXPON,,,4   Priority 4 jobs
32   11       ASSIGN         4,4                  P4 is the job priority
33   12       TRANSFER       ,NEXT
34
35   13 NEXT ASSIGN         1,FN$JOBTPE          P1 ⟵ job type.
36   14       ASSIGN         2,FN$MST             P2 ⟵ mean CPU time.
37   15       ASSIGN         3,FN$SDST            P3 ⟵ st. dev. of CPU time.
38   16       QUEUE          5                    Place job in overall queue.
39   17       QUEUE          P4                   Place job in queue stipulated
40   18       SEIZE          CPU                  by priority.
41   19       DEPART         P4
42   20       DEPART         5
43   21       ADVANCE        V$STME               Obtain CPU service.
44   22       RELEASE        CPU
45   23       TERMINATE
46
47   24       GENERATE       10800000
48   25       TERMINATE      1
49            START          1
```

Figure 7.1–2. A modified program for the batch computing system.

FRj, FR$sn: The utilization of the facility given in parts per 1000. For instance, a value of 345 indicates 0.345, or 34.5% utilization.

FTj, FT$sn: The average length of time the facility is used. This SNA is truncated to an integer value.

The truncation in the case of FT and the scaling to parts per thousand of FR make all facility-related SNAs integer-valued.

7.1.2 Queues

There are seven SNAs describing various aspects of queue entities:

Qj, Q$sn: Number of transactions in the queue.

QAj, QA$sn: Time-weighted average of the number of transactions in the queue since the start of execution. This SNA is truncated to an integer value.

QCj, QC$sn: The number of entries into the queue. This count is obtained by keeping a running total of the B operands of QUEUE blocks that reference this queue. Recall that the default value of the B operand of a QUEUE block is 1.

QMj, QM$sn: Maximum queue contents; equivalently the maximum value of the SNA Q since the start of execution.

QTj, QT$sn: The average of the times that transactions have stayed in the queue. This SNA is truncated to an integer value.

QXj, QX$sn: The average of the times that nonzero transactions (those with positive waiting times) have stayed in the queue. As in the case of QT, this SNA is also truncated.

QZj, QZ$sn: Number of zero-entry transactions (ones that departed from the queue in zero time) that have entered the queue.

7.1.3 Storages

There are seven SNAs related to storages and their use:

Rj, R$sn: The number of storage units available (i.e., not in use by transactions).

Sj, S$sn: The number of storage units currently in use.

SAj, SA$sn: Truncated portion of the average number of storage items that have been in use.

SCj, SC$sn: Total number of entries for the storage. This is the running total of the B operands of all ENTER blocks referencing this storage.

SMj, SM$sn: Maximum number of units that have been in use; equivalently the maximum value of the SNA S since the start of execution.

SRj, SR$sn: The utilization of the storage, given in parts per thousand.

STj, ST$sn: Truncated portion of the average of the times that transactions use the storage.

7.1.4   Blocks

The two SNAs in this area give the current and total counts of the blocks:

Nj, N$sn: The total number of transactions that have entered the block.

Wj, W$sn: The number of transactions that are currently in the block.

7.1.5   Transactions

The SNAs describing transaction attributes are

Pj: The value of the j-th transaction parameter.

PR: The priority of the transaction.

M1: The difference between the current absolute clock and the value of the absolute clock when the transaction entered the simulation. Like the transaction parameters and the transaction priority, every transaction has an M1 value uniquely associated with it. When M1 is used in the model, the value of M1 belonging to the active transaction is accessed.

MPj: The difference between the current absolute clock and the value of the absolute clock when the transaction last moved into a MARK block. The MARK block and the use of the MP SNA will be described more fully in Section 7.7.

7.1.6   The Clock

Early versions of GPSS had only one clock SNA, representing the value of the relative clock and designated by C1. Later versions (GPSS V, GPSS/PC, GPSS/H) added the SNA AC1 for the absolute clock.

7.1.7   Other SNAs

In addition to the SNAs listed in the preceding sections and the FN, V, and RN SNAs, which have already been discussed, GPSS has a number of SNAs dealing with various model entities. These will be considered, along with their associated model entities, in later sections.

7.2   Savevalues, the INITIAL Statement, and the SAVEVALUE Block

Data items that are uniquely associated with a transaction are stored as parameters of that transaction. The limitation of this form of data storage is that a transaction has access only to its own parameter values. In this sense, transaction parameters act as local variables. To be able to store and manipulate data items that can be accessed by all transactions, GPSS uses savevalue entities. Savevalues are similar to global variables—they can be initialized by the INITIAL statement and modified by any transaction, through the SAVEVALUE block. Consistent with the convention described in Section 7.1, savevalues are referenced by Xj or X$sn. Depending on the particular version of GPSS being used, the family names XF, XH, XB, and XL may also be available to designate fullword (32-bit), halfword (16-bit), byte (8-bit), and floating-point savevalues, respectively. Only Xj and X$sn, which in all versions of GPSS refer to fullword integer values, will be considered here.

7.2.1   The INITIAL Statement

To override the default initialization of all savevalues to 0, the INITIAL statement can be used to supply nonzero initial values. An INITIAL statement consists of

1. The word INITIAL in the operation field.

2. Name and initial value pairs, with a comma separating the name and initial value and a / separating the pairs. These pairs are listed in the operand field.

The INITIAL statement

```
INITIAL    X$RATE,5/X$COST,1000/X3,4
```

initializes savevalues RATE and COST to 5 and 1000, respectively, and the third savevalue to 4. When numeric reference is being made to savevalues, it is possible to have a number of successive savevalues assume a specific value. For example,

INITIAL   X$RATE,5/X5-X12,32

will initialize savevalue RATE to 5 and the 5th through 12th savevalues to 32.

If, through the use of multiple START statements, the program is to execute more than once, then the modeler should be able to stipulate the content of savevalues as each execution starts. Since almost all such situations involve the use of the RESET and CLEAR statements, the impact of these statements on savevalues needs to be understood. **The RESET statement does not alter any savevalues.** Thus, if a program is reexecuted following a RESET statement, the initial content of all savevalues will be what it was at the end of the previous execution. **The CLEAR statement**, on the other hand, **reinitializes all savevalues to zero.** Since a modeler may want to reinitialize some (but not all) savevalues to 0, GPSS provides a special use of the CLEAR statement for this purpose. If a list of savevalues, referenced numerically or by name, is included in the operand field of a CLEAR statement, the savevalues in that list are excluded from reinitialization. For example, the CLEAR statement

CLEAR   X1-X4,X$RATE

will keep the values of X1 through X4 and the savevalue RATE intact from the previous execution of the program. It will also zero all other savevalues, if there are others in the model.

Since the GPSS processor schedules an arrival through each GENERATE block before the start of execution, it is important that the initialization of all savevalues that are used as operands in GENERATE blocks physically precede the GENERATE blocks in the program. Otherwise, GPSS will use the default zero value for these savevalues as it schedules initial transactions for the GENERATE blocks. In a situation where savevalues are being used as operands of GENERATE blocks, it might be tempting to use the sequence shown in Figure 7.2–1.

At first glance, it may seem as if the GENERATE block

GENERATE   40,30

is operative during the first execution of the statements in Figure 7.2–1, and

```
          ⋮
INITIAL        X1,30
          ⋮
GENERATE       40,X1
          ⋮
START
CLEAR
INITIAL        X1,20
START
END
```

Figure 7.2–1. The CLEAR and INITIAL statements in the wrong order.

```
          GENERATE    40,20
```

is operative during the second execution. This is not quite true, since the CLEAR statement not only initializes X1 to 0, but also schedules the first arrival at every GENERATE block before execution is restarted. Therefore, the first transaction of the second execution enters the system through the GENERATE block

```
          GENERATE    40,30
```

and not through

```
          GENERATE    40,20
```

as intended. The correct way to use

```
          GENERATE    40,20
```

in the second run is given in Figure 7.2–2. This approach reinitializes X1 to 20, clears all savevalues and model statistics except for X1, and restarts execution.

7.2.2   The SAVEVALUE Block

The SAVEVALUE block is used to modify the content of savevalues during execution. This block has the following three operands:

Operand A: This operand is the numeric or user-defined name of the savevalue that is to be modified; it cannot be defaulted.

```
                    ⋮
        INITIAL        X1,30
                    ⋮
        GENERATE       40,X1
                    ⋮
        START
        INITIAL        X1,20
        CLEAR          X1
        START
```

Figure 7.2–2. The CLEAR and INITIAL statements in the proper order.

Operand B: This operand specifies the value that will be assigned to the savevalue indicated by operand A. This operand also cannot be defaulted.

Operand C: Operand C is the savevalue type. The default type is fullword. In the case of halfword savevalues, operand C is literally H.

In its simplest form, a SAVEVALUE block changes the previous content of a savevalue, as in

                    SAVEVALUE   CREWNUM,5

where the savevalue with a user-defined name CREWNUM is assigned value 5. In

                    SAVEVALUE   CREWNUM,X$NUM

savevalue CREWNUM is set to the same value as savevalue NUM. Note that the X$ designation is inappropriate for operand A, since operand A is supposed to be a savevalue. By contrast, the B operand can be any SNA; therefore, the X$ designation is needed to identify NUM as a savevalue.

Some versions of GPSS also allow indirect addressing of savevalues by having X∗Pj specify the savevalue whose number is in parameter j of the active transaction. Thus, if the third parameter of a transaction has value 7 and that transaction enters the block

                    SAVEVALUE   X$∗P3,0

the 7th savevalue will be changed to 0. The indirect addressing scheme discussed here is not limited to savevalues. For instance, Q*P4 refers to the length of the queue (see the Q SNA in Section 7.1.2) whose number is in parameter 4. There are GPSS implementations that use variations of the indirect addressing scheme given above. For example, in GPSS/PC indirect addressing can only be done through parameters; hence, the parameter specification is not necessary and Q*4 is used to refer to the length of the queue whose number is in parameter 4.

The SAVEVALUE block can also be used in increment and decrement modes in the same way that ASSIGN blocks can. The block

$$\text{SAVEVALUE} \quad \text{DAYCOUNT+,1}$$

increments the content of the savevalue DAYCOUNT by 1 and

$$\text{SAVEVALUE} \quad \text{STORE-,P3}$$

decrements the savevalue STORE by an amount equal to the third parameter of the transaction entering the SAVEVALUE block.

7.3   Example: A Repair Service Operation

A recently established appliance repair service in an urban area receives requests for repairs by telephone throughout an 8-hour work day. The repair service employs three repairmen and supplies each of them with a van equipped with appropriate tools and replacement parts to respond to service calls. The charge for a service call consists of a $15 fee plus a $30 per hour charge for travel and service time.

The agreement between the owner of the repair service and the repairmen is that if a repairman becomes available any time during the working day following a service call, that call must receive service, even if a repairman must work beyond the 8-hour day. If all repairmen remain busy for the rest of the 8-hour period following a call, that call is considered a lost business opportunity. The operation of the repair service is simulated to determine what the expected profits should be under present operations and to determine alternative modes of operation for a higher profit margin. The simulation program given in Figure 7.3–1, with a portion of its output shown in Figure 7.3–2, is based on the following observations:

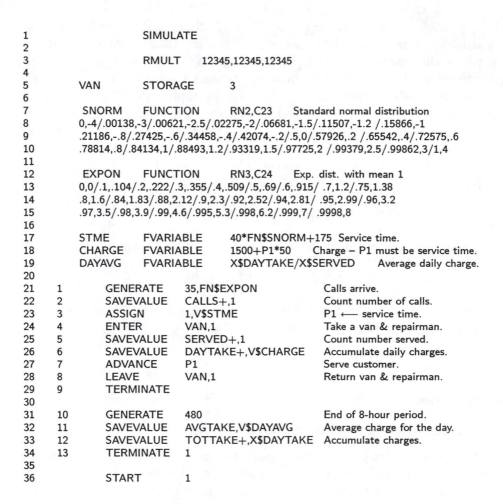

```
1                  SIMULATE
2
3                  RMULT      12345,12345,12345
4
5       VAN        STORAGE       3
6
7       SNORM      FUNCTION      RN2,C23     Standard normal distribution
8       0,-4/.00138,-3/.00621,-2.5/.02275,-2/.06681,-1.5/.11507,-1.2 /.15866,-1
9       .21186,-.8/.27425,-.6/.34458,-.4/.42074,-.2/.5,0/.57926,.2 /.65542,.4/.72575,.6
10      .78814,.8/.84134,1/.88493,1.2/.93319,1.5/.97725,2 /.99379,2.5/.99862,3/1,4
11
12      EXPON      FUNCTION      RN3,C24     Exp. dist. with mean 1
13      0,0/.1,.104/.2,.222/.3,.355/.4,.509/.5,.69/.6,.915/ .7,1.2/.75,1.38
14      .8,1.6/.84,1.83/.88,2.12/.9,2.3/.92,2.52/.94,2.81/ .95,2.99/.96,3.2
15      .97,3.5/.98,3.9/.99,4.6/.995,5.3/.998,6.2/.999,7/ .9998,8
16
17      STME       FVARIABLE     40*FN$SNORM+175  Service time.
18      CHARGE     FVARIABLE     1500+P1*50      Charge – P1 must be service time.
19      DAYAVG     FVARIABLE     X$DAYTAKE/X$SERVED      Average daily charge.
20
21   1     GENERATE     35,FN$EXPON           Calls arrive.
22   2     SAVEVALUE    CALLS+,1              Count number of calls.
23   3     ASSIGN       1,V$STME              P1 ←— service time.
24   4     ENTER        VAN,1                 Take a van & repairman.
25   5     SAVEVALUE    SERVED+,1             Count number served.
26   6     SAVEVALUE    DAYTAKE+,V$CHARGE     Accumulate daily charges.
27   7     ADVANCE      P1                    Serve customer.
28   8     LEAVE        VAN,1                 Return van & repairman.
29   9     TERMINATE
30
31   10    GENERATE     480                   End of 8-hour period.
32   11    SAVEVALUE    AVGTAKE,V$DAYAVG      Average charge for the day.
33   12    SAVEVALUE    TOTTAKE+,X$DAYTAKE    Accumulate charges.
34   13    TERMINATE    1
35
36         START        1
```

Figure 7.3–1. A program for repair service operation.

- The interarrival times between service calls are exponential with a mean of 35 minutes, and service times for travel and repairs are normally distributed with a mean of 175 minutes and a standard deviation of 40 minutes.

- The total compensation (salary, benefits, indirect costs) of each repairman is \$120 per day, regardless of the number of hours worked.

- The cost of operating a van, including amortization, is \$35 per work day, and the overhead of running the business is \$215 per work day.

RELATIVE CLOCK 480   ABSOLUTE CLOCK 480

BLOCK	CURRENT	TOTAL
1	1	14
2	0	13
3	5	13
4	0	8
5	0	8
6	0	8
7	3	8
8	0	5
9	0	5
10	1	2

BLOCK	CURRENT	TOTAL
11	0	1
12	0	1
13	0	1

STORAGE	CAPACITY	AVERAGE CONTENTS	TOTAL ENTRIES	AVERAGE TIME/TRANS	AVERAGE UTILIZ.
VAN	3	2.608	8	156.500	0.869

		CURRENT STATUS	FRACTION AVAIL.	CURRENT CONTENTS	MAXIMUM CONTENTS
		A	1.000	3	3

NON-ZERO FULLWORD SAVEVALUES

SAVEX	VALUE	SAVEX	VALUE	SAVEX	VALUE
AVGTAKE	8480	CALLS	13	DAYTAKE	67840
SERVED	8	TOTTAKE	67840		

37	CLEAR	X$TOTTAKE,X$CALLS,X$SERVED
38	START	1

⋮   ⋮   ⋮

RELATIVE CLOCK 480   ABSOLUTE CLOCK 480

BLOCK	CURRENT	TOTAL
1	1	13
2	0	12
3	4	12
4	0	8
5	0	8
6	0	8
7	3	8
8	0	5
9	0	5
10	1	2

BLOCK	CURRENT	TOTAL
11	0	1
12	0	1
13	0	1

STORAGE	CAPACITY	AVERAGE CONTENTS	TOTAL ENTRIES	AVERAGE TIME/TRANS	AVERAGE UTILIZ.
VAN	3	2.621	8	157.250	0.874
		CURRENT STATUS	FRACTION AVAIL.	CURRENT CONTENTS	MAXIMUM CONTENTS
		A	1.000	3	3

Figure 7.3–2. The output for the repair service simulation (continues).

```
NON-ZERO FULLWORD SAVEVALUES
SAVEX      VALUE        SAVEX     VALUE         SAVEX     VALUE
AVGTAKE    881          CALLS     136           DAYTAKE   74960
SERVED     85           TOTTAKE   739740
```

Figure 7.3–2. The output for the repair service simulation (concluded).

- It is projected that if repairmen and vans are added to the operation, there would be an overhead increase of $25 (per repairman-van combination, per work day) beyond the direct costs incurred.

Minutes are chosen as the time unit for the simulation, and charges are made on a per-minute rather than an hourly basis. The program uses CALLS and SERVED to keep a record of the number of service calls received and the number of customers served. The variables STME and CHARGE compute the service time (including time for travel) and the charge to the customer. The savevalue DAYTAKE accumulates the charges made throughout the day. Notice that charges are made only if a transaction (service call) succeeds in obtaining a repairman and a van at the ENTER block before the timer segment shuts off the simulation at the end of the 8-hour period.

In the timer segment (lines 31–34), the savevalue AVGTAKE captures the average charge for the day as the quotient of the savevalues DAYTAKE and SERVED. The savevalue TOTTAKE, which accumulates the day's gross receipts, is not necessary if the simulation is to be run for only one day. However, with the proper use of control statements, the simulation can be run for successive days, in which case the gross receipts over the entire period would be accumulated in TOTTAKE. A repetition of the control statements

```
CLEAR      X$TOTTAKE, X$CALLS, X$SERVED
START      1
```

can be used to simulate the repair service operation for any number of days.

After storing the service time for a call in P1, the CHARGE variable determines a fee of $15 plus $30 per hour of service. To avoid fractions altogether, the fee is computed in cents rather than dollars, as 1500 plus 50 cents per minute. The portion of the output of a 10-day simulation given in Figure 7.3–2 shows that 85 of the 136 service calls taken received service and the gross income during the 10-day period simulated was $7397.40. The expenses during the same period would be $120 × 3 × 10 = $3600 for compensation for three repairmen for 10 days, $35 × 3 × 10 = $1050 for the operation of the three vans, and $215 × 10 = $2150 for general overhead, for a total of

$6800. Thus, for the 10-day period simulated, the repair service operated at a $597.40 profit, or just under $60 profit per day.

The fact that 51 calls were lost during the 10-day simulation suggests that a combination of four, perhaps even five, repairmen and vans should also be investigated to see if a reduction in lost business opportunities would result in greater profits. Although the output from these simulations is not included here, their results are summarized in Table 7.3–3.

Table 7.3–3. Receipts and profits of the repair service operation.

	NUMBER OF REPAIRMEN AND VANS		
	3	4	5
10-DAY GROSS RECEIPTS	$7397.40	$9831.00	$10,910.40
10-DAY EXPENSES	$6800.00	$8600.00	$10,400.00
PER-DAY PROFITS	$59.74	$123.10	$51.04

7.4 The PRINT Block

Although the PRINT block is part of GPSS V, it is not included in some GPSS implementations (e.g., GPSS/PC). The GPSS/PC model (included on the disk) corresponding to the program of Figure 7.5–1 of the next section illustrates how other features of GPSS/PC can be used to compensate for the absence of the PRINT block. This PRINT block allows users to include in the program output specific information produced at designated points during the simulation. This special output, resulting from the use of a PRINT block, is generated when a transaction enters the PRINT block. The PRINT block has four operands:

Operands A and B: These are the lower and upper limits, respectively, of the entity numbers for which output is to be generated. If defaulted, output will be generated for the entire range of the chosen entity.

Operand C: This is the mnemonic of the entity for which output is to be generated. If defaulted, the savevalue entity (with the mnemonic X) will be assumed.

Operand D: If this operand is defaulted, the output associated with the
    PRINT block will appear at the top of a new page. If any alphabetic
    character is used for operand D, the form feeds prior to the output will
    be suppressed.

Table 7.4–1 gives the choices that are available for the C operand of the
PRINT block. The rightmost column in the table gives the section in this
book in which the relevant SNA or entity is discussed. As can be seen, some
of the choices for operand C involve GPSS concepts that have not yet been
covered. The SNAs and entities listed in the portion of the table below the
heavy line do not have numeric references. In these cases, therefore, the A
and B operands are omitted.

Table 7.4–1. The C operand of the PRINT block.

Mnemonic	Output	Section
F	Facilities	7.1.1
Q	Queues	7.1.2
S	Storages	7.1.3
X,XF,default	Fullword savevalues	7.2
XH	Halfword savevalues	†
MX	Fullword matrix savevalues	†
MH	Halfword matrix savevalues	†
LG	Logic switches	7.14
G	Members of groups	†
T	Tables	7.8
U	User-chain statistics	7.17
CHA	User-chain listings	7.17
C	Absolute and relative clock	7.1.6
B	Current and total block counts	7.1.4
N	Total block counts	7.1.4
W	Current block counts	7.1.4
MOV	Current events chain	5.1
FUT	Future events chain	5.1
I	Interrupt chain	7.9
MAT	Matching-status chain	†

†Not discussed in this text.

When a transaction enters the PRINT block

                    PRINT   3,7,Q

the usual statistics related to queues 3 through 7 will be produced. A transaction that enters the blocks

```
PRINT    ,,F
PRINT    ,,FUT
PRINT    ,,C
```

will generate output associated with all the facilities in the model, the usual output for the FEC, and the output giving the values of the absolute and relative simulation times, respectively.

As one use of the PRINT block, consider the program of Section 7.3 (Figure 7.3–1). Only the savevalue component of the voluminous output provided by GPSS is of interest here. To obtain what is necessary without getting lost in many pages of output, the NP designation for operand B of the START statement could be used to suppress all output. The PRINT statement

```
PRINT    ,,X,A
```

can then be used in the timer segment to produce the savevalue output at the end of each simulated day. Figure 7.4–2 gives the initial and final portions of the output of the modified program.

```
PRINT BLOCK 13 AT CLOCK TIME 480
NON-ZERO FULLWORD SAVEVALUES
```

SAVEX	VALUE	SAVEX	VALUE	SAVEX	VALUE
AVGTAKE	8480	CALLS	13	DAYTAKE	67840
SERVED	8	TOTTAKE	67840		

$\vdots \qquad\qquad\qquad \vdots \qquad\qquad\qquad \vdots$

```
64                    CLEAR        X$TOTTAKE,X$CALLS,X$SERVED
65                    START        1,NP
```

```
PRINT BLOCK 13 AT CLOCK TIME 480
NON-ZERO FULLWORD SAVEVALUES
```

SAVEX	VALUE	SAVEX	VALUE	SAVEX	VALUE
AVGTAKE	881	CALLS	136	DAYTAKE	74960
SERVED	85	TOTTAKE	739740		

```
66                    END
```

Figure 7.4–2. The savevalue output of the program in Figure 7.3–1.

7.5   Example: Demand Estimation

To illustrate the use of savevalues, we consider the simulation of the problem of Example 6.1.2, given in Figure 7.5–1. This problem deals with the operation of a fire station under the following constraints:

1. The number of fires per day follows a Poisson distribution with a mean of 4 fires per day.

2. Only one response unit, a fire truck, is required in 75% of the fires, and in these cases the time to service the fire is normally distributed with a mean of 3 hours and a standard deviation of 0.5 hour.

3. The remaining 25% of fires require two response units, and the time to service these fires is normally distributed with a mean of 4 hours and a standard deviation of 1 hour.

The central concern of Example 6.1.2 was to determine the number of minutes of unit time used per day to within 60 minutes with 90% confidence. The X_is (number of minutes of unit time for the i-th simulated day) used in the example were obtained through the GPSS model given in Figure 7.5–1. The simulation time units are minutes, and the TIME1 and TIME2 variables give durations of 1-unit and 2-unit fires, respectively. The overall logic of the program creates transactions (fires) that branch to one of the locations ONE or TWO, depending on whether we have a 1-unit or a 2-unit fire. The savevalue DAYTOT is incremented, once for 1-unit fires and twice for 2-unit fires, to accumulate the simulated day's requests. The timer segment of the model (lines 43 through 47) produces a transaction at the end of each day (1440 minutes) that keeps a running total of accumulated service times in the savevalue TOTAL. This segment also reinitializes savevalue DAYTOT to 0 prior to the start of a new day.

The program will run for as many days as stipulated by the A operand of the START statement. Without the PRINT block, the 10-day results used in Example 6.1.2 can be obtained through 10 successive START statements, each with its A operand set to 1. $X_1, \ldots, X_{10}$ could then be derived as successive differences of TOTAL. Figure 7.5–2 gives a portion of the output that results from the PRINT block. The mean (989 minutes) and standard deviation (425.07 minutes) are calculated (by hand) from these results. Through the use of the TABLE and TABULATE statements of GPSS (Section 7.8), it is possible to have GPSS provide means and standard deviations.

```
1
2
3                      SIMULATE
4
5                      RMULT       12345,12345,12345
6
7       EXPON    FUNCTION     RN3,C24    Exponential distribution
8       0,0/.1,.104/.2,.222/.3,.355/.4,.509/.5,.69/.6,.915
9       .7,1.2/.75,1.38/.8,1.6/.84,1.83/.88,2.12/.9,2.3
10      .92,2.52/.94,2.81/.95,2.99/.96,3.2/.97,3.5/.98,3.9
11      .99,4.6/.995,5.3/.998,6.2/.999,7/.9998,8
12
13      SNORM    FUNCTION     RN2,C23    Standard normal distribution
14      0,-4/.00138,-3/.00621,-2.5/.02275,-2/.06681,-1.5
15      .11507,-1.2/.15866,-1/.21186,-.8/.27425,-.6
16      .34458,-.4/.42074,-.2/.5,0/.57926,.2/.65542,.4
17      .72575,.6/.78814,.8/.84134,1/.88493,1.2/.93319,1.5
18      .97725,2/.99379,2.5/.99862,3/1,4
19
20      TIME1    FVARIABLE    30*FN$SNORM+180     Duration of 1-unit fires
21      TIME2    FVARIABLE    60*FN$SNORM+240     Duration of 2-unit fires
22
23      UNIT   STORAGE        10                  10 units should suffice.
24
25   1         GENERATE       360,FN$EXPON        Create fires.
26   2         TRANSFER       .25,ONE,TWO         Branch to 1- or 2-unit fires.
27
28   3   ONE   ENTER          UNIT,1              Get a unit.
29   4         ASSIGN         1,V$TIME1           P1 ⟵ Time to put out fire.
30   5         SAVEVALUE      DAYTOT+,P1          Accumulate times for the day.
31   6         ADVANCE        P1                  Service the fire.
32   7         LEAVE          UNIT,1              Return the unit.
33   8         TERMINATE
34
35   9   TWO   ENTER          UNIT,2              Get 2 units.
36   10        ASSIGN         1,V$TIME2           P1 ⟵ Time to put out fire.
37   11        SAVEVALUE      DAYTOT+,P1          Accumulate times for the day,
38   12        SAVEVALUE      DAYTOT+,P1          − for each unit.
39   13        ADVANCE        P1                  Service the fire.
40   14        LEAVE          UNIT,2              Return both units.
41   15        TERMINATE
42
43   16        GENERATE       1440                At the end of each day,
44   17        SAVEVALUE      TOTAL+,X$DAYTOT     accumulate times,
45   18        PRINT          ,,X                 obtain output, and
46   19        SAVEVALUE      DAYTOT,0            zero total for day.
47   20        TERMINATE      1
48
49             START          10,NP
```

Figure 7.5–1. A simulation of emergency service delivery.

```
PRINT BLOCK 18 AT CLOCK TIME 1440
NON-ZERO FULLWORD SAVEVALUES
SAVEX       VALUE          SAVEX       VALUE
DAYTOT      1505           TOTAL       1505


PRINT BLOCK 18 AT CLOCK TIME 2880
NON-ZERO FULLWORD SAVEVALUES
SAVEX       VALUE          SAVEX       VALUE
DAYTOT      804            TOTAL       2309
   .            .              .
   .            .              .


PRINT BLOCK 18 AT CLOCK TIME 14400
NON-ZERO FULLWORD SAVEVALUES
SAVEX       VALUE          SAVEX       VALUE
DAYTOT      603            TOTAL       9890
50          END
```

Figure 7.5–2. The output of the simulation of emergency
service delivery in Figure 7.5–1.

The standard output from 10 successive executions of this simulation
would produce considerable output. At the model development stage, all
the output should be studied carefully to make sure that the program is free
of logical errors. However, when the simulation is extended to 169 days (recall that we needed to do this in the example of Section 6.1.2), the standard
GPSS output becomes excessive. For this reason, we use the PRINT block
(line 45) to obtain the output of Figure 7.5–2. The successive values of DAY-
TOT give $X_1 = 1505$, $X_2 = 804, \ldots, X_{10} = 603$. The values of X$DAYTOT
are accumulated in X$TOTAL. At time 14,400 (end of the 10-day simulation),
9890 (value of X$TOTAL) divided by 10 gives the daily average.

7.6   User-Defined Random Number Generators

Users often place a great deal of confidence in algorithms that are bundled with standard software packages. This is particularly true for random
number generating algorithms. We saw in Section 3.5.1 that some random
number generators come with the operating system of the computer (e.g.,
URN22 and, to a lesser extent, URN07, URN08, and URN09); other generators (e.g., URN04 and URN10) are bundled with software packages; and
still others (e.g., URN27 and URN30) are part of simulation languages.

Unfortunately, the fact that a generator may be incorporated within a larger software environment does not provide any assurance of its quality. Of the eight random number generators just mentioned, only two (URN22 and URN30) had satisfactory performance when tested (see Section 3.5.2). The quality of random number generators that come as part of GPSS varies from one implementation to another. For example, GPSS/VX, as implemented on VAX/VMS systems, uses URN22, which comes with the VMS operating system. This was one of the best generators considered in Section 3.5.2 (see Table 3.5–1). By contrast, URN27, which comes with GPSS/H release 1, does poorly on the tests discussed in Section 3.3.

Figure 7.6–1 shows how the issue of the quality of a GPSS random number generator can be circumvented altogether through the insertion of a known high-quality generator into a program. This can be done rather simply by using variables and savevalues. Figure 7.6–1 shows a program, with its output, that uses the random number generator described in Problem 1.1 to produce interarrival times. This generator (see URN38 of Section 3.7) is being used here for illustration only; we found it to be unsuitable for use in Section 3.7 and we do **not** recommend its use. In the program in Figure 7.6–1, the variable RANDINT describes the congruential relation of URN38, namely

$$X_i = 1220703125 \times X_{i-1} \bmod 32768. \tag{7.6.1}$$

When numbers of the magnitude of 1,220,703,125 are used, we need to be concerned with the hardware and software environment in which GPSS is executing. If GPSS is running on a computer that uses 32-bit architecture, then 1,220,703,125 can be easily represented. But even in this case, we need to make sure that the overflow that may result by multiplying 1,220,703,125 by X_{i-1} is handled properly by the software. If there are integer overflow problems, line 2 of the program in Figure 7.6–1 could be changed to

RANDINT   5*(15625*(15625*X$SEED@32768)@32768) @32768,

reflecting the relationship (see equation (1.2.3))

$$X_i = 5(15625(15625 \times X_{i-1} \bmod 32768) \bmod 32768) \bmod 32768. \tag{7.6.2}$$

The variable IAT uses X$SEED (successively $X_0, X_1, \ldots$) to compute the random numbers X$SEED/32768 and eventually converts them to integer values in the range of 9 to 15. The only purpose of lines 6 through 10 is to update $X_0 = 32767$, ensuring that the program will use $X_1, X_2, \ldots$ during its run. If this is not of concern, this portion of the program can be deleted.

```
1                          SIMULATE
2            RANDINT       VARIABLE     (1220703125*X$SEED)@32768)
3            IAT           FVARIABLE    9+(7*(X$SEED/32768))
4                          INITIAL      X$SEED,32767
5
6     1                    GENERATE     0,,,1
7     2                    SAVEVALUE    SEED,V$RANDINT
8     3                    TERMINATE    1
9                          START        1,NP
10                         RESET
11
12    4                    GENERATE     V$IAT
13    5                    SAVEVALUE    SEED,V$RANDINT
14    6                    QUEUE        WAIT
15    7                    SEIZE        SERVER
16    8                    DEPART       WAIT
17    9                    ADVANCE      11
18    10                   RELEASE      SERVER
19    11                   TERMINATE    1
20                         START        10
```

RELATIVE CLOCK 138			ABSOLUTE CLOCK 139		
BLOCK	CURRENT	TOTAL	BLOCK	CURRENT	TOTAL
1	0	0	11	0	10
2	0	0			
3	0	0			
4	1	12			
5	0	11			
6	1	11			
7	0	10			
8	0	10			
9	0	10			
10	0	10			

QUEUE	MAXIMUM CONTENTS	AVERAGE CONTENTS	TOTAL ENTRIES	ZERO ENTRIES	PERCENT ZEROS
WAIT	1	0.029	11	9	81.818

	AVERAGE TIME/TRANS	$AVERAGE TIME/TRANS	TABLE NUMBER	CURRENT CONTENTS
	0.364	2.000	1	

FACILITY	AVERAGE UTILIZATION	NUMBER ENTRIES	AVERAGE TIME/TRANS	SEIZING TRANS. NO.	PREEMPTING TRANS. NO.
SERVER	0.797	10	11.000		

NON-ZERO FULLWORD SAVEVALUES

SAVEX	VALUE	SAVEX	VALUE	SAVEX	VALUE
SEED	15343				
23		END			

Figure 7.6–1. A user-defined random number generator in GPSS.

Lines 12 and 13 use the random number generator we have supplied to produce interarrival times from the uniform distribution on the interval from 9 to 15. Note that at line 13, the savevalue SEED has to be updated in order to produce a new random number with each successive invocation of V$IAT.

7.7 Transit Time and the MARK Block

The need to record the transit times of transactions through an entire model, or through subsections of a model, has come up several times. In the manufacturing shop example in Section 5.6, the turnaround time of jobs (i.e., transit times through the entire model) was needed, and in the batch computer system example in Section 5.11, the waiting times of jobs prior to the capture of the CPU were of interest. One way to determine the average transit time over a subsection of a model is to bracket that subsection with QUEUE and DEPART blocks and capture the average transit time as the average queue residence time.

The use of QUEUE and DEPART blocks could pose a problem in situations where relatively few transactions move through the subsection. In such cases there could be a significant difference between the average queue residence time and the average transit time, since transactions that have partially moved through the subsection are included in the queue residence computation. To see this discrepancy, consider the program in Figure 7.7–1. Parameter 1 of the transactions moving through the first segment of the program contains the simulated times when the transactions enter the ASSIGN block. These are the arrival times of the transactions. The variable TRTM computes the transit time for each transaction completing its progress through the model. The savevalues COUNT and TOTTME accumulate the total number of transactions and transit times of those transactions that move through the entire first segment. From the output of this program, given in Figure 7.7–2, it can be seen that the true average transit time for this simulation is 148 units (following truncation), whereas the average queue residence time is 141.8 simulation-time units.

Since the use of savevalues to compute transit times is somewhat cumbersome, GPSS provides the MARK block and the MPj SNA (see Section 7.1.5) to make such computations simpler. The MARK block records the absolute clock when a transaction enters it and later, when the MPj SNA is accessed, its value becomes the current absolute clock value minus the time previously recorded by the MARK block. The MARK block has one operand:

```
1
2                    SIMULATE
3
4       TRTM  VARIABLE        C1-P1                      Transit Time; P1=arriv. time.
5       AVG   FVARIABLE       X$TOTTME/X$COUNT           Average transit time.
6
7    1         GENERATE       45,30                      Transactions arrive.
8    2         QUEUE          LINE                       Enter waiting LINE.
9    3         ASSIGN         1,C1                       P1=arrival time.
10   4         SEIZE          FAC                        Get facility.
11   5         ADVANCE        50,30                      Get service.
12   6         RELEASE        FAC                        Leave facility.
13   7         DEPART         LINE                       Leave waiting LINE.
14   8         SAVEVALUE      COUNT+,1                   COUNT=No. of trans. serviced.
15   9         SAVEVALUE      TOTTME+,V$TRTM             TOTTME=Total transit time.
16   10        TERMINATE
17
18   11        GENERATE       1000                       Timer arrives.
19   12        SAVEVALUE      AVGTME,V$AVG               Computes avg. transit time.
20   13        PRINT          ,,Q                        Prints queue info.
21   14        PRINT          ,,X                        Prints savevalue info.
22   15        TERMINATE      1
23
24             START          1,NP                       Suppress all other output.
```

Figure 7.7–1. A program to compute average transit time.

```
PRINT BLOCK 13 AT CLOCK TIME 1000
QUEUE      MAXIMUM        AVERAGE        TOTAL      ZERO       PERCENT
           CONTENTS       CONTENTS       ENTRIES    ENTRIES    ZEROS
LINE       5              2.978          21         0          0.000

                          AVERAGE        $AVERAGE   TABLE      CURRENT
                          TIME/TRANS     TIME/TRANS NUMBER     CONTENTS
                          141.810        141.810               3

PRINT BLOCK 14 AT CLOCK TIME 1000
NON-ZERO FULLWORD SAVEVALUES
SAVEX      VALUE          SAVEX          VALUE      SAVEX      VALUE
AVGTME     148            COUNT          18         TOTTME     2678
25                        END
```

Figure 7.7–2. The output of the program in Figure 7.7–1.

Operand A: This operand is the parameter number into which the current
          value of the absolute clock will be recorded. If the operand is defaulted,
          the value of the absolute clock is copied into the transaction's mark-
          time area, destroying the previous content of that area.

With the use of the A operand, the sequence

$$\vdots$$

X   MARK   5

$$\vdots$$

Y   ASSIGN   2,MP5

will cause the absolute clock to be stored in parameter 5 at point X in the program. At point Y, MP5, the difference between the current absolute clock value and the value of parameter 5 will be stored in parameter 2. Thus, parameter 2 will contain the transit time between points X and Y. If the A operand of the MARK block is defaulted and the ASSIGN block is changed to

Y   ASSIGN   2,M1

then the absolute clock will be stored in the mark-time area of the transaction at location X. At location Y, M1, the SNA designating the difference between the current absolute clock value and the mark-time value will be stored in parameter 2. Again, parameter 2 will represent the transit time between points X and Y. Figure 7.7–3 shows a modification of the program in Figure 7.7–1, using the MARK block. The truncated average transit time AVGTME will be 148, as was the case in the earlier version.

```
1                SIMULATE
2      TRTM  VARIABLE    C1-P1                     Transit Time; P1=arriv. time.
3      AVG   FVARIABLE   X$TOTTME/X$COUNT          Average transit time.
4
5      1     GENERATE    45,30                     Transactions arrive.
6      2     QUEUE       LINE                      Enter waiting LINE.
7      3     MARK        7                         Capture arrival time.
8      4     SEIZE       FAC                       Get facility.
9      5     ADVANCE     50,30                     Get service.
10     6     RELEASE     FAC                       Leave facility.
11     7     DEPART      LINE                      Leave waiting LINE.
12     8     SAVEVALUE   COUNT+,1                  COUNT=No. of trans. serviced.
13     9     SAVEVALUE   TOTTME+,MP7               TOTTME=Total transit time.
14     10    TERMINATE
15
16     11    GENERATE    1000                      Timer arrives.
17     12    SAVEVALUE   AVGTME,V$AVG              Computes avg. transit time.
18     13    PRINT       ,,Q                       Prints queue info.
19     14    PRINT       ,,X                       Prints savevalue info.
20     15    TERMINATE   1
21           START       1,NP                      Suppress all other output.
```

Figure 7.7–3. The use of the MARK block to capture transit times.

7.8   The TABLE Statement and the TABULATE Block

The default output given by GPSS may not provide sufficiently detailed information about a model entity. For example, the modeler may want to know more than the average queue residence time and the percentage of zero entries associated with a certain queue. Additional statistics such as the standard deviation or a frequency distribution of queue residence times could be very useful. The TABLE statement and the TABULATE block can be used to compute and print the means, standard deviations, and frequency and cumulative distributions of specified SNAs. The variations of the TABLE block (Section 7.8.3), enable users to capture information on interarrival times, on queue residence times, and on rates at which certain events occur during the simulation.

The primary function of the TABLE and TABULATE combination is to construct a frequency table (or, equivalently, a histogram) for a chosen SNA. The TABLE statement specifies the SNA and the class interval structure for the frequency table. The TABULATE block causes the entry or tallying of the SNA into the frequency table. Since the interval structure must be known at the time the TABULATE block is executed, the TABLE statement should precede the executable portion of the program.

7.8.1   The TABLE Statement

In addition to indicating the SNA that is to be tabulated, the TABLE statement sets up all the intervals over which the data values of the SNA are to be tallied. The TABLE statement consists of a numeric or symbolic table name in the label field, the word TABLE in the operator field, and operands in the operand field.

The TABLE statement is generally used with operands A, B, C, and D; a fifth operand, E, is also available in specific situations.

Operand A: The name (mnemonic or numeric) of the SNA whose values are to be tabulated.

Operand B: The leftmost boundary point of the intervals over which the SNA is to be tallied.

Operand C: The length of the **equal sized** intervals (except for the first and last) to be used for tallying the SNA.

Operand D: The total number of intervals, including the first and last ones.

The B, C, and D operands of the TABLE statement produce the tabulation intervals shown in Figure 7.8–1.

Interval 1	Interval 2	Interval 3	...	Interval D
B	B+C	B+2C	... B+(D–2)C	

Figure 7.8–1. The intervals specified by the B, C, and D
operands of the TABLE statement.

7.8.2 The TABULATE Block

The TABULATE block's only operand references the name (mnemonic or numeric) of a TABLE statement that specifies the SNA and the interval structure over which that SNA is to be tallied. When a transaction enters a TABULATE block, the SNA indicated by the associated TABLE statement is classified into one of the classes stipulated by the B, C, and D operands of the TABLE statement, and the frequency count of that class is incremented by 1.

Operand A: The table name (numeric or mnemonic) into which the SNA is to be tallied.

The TABLE and TABULATE combination

```
        T       TABLE       P3,100,10,3

        ⋮         ⋮             ⋮

                TABULATE    T
```

will enter the value of the third parameter of each transaction entering the TABULATE block into table T. The simple program and its output, given in Figure 7.8–2, illustrate the use of the TABLE and TABULATE blocks to produce a frequency table of arrival times. Note that the six values at or below 600 are collected in the first interval and all 23 values above 7800 are collected into the last interval.

The headings and contents of the first five columns are self-explanatory. The MULT. OF MEAN column is obtained by dividing the upper interval limit by the mean. Thus, the 0.118 in row 1 of column 5 is 600/5082.608. The last column is the standardized value of the upper interval limit (often called the z-value), obtained by $z = (x-\overline{x})/s$ where $\overline{x}$ and s are the mean and standard deviations, respectively. In row 1, $-1.544 = (600 - 5082.608)/2904.146$.

```
1                      SIMULATE
2
3          ARRIV      TABLE        C1,600,1200,8
4
5      1              GENERATE     100,50
6      2              TABULATE     ARRIV
7      3              SEIZE        FAC
8      4              ADVANCE      90,50
9      5              RELEASE      FAC
10     6              TERMINATE
11
12     7              GENERATE     10000
13     8              PRINT        ,,T
14     9              TERMINATE    1
15
16                    START        1,NP
```

PRINT BLOCK 8 AT CLOCK TIME 10000

TABLE	ARRIV					
ENTRIES	MEAN ARGUMENT	ST. DEV.		SUM OF ARGUMENTS		
97	5082.608	2904.146		493013.000		NON-WEIGHTED

UPPER LIMIT	OBS. FREQ.	PER CENT OF TOTAL	CUM. PERCENT	CUM. REMAINDER	MULT. OF MEAN	DEVIATION FROM MEAN
600	6	6.19	6.19	93.81	0.118	-1.544
1800	12	12.37	18.56	81.44	0.354	-1.130
3000	9	9.28	27.84	72.16	0.590	-0.717
4200	12	12.37	40.21	59.79	0.826	-0.304
5400	12	12.37	52.58	47.42	1.062	0.109
6600	12	12.37	64.95	35.05	1.299	0.522
7800	11	11.34	76.29	23.71	1.535	0.936

AVERAGE VALUE OF THE 23 OVERFLOW ITEMS IS 8830.174

```
18                           END
```

Figure 7.8–2. Use of the TABLE and TABULATE blocks to produce a frequency table of arrival times.

7.8.3 Variations of the TABLE Statement

GPSS allows three variations of the TABLE statement for the accumulation of special information into tables. The three variations, described below, are called modes.

IA (Interarrival) Mode: If the A operand of the TABLE statement is literally IA, then when transactions enter a TABULATE block associated with the TABLE statement, the interarrival times between successive transactions at the TABULATE block are entered into the table.

QTABLE (Queue TABLE) Mode: The statement mnemonic for this mode is QTABLE instead of TABLE, with the A operand specifying the name of a queue. Since QTABLE by definition refers to queue residence times, no TABULATE block is needed in this situation. The residence times of the queue indicated by the QTABLE block are tallied in the table according to the usual specifications of the B, C, and D operands.

RT (Rate) Mode: The A operand of the TABLE block must be the letters RT. In addition to the B, C, and D operands, an E operand specifying a time span must be present. The rate at which transactions arrive at a corresponding TABULATE block, (or, equivalently, the number of transactions passing through the TABULATE block), in time spans given by the E operand, are entered into the table.

For example,

TABLE   RT,10,15,15,100

and an associated TABULATE block will capture information on the rate (per 100 units of simulated time) at which transactions enter the TABULATE block. GPSS uses a special counter to gather the rate information mandated by a TABLE statement in the RT mode. This counter is incremented by 1 every time a transaction enters the TABULATE block. After a time lapse equal to the value specified by the E operand, the value of the counter is entered into the table, the counter is reset to 0, and the sequence is repeated. The program and output given in Figure 7.8–3 illustrate the use of tables in the IA and QTABLE modes. Notice that a numeric reference is chosen for the first table, in the IA mode. This table shows an average interarrival time of 102.875, which is close to the ideal of 100 stipulated by the GENERATE block. The second table, WAIT, gives a frequency distribution of the queue residence times.

```
1
2                    SIMULATE
3
4        1          TABLE        IA,50,25,6
5        WAIT       QTABLE       LINE,10,15,7
6
7        1          GENERATE     100,50
8        2          TABULATE     1
9        3          QUEUE        LINE
10       4          SEIZE        FAC
11       5          DEPART       LINE
12       6          ADVANCE      90,50
13       7          RELEASE      FAC
14       8          TERMINATE
15
16       9          GENERATE     10000
17       10         PRINT        ,,T
18       11         TERMINATE    1
19
20                  START        1,NP
```

PRINT BLOCK 10 AT CLOCK TIME 10000

TABLE	1					
ENTRIES	MEAN ARGUMENT		ST. DEV.	SUM OF ARGUMENTS		
96	102.875		29.372	9876.000		NON-WEIGHTED

UPPER LIMIT	OBS. FREQ.	PER CENT OF TOTAL	CUM. PERCENT	CUM. REMAINDER	MULT. OF MEAN	DEVIATION FROM MEAN
50	1	1.04	1.04	98.96	0.486	-1.800
75	18	18.75	19.79	80.21	0.729	-0.949
100	28	29.17	48.96	51.04	0.972	-0.089
125	23	23.96	72.92	27.08	1.215	0.753
150	26	27.08	100.00	0.00	1.458	1.604

TABLE	WAIT					
ENTRIES	MEAN ARGUMENT		ST. DEV.	SUM OF ARGUMENTS		
97	25.742		35.904	2497.000		NON-WEIGHTED

UPPER LIMIT	OBS. FREQ.	PER CENT OF TOTAL	CUM. PERCENT	CUM. REMAINDER	MULT. OF MEAN	DEVIATION FROM MEAN
10	55	56.70	56.70	43.30	0.388	-0.438
25	9	9.28	65.98	34.02	0.971	-0.021
40	5	5.15	71.13	28.87	1.554	0.397
55	6	6.19	77.32	22.68	2.137	0.815
70	8	8.25	85.57	14.43	2.719	1.233
85	6	6.19	91.75	8.25	3.302	1.650

AVERAGE VALUE OF THE 8 OVERFLOW ITEMS IS   110.125

```
21       END
```

Figure 7.8–3. The use of IA and QTABLE modes.

7.9 The PRIORITY, PREEMPT, and RETURN Blocks

In many applications, certain transactions must receive preferential treatment when competing with other transactions for the use of a facility. For example, if computational processes are modeled as transactions in the simulation of a time-sharing computer system, some of the critical operating system functions—those responsible for memory management or resource allocation, for instance—should take precedence in obtaining the services of the system.

Since GPSS orders the CEC by transaction priority and then by waiting time within priority levels, the order in which transactions are activated from the CEC can be influenced by assigning suitable priorities to transactions. So far, the only way to assign priority that has been described is through the E operand of the GENERATE block (Section 2.3). This allows the user to assign an integer from 0 to 127 as the priority of a transaction entering the simulation. The PRIORITY block provides the additional flexibility of altering the priority of a transaction as the transaction moves through the model blocks. Through the PREEMPT block, a transaction can capture a facility even if it is in use by another transaction with a lower priority than the preempting transaction. When a facility is captured by preemption, it must be relinquished through the RETURN block instead of the usual RELEASE block.

7.9.1 The PRIORITY Block

If a priority is not explicitly assigned to a transaction by the E operand of a GENERATE block, the transaction enters the simulation with a default priority of 0. The PRIORITY block allows the analyst to modify the priority of a transaction when the simulation is in progress. The PRIORITY block has two operands:

Operand A: The priority that the transaction will assume by entering the PRIORITY block. It must be an SNA with an integer value from 0 to 127 and cannot be defaulted.

Operand B: When defaulted, this operand will limit the action of the PRIORITY block to the reassignment of the transaction priority. If operand B is not defaulted, it must be the word BUFFER. The presence of this operand will cause the GPSS processor to stop the progress of the transaction that is in the PRIORITY block after adjusting its priority as dictated by operand A. GPSS then places the transaction, with its new priority, on the CEC and restarts scanning the CEC for a transaction to be moved forward.

The use of the B operand of the PRIORITY block is awkward because, with the B operand option, the PRIORITY block is actually a compaction of the same PRIORITY block, with the B operand defaulted, and the BUFFER block. The BUFFER block (which does not have any operands), when used by itself, causes the suspension of the active transaction and a rescanning of the CEC.

If processes are competing for the CPU of a time-sharing system, it may be appropriate to lower the priority of a process after it has received a certain amount of service from the CPU, simply by using the block

PRIORITY   0

or, if appropriate,

PRIORITY   0,BUFFER

7.9.2   The PREEMPT and RETURN Blocks

The SEIZE and RELEASE blocks, which were first encountered in Section 2.4, allowed transactions to capture a facility when it became available and to vacate the facility following some time lapse. By contrast, the PREEMPT and RETURN blocks allow transactions, under certain conditions, to capture a facility even when that facility is in use. Facility capture in the case of preemption is more complicated because, in addition to describing the progress of the preempting transaction, the analyst must also specify what is to happen to the preempted transaction. The details of preemption are given by the five operands of the PREEMPT block:

Operand A: This operand, which may not be defaulted, gives the name of the facility that is to be preempted.

Operand B: If this operand is defaulted, preemption will occur only if the current occupant of the facility is not itself a preemptor. If operand B is not defaulted, it must be the letters PR. In this case, preemption occurs only if the transaction attempting to preempt has a higher priority than the transaction (if any) using the facility.

Operand C: This operand is the label specifying the location to which the preempted transaction is to be sent.

Operand D: If preemption occurs, the amount of service time that is yet to be completed for the preempted transaction is saved as one of the parameters of the preempted transaction. Operand D gives the parameter number of the preempted transaction in which this information is stored.

Operand E: If operand E is defaulted, the preempted transaction continues to remain in contention for the facility from which it was removed. If this operand is the letters RE, the preempted transaction aborts its use of the facility and is removed from contention for that facility.

Section 7.9.3 contains a more detailed discussion of what happens to a preempted transaction and the options available to it. Following a preemptive capture, a facility is relinquished through the RETURN block so that the GPSS processor may take appropriate action associated with the possible reallocation of the facility to a previously preempted transaction. The RETURN block has a single A operand that specifies the name of the facility to be relinquished.

As an illustration of the use of the PREEMPT and RETURN blocks, consider the case of a chemical-processing plant that occasionally is interrupted because of equipment failure. Suppose that, due to the nature of the chemical process, after a failure is detected and corrected, the accumulated chemicals must be discarded and the chemical reactions restarted. To make the example more specific, assume that the time between equipment failures, the equipment repair times, and the time required for the chemical process are all uniformly distributed on the intervals 1000 ± 500, 60 ± 30, and 300 ± 10 minutes, respectively. The program structure given in Figure 7.9–1 assumes that transactions in the chemical process segment of the model have priority 0 and uses the PREEMPT and RETURN blocks to simulate such a situation.

For another use of the PREEMPT and RETURN blocks, consider a computing system in which executing programs are interrupted by a higher priority operating system process. For simplicity, assume that the interrupting process is a "timer interrupt," which interrupts the system at fixed intervals of 100 milliseconds. In such a situation, the interrupted executing program must return to the CPU to complete its necessary computation time. A program structure that accomplishes this is given in Figure 7.9–2.

7.9.3 Preempted Transactions

When a preemption is attempted, one of the following three conditions must prevail:

```
*               MODEL SEGMENT FOR FAILURES

                GENERATE        1000,500,,,1
                PREEMPT         EQUIP,PR,RESTART,,RE
                ADVANCE         60,30
                RETURN          EQUIP
                TERMINATE


*               MODEL SEGMENT FOR CHEMICAL PROCESS
                   ⋮
RESTART         SEIZE           EQUIP
                ADVANCE         300,10
                RELEASE         EQUIP

                   ⋮
```

Figure 7.9–1. Preemption where service is restarted.

```
*               TIMER-INTERRUPT SEGMENT

                GENERATE        100,,,1,1
                PREEMPT         CPU,PR,GETCPU,2
                ADVANCE         20,5
                RETURN          CPU
                TERMINATE


*               EXECUTING PROGRAM SEGMENT
                   ⋮
                SEIZE           CPU
                ADVANCE         800,200
                TRANSFER        DONE
GETCPU          ADVANCE         P2
DONE            RELEASE         CPU

                   ⋮
```

Figure 7.9–2. Preemption where service is continued.

1. The facility is idle.

2. The facility is in use with its occupant on the CEC, which can happen only if the preemption and the release of the facility by its current user are to occur at the same time.

3. The facility is in use with its occupant on the FEC and in an ADVANCE block waiting to complete its use of the facility.

In the first case, the preempting transaction is simply allowed to capture the facility. In the second case, the GPSS processor moves the transaction that has just completed its use of the facility out of the facility and allows the preempting transaction to capture the facility. Note that GPSS makes sure that this sequence is followed, even if the initial ordering on the CEC stipulated that preemption must occur prior to the release of the facility.

The third case requires something special to be done about the preempted transaction. To make possible the return of the transaction to the facility from which it was preempted, the transaction is removed from the FEC and placed on a special transaction list called an **Interrupt Chain** (IC). The IC is operated as a stack (last in, first out) rather than as a queue. Since any facility in the model may be preempted, several ICs may be needed, each uniquely identified with a single facility. The precise actions taken to move transactions from one chain to another depend on which of the following conditions prevail:

1. The B operand of the PREEMPT block is defaulted. In this situation, GPSS ignores the C, D, and E operands, moves the preempted transaction from the FEC to the IC, and keeps a record of the remaining service time that the preempted transaction needed to complete its residence at the facility. As soon as the preempting transaction is done with the facility, GPSS allocates the facility to the preempted transaction by setting the facility condition to busy and moving the preempted transaction from the IC to the FEC, with a departure time set to the current simulated time plus the remaining service time. Note that in this situation, facility interruption cannot extend beyond a single level because a preemptor cannot itself be preempted.

2. The B operand is not defaulted but the C operand is. The only difference between this case and the preceding one is that now preempting transactions may themselves be preempted by higher priority transactions. This makes multiple-leveled (up to 128 levels, since priorities range from 0 to 127) facility interruptions possible.

3. The B, C, and E operands are not defaulted, but the D operand is. This situation requires that the preempted transaction be redirected to a different location in the model, with no intention of finishing its service time in the facility from which it was preempted. The GPSS processor removes the preempted transaction from the FEC and places it on the CEC, as the last member within its priority group. When the progress of the preempting transaction is suspended, the CEC is rescanned. In this or some future rescanning of the CEC, the preempted transaction is set in motion starting with the block designated by the C operand.

4. The B, C, and D operands are not defaulted, but the E operand is. In this case, the preempted transaction must be returned to the facility from which it was preempted to complete its service in that facility. The GPSS processor moves the preempted transaction from the FEC and the ADVANCE block (designated here by ADV1) to the CEC as the last entry of its priority group and directs it to the block specified by the C operand. The preempted transaction moves through the blocks mandated by the program until it enters an ADVANCE block (ADV2, not necessarily the same ADVANCE block from which it was removed). It is at this point that the transaction is placed on the IC. When the facility from which the transaction was preempted becomes available, GPSS moves the preempted transaction from the IC to the FEC and into the ADVANCE block designated by ADV2, with a scheduled delay time equal to the remaining service time stored in the parameter of the transaction specified by the D operand of the original PREEMPT block.

7.10   Example: Emergency Medical System

An agency that is responsible for providing emergency medical assistance in the event of a natural disaster wishes to investigate the consequences of emergency medical delivery models in a variety of situations. The delivery system of particular interest is one in which a small team of medical personnel and the resources for setting up a minimal medical facility are dispatched to the disaster scene immediately. When injured individuals are brought in for treatment, it is first decided whether they are critically injured and need attention as soon as possible, or their injury is not critical and they can wait almost indefinitely for medical treatment. This form of medical service continues until additional help arrives or injured individuals can be evacuated to other medical centers.

As a first step in this study, the model in Figure 7.10–1 is developed. However, because of inaccuracies in the output of this program, a refinement of the program and its output are given in Figures 7.10–2 and 7.10–3, respectively. This study is based on the following worst-case assumptions:

1. The interarrival times between critically injured persons are exponentially distributed, with a mean of 60 minutes; the treatment times for these individuals are also exponentially distributed, with a mean of 45 minutes.

2. The interarrival times and treatment times of noncritically injured patients are exponentially distributed, with means of 10 and 7 minutes, respectively.

3. In 5% of the noncritical cases, after treatment has started, it is discovered that the injury is critical enough that the patient must be reclassified into the critical category, with a corresponding change in the patient's treatment time.

4. It may take as long as 18 hours before additional help arrives or patients can be evacuated to other medical facilities.

Waiting times, particularly for the critically injured, are of primary interest in this study. The second segment of the model given in Figure 7.10–1 describes the activities associated with critically injured arrivals and is straightforward. In this segment, critically injured patients arrive, acquire the medical facility through preemption, receive treatment, and leave. The purpose of the CRITICAL queue is to gather waiting-time statistics.

The first segment, dealing with arrivals that are initially labeled noncritical, is more complicated. Following entry into the NONCRIT queue, and eventually capturing the medical facility, the TYPE function (which will have value 0 with 95% probability and value 1 with 5% probability) is used to determine if the patient should be reclassified to critical status. The priority of the patient is adjusted, and the appropriate average treatment time (7 minutes if the classification is noncritical, 45 minutes if it is critical), which is given by the function AVGST, is used in the ADVANCE block. The standard output resulting from the model in Figure 7.10–1 gives the average queue residence times for the CRITICAL and NONCRIT queues.

There are two reasons that these averages, by themselves, do not provide the type of information that is needed in this situation. First, since the num-

```
1                        SIMULATE
2                        RMULT          54321,54321,54321
3
4        EXPON      FUNCTION       RN3,C24      Exp. dist. with mean 1
5        0,0/.1,.104/.2,.222/.3,.355/.4,.509/.5,.69/.6,.915/.7,1.2
6        .75,1.38/.8,1.6/.84,1.83/.88,2.12/.9,2.3/.92,2.52/.94,2.81/.95,2.99
7        .96,3.2/.97,3.5/.98,3.9/.99,4.6/.995,5.3/.998,6.2/.999,7/.9998,8
8
9        TYPE       FUNCTION       RN2,D2       95% remain priority 0; 5% change to 1.
10       0.95,0/1.0,1
11       AVGST      FUNCTION       P1,D2        Average service time, dependent on P1
12       1,7/2,45
13
14   1       GENERATE    10,FN$EXPON,,,0        Noncritical patients arrive.
15   2       QUEUE       NONCRIT                Get in line.
16   3       SEIZE       FAC                    Get the medical facility.
17   4       DEPART      NONCRIT                Leave the waiting line.
18   5       ASSIGN      1,FN$TYPE              Find out if they are really noncrit.
19   6       PRIORITY    P1                     and adjust their priority accordingly.
20   7       ADVANCE     FN$AVGST,FN$EXPON      Attention for avg. of 7 or 45 min.
21   8       RELEASE     FAC                    Leave the medical facility.
22   9       TERMINATE
23
24   10      GENERATE    60,FN$EXPON,,,1        Critically injured patients arrive.
25   11      QUEUE       CRITICAL               Get in line.
26   12      PREEMPT     FAC,PR                 Get facility, by preemption.
27   13      DEPART      CRITICAL               Leave the waiting line.
28   14      ADVANCE     45,FN$EXPON            Receive medical attention.
29   15      RETURN      FAC                    Leave the medical facility.
30   16      TERMINATE
31
32   17      GENERATE    1080                   18 hours.
33   18      TERMINATE   1
34
35           START       1
```

Figure 7.10–1. A simulation of the emergency medical delivery system.

ber of arrivals is rather small, particularly through the first segment, there is likely to be a significant disparity between the average waiting times and the average queue residence times. Second, regardless of what the average waiting time may be, it is important to know if some critically injured patients had to wait as long as an hour before receiving treatment. For these reasons, the program given in Figure 7.10–2 (with output shown in Figure 7.10–3) includes two QTABLE statements to gather waiting-time information of those with initial critical and noncritical classifications.

```
1                    SIMULATE
2
3                    RMULT          54321,54321,54321
4
5        EXPON       FUNCTION       RN3,C24     Exp. dist. with mean 1
6        0,0/.1,.104/.2,.222/.3,.355/.4,.509/.5,.69/.6,.915/.7,1.2
7        .75,1.38/.8,1.6/.84,1.83/.88,2.12/.9,2.3/.92,2.52/.94,2.81/.95,2.99
8        .96,3.2/.97,3.5/.98,3.9/.99,4.6/.995,5.3/.998,6.2/.999,7/.9998,8
9
10       TYPE        FUNCTION       RN2,D2      95% remain priority 0; 5% change to 1.
11       0.95,0/1.0,1
12       AVGST       FUNCTION       P1,D2       Average service time, dependent on P1
13       1,7/2,45
14
15       CWAIT       QTABLE CRITICAL,5,10,10              Times for critically injured.
16       NCWAIT      QTABLE NONCRIT,15,45,10              Times noncrit. injured.
17
18   1   GENERATE    10,FN$EXPON,,,0      Noncritical patients arrive.
19   2   QUEUE       NONCRIT              Get in line.
20   3   SEIZE       FAC                  Get the medical facility.
21   4   DEPART      NONCRIT              Leave the waiting line.
22   5   ASSIGN      1,FN$TYPE            Find out if they are really noncrit.
23   6   PRIORITY    P1                   and adjust their priority accordingly.
24   7   ADVANCE     FN$AVGST,FN$EXPON    Attention for 7 or 45 min.
25   8   RELEASE     FAC                  Leave the medical facility.
26   9   TERMINATE
27
28   10  GENERATE    60,FN$EXPON,,,1      Critically injured patients arrive.
29   11  QUEUE       CRITICAL             Get in line.
30   12  PREEMPT     FAC,PR               Get facility, by preemption.
31   13  DEPART      CRITICAL             Leave the waiting line.
32   14  ADVANCE     45,FN$EXPON          Receive medical attention.
33   15  RETURN      FAC                  Leave the medical facility.
34   16  TERMINATE
35
36   17  GENERATE    1080                 18 hours.
37   18  TERMINATE   1
38
39       START       1
```

Figure 7.10–2. A modification of the program in Figure 7.10–1 to capture detailed output on waiting times.

It is clear from the output shown in Figure 7.10–3 that the queue lengths are generally increasing over the 18 hours of simulated time. To develop further insight into the increasing waiting times as the simulation progresses, the GENERATE block of the timer segment and the START statement could be changed to

RELATIVE CLOCK 1080   ABSOLUTE CLOCK 1080

BLOCK	CURRENT	TOTAL		BLOCK	CURRENT	TOTAL
1	1	139		11	0	16
2	88	138		12	0	16
3	0	50		13	0	16
4	0	50		14	0	16
5	0	50		15	0	16
6	0	50		16	0	16
7	1	50		17	0	1
8	0	49		18	0	1
9	0	49				
10	1	17				

QUEUE	MAXIMUM CONTENTS	AVERAGE CONTENTS	TOTAL ENTRIES	ZERO ENTRIES	PERCENT ZEROS
CRITICAL	2	0.492	16	6	37.500
NONCRIT	91	33.606	138	4	2.899

	AVERAGE TIME/TRANS	$AVERAGE TIME/TRANS	TABLE NUMBER	CURRENT CONTENTS
	33.188	53.100	1	0
	263.007	270.858	2	88

FACILITY	AVERAGE UTILIZATION	NUMBER ENTRIES	AVERAGE TIME/TRANS	SEIZING TRANS. NO.	PREEMPTING TRANS. NO.
FAC	0.944	66	15.455	18	

TABLE CWAIT ENTRIES	MEAN ARGUMENT	ST. DEV.	SUM OF ARGUMENTS	
16	33.188	48.115	531.000	NON-WEIGHTED

UPPER LIMIT	OBS. FREQ.	PER CENT OF TOTAL	CUM. PERCENT	CUM. REMAINDER	MULT. OF MEAN	DEVIATION FROM MEAN
5	6	37.50	37.50	62.50	0.151	-0.586
15	3	18.75	56.25	43.75	0.452	-0.378
25	1	6.25	62.50	37.50	0.753	-0.170
35	1	6.25	68.75	31.25	1.055	0.038
45	1	6.25	75.00	25.00	1.356	0.246
55	1	6.25	81.25	18.75	1.657	0.453
65	0	0.00	81.25	18.75	1.959	0.661
75	1	6.25	87.50	12.50	2.260	0.869
85	0	0.00	87.50	12.50	2.561	1.077

AVERAGE VALUE OF THE 2 OVERFLOW ITEMS IS   141.500

TABLE NCWAIT ENTRIES	MEAN ARGUMENT	ST. DEV.	SUM OF ARGUMENTS	
50	206.820	269.456	10341.000	NON-WEIGHTED

Figure 7.10–3. The output of the program in Figure 7.10–2 (continues).

UPPER LIMIT	OBS. FREQ.	PER CENT OF TOTAL	CUM. PERCENT	CUM. REMAINDER	MULT. OF MEAN	DEVIATION FROM MEAN
15	15	30.00	30.00	70.00	0.073	-0.712
60	12	24.00	54.00	46.00	0.290	-0.545
105	0	0.00	54.00	46.00	0.508	-0.378
150	0	0.00	54.00	46.00	0.725	-0.211
195	10	20.00	74.00	26.00	0.943	-0.044
240	1	2.00	76.00	24.00	1.160	0.123
285	0	0.00	76.00	24.00	1.378	0.290
330	0	0.00	76.00	24.00	1.596	0.457
375	1	2.00	78.00	22.00	1.813	0.624

AVERAGE VALUE OF THE 11 OVERFLOW ITEMS IS   686.000

48   END

Figure 7.10–3. The output of the program in Figure 7.10–2 (concluded).

GENERATE   180

START   6,,1

respectively, to obtain intermediate output at simulated times of 180, 360, 540, ...

7.11   SNA Comparisons and the TEST Block

The purpose of the TEST block is to compare the values of two SNAs and take appropriate action, depending on the outcome of the comparison. The six types of comparisons that can be made are designated by an auxiliary operator that is set to one of E, G, GE, L, LE, and NE, representing, respectively, the relational operators equal to, greater than, greater than or equal to, less than, less than or equal to, and not equal to. In addition to operands A, B, and C, the TEST block has an auxiliary operator X that indicates which relation is to be used in connection with the TEST block.

Operand A: The first SNA to be used in the comparison.

Operand B: The second SNA to be used in the comparison.

Operand C: The location to which the transaction should move if the test result is false. The test result and the default option of the C operand are discussed below.

Auxiliary Operand X: This operand must be one of E, G, GE, L, LE, or NE.

When the TEST block executes, first the result of the implied test is determined. The implied test is "does the A operand have the prescribed relationship to the B operand?" The result of this test is either true or false. For example,

TEST G   5,Q$WAIT

will produce a true test result if 5 is greater than the current contents of the queue called WAIT; otherwise, the result of the test will be false.

When the C operand is used, the test is conducted in **alternate exit mode**: if the test result is true, the transaction continues its sequential path through the model; if the test result is false, the transaction moves to the block specified by the C operand. When the C operand is defaulted, the test is conducted in **refusal mode**. In this case, the test is conducted **before** the transaction enters the TEST block. If the test result is true, the transaction moves into the TEST block and proceeds sequentially through the model. If the test result is false, the transaction is refused entry into the test block and is kept in its current block location. Whenever the CEC is scanned, the test is performed again to see if a true test result could be obtained so that the transaction can move on. In the example of the TEST block

TEST G   5,Q$WAIT

if 5 is greater than the queue contents, then the transaction will move through the TEST block. If, however, the queue contains 5 or more entries, the transaction will be held up in its current block until the queue contents are less than 5 when the CEC is scanned.

The program given in Figure 7.11–1 simulates a telephone switchboard with eight lines set up to receive orders and inquiries during an advertising campaign, under the following conditions:

1. New calls come in with exponentially distributed interarrival times with an average of 2 minutes.

2. The conversation time for calls is normally distributed with mean and standard deviation of 2 minutes and 1/2 minute, respectively.

3. The 25% of callers who do not get through wait for some time (this waiting time is exponential with mean of 10 minutes) and attempt to call again.

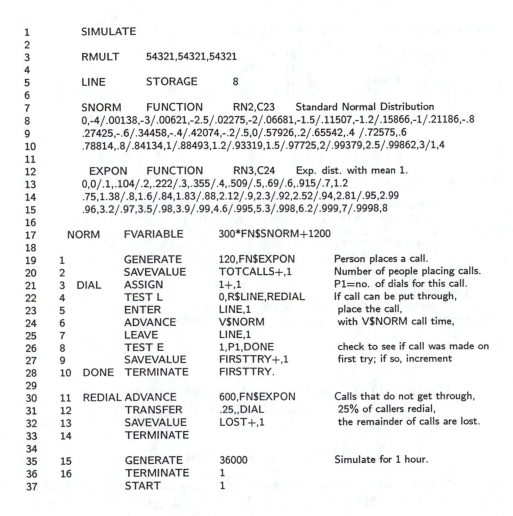

```
1         SIMULATE
2
3         RMULT       54321,54321,54321
4
5         LINE        STORAGE     8
6
7         SNORM       FUNCTION    RN2,C23    Standard Normal Distribution
8         0,-4/.00138,-3/.00621,-2.5/.02275,-2/.06681,-1.5/.11507,-1.2/.15866,-1/.21186,-.8
9         .27425,-.6/.34458,-.4/.42074,-.2/.5,0/.57926,.2/.65542,.4 /.72575,.6
10        .78814,.8/.84134,1/.88493,1.2/.93319,1.5/.97725,2/.99379,2.5/.99862,3/1,4
11
12        EXPON       FUNCTION    RN3,C24    Exp. dist. with mean 1.
13        0,0/.1,.104/.2,.222/.3,.355/.4,.509/.5,.69/.6,.915/.7,1.2
14        .75,1.38/.8,1.6/.84,1.83/.88,2.12/.9,2.3/.92,2.52/.94,2.81/.95,2.99
15        .96,3.2/.97,3.5/.98,3.9/.99,4.6/.995,5.3/.998,6.2/.999,7/.9998,8
16
17   NORM        FVARIABLE    300*FN$SNORM+1200
18
19   1          GENERATE     120,FN$EXPON     Person places a call.
20   2          SAVEVALUE    TOTCALLS+,1      Number of people placing calls.
21   3   DIAL   ASSIGN       1+,1             P1=no. of dials for this call.
22   4          TEST L       0,R$LINE,REDIAL  If call can be put through,
23   5          ENTER        LINE,1            place the call,
24   6          ADVANCE      V$NORM            with V$NORM call time,
25   7          LEAVE        LINE,1
26   8          TEST E       1,P1,DONE         check to see if call was made on
27   9          SAVEVALUE    FIRSTTRY+,1      first try; if so, increment
28   10  DONE   TERMINATE    FIRSTTRY.
29
30   11  REDIAL ADVANCE      600,FN$EXPON     Calls that do not get through,
31   12         TRANSFER     .25,,DIAL         25% of callers redial,
32   13         SAVEVALUE    LOST+,1           the remainder of calls are lost.
33   14         TERMINATE
34
35   15         GENERATE     36000            Simulate for 1 hour.
36   16         TERMINATE    1
37              START        1
```

Figure 7.11–1. A simulation of a telephone switchboard.

The output of the program (Figure 7.11–2) gives TOTCALLS, FIRSTTRY, and LOST, which represent the total number of callers, the number of callers who got through on their first attempt, and the number of calls that were lost, respectively. Tracing the path of transactions (calls), it can be seen that as a transaction enters the system, its first parameter is incremented to 1 to indicate that the call to be made will be a first call attempt. The TEST block

```
RELATIVE CLOCK 36000        ABSOLUTE CLOCK 36000

BLOCK   CURRENT   TOTAL        BLOCK   CURRENT   TOTAL
1       1         314          11      0         142
2       0         313          12      0         142
3       0         348          13      0         107
4       0         348          14      0         107
5       0         206          15      0         1
6       5         206          16      0         1
7       0         201
8       0         201
9       0         177
10      0         201
```

STORAGE	CAPACITY	AVERAGE	TOTAL	AVERAGE	AVERAGE
	CONTENTS	ENTRIES	TIME/TRANS	UTILIZ.	
LINE	8	6.585	206	1150.806	0.823
	CURRENT	FRACTION	CURRENT	MAXIMUM	
	STATUS	AVAIL.	CONTENTS	CONTENTS	
	A	1.000	5	8	

```
NON-ZERO FULLWORD SAVEVALUES
SAVEX      VALUE      SAVEX      VALUE         SAVEX      VALUE
FIRSTTRY   177        LOST       107           TOTCALLS   313

45                               END
```

Figure 7.11–2. The output of the telephone switchboard simulation in Figure 7.11–1.

TEST L   0,R$LINE,REDIAL

at line 22 checks to see if the number of telephone lines available (R$LINE) is positive. If not, the transaction is directed to location REDIAL where, following a waiting period, it is transferred back to location DIAL with a probability of 0.25. Each time a call is repeated, P1 is incremented by 1. Hence, P1 represents the number of attempts made by the caller to get through to the switchboard.

Once a call is made, the TEST block

TEST E   1,P1,DONE

at line 26 checks if it was a call that got through on its first attempt (the value of P1 would be 1 for all such calls) and increments savevalue FIRSTTRY if that was the case. The results of the 10-hour simulation, given in Figure 7.11–2, show that, of 313 total calls, 177 got through on their first attempts and 107 were lost.

7.12 Example: A Time-Shared Computer System

In this section, we reconsider the time-shared computer system that was analyzed in Section 6.4.1. In this system, n terminals are attached to a single CPU (central processing unit) and users at the terminals interact with the computer through cycles of thinking and making CPU requests. The thinking times and CPU requests follow exponential distributions with means of 25 seconds and 0.8 second, respectively. The CPU functions by placing all requests in a queue and serving them on a FIFO basis for up to 0.1 second plus 0.015 second that is used up internally for such things as queue maintenance. If a job that has captured the CPU is finished, a response is sent to the terminal; otherwise, it is placed at the end of the queue with its CPU need reduced by 0.1 second. In Section 6.4 we estimated the number of terminals that could be activated under the stipulation that the average response time would remain ≤ 6.

The average response times of 1.637, 3.125, 7.613, and 16.518 seconds (for 15, 25, 35 and 45 terminals, respectively) used in Section 6.4 were obtained through executing the program given in Figure 7.12–1 (see Figure 7.12–2 for the output). The program uses separate streams of exponential random variables to generate thinking times and CPU requests. The EXPON1 function is used by the variable CPUTME to produce exponentially distributed CPU requests with a mean of 0.8 second or 800 ms (milliseconds). At line 26, the EXPON2 function is used to generate exponentially distributed thinking times. The thinking time and CPU request patterns will be identical in repeated executions of the program with varying numbers of terminals.

The time unit used in savevalue initializations and throughout the simulation is milliseconds. We use the savevalues SLICE (the 100 ms of time allocated by the CPU at each pass), AVG (the 800 ms of average thinking time), SWAP (115 ms for CPU time allocation plus 15 ms of overhead required at each pass), and NOTERM (the number of active terminals on the system). Through the use of these savevalues, we can model different configurations of this system easily. Two additional savevalues, CTR and NOJOBS, are used to count the number of completed requests and the number of jobs (requests) to be completed before the end of the simulation.

The transactions in the main part of the program represent active terminals. The X$NOTERM transactions created at the start of the simulation immediately enter the thinking state at the ADVANCE block (line 26). The two ASSIGN blocks (lines 27 and 28) place the CPU time demand from the terminal into parameters 1 and 2 of the transaction. P2 will continue to

```
1                       SIMULATE
2
3       EXPON1   FUNCTION                    RN1,C24      Exp. dist. with mean 1.
4       0,0/.1,.104/.2,.222/.3,.355/.4,.509/.5,.69/.6,.915/.7,1.2
5       .75,1.38/.8,1.6/.84,1.83/.88,2.12/.9,2.3/.92,2.52/.94,2.81/.95,2.99
6       .96,3.2/.97,3.5/.98,3.9/.99,4.6/.995,5.3/.998,6.2/.999,7/.9998,8
7
8       EXPON2   FUNCTION                    RN2,C24      Exp. dist. with mean 1.
9       0,0/.1,.104/.2,.222/.3,.355/.4,.509/.5,.69/.6,.915/.7,1.2
10      .75,1.38/.8,1.6/.84,1.83/.88,2.12/.9,2.3/.92,2.52/.94,2.81/.95,2.99
11      .96,3.2/.97,3.5/.98,3.9/.99,4.6/.995,5.3/.998,6.2/.999,7/.9998,8
12
13      INITIAL      X$OHEAD,15/X$SLICE,100   Overhead/CPU slice in ms.
14      INITIAL      X$AVG,800/X$SWAP,115     Avg. CPU request/Swap cycle.
15      INITIAL      X$NOTERM,32              Number of terminals.
16      INITIAL      X$CTR,0/X$NOJOBS,5000    Counter for jobs/total jobs.
17
18      CPUTME   FVARIABLE     X$AVG*FN$EXPON1    CPU time required.
19      LSTTME   FVARIABLE     P1+X$OHEAD         CPU time on last cycle.
20
21      RSPTME   TABLE         M1,600,600,25      Table for response times.
22
23      ******** MODEL LOGIC SEGMENT
24
25   1          GENERATE     ,,,X$NOTERM           Start all terminals.
26   2  STRT    ADVANCE      25000,FN$EXPON2       Terminals in think state.
27   3          ASSIGN       2,V$CPUTME            P2 = total CPU time needed.
28   4          ASSIGN       1,P2                  P1 = CPU time left on job.
29   5          MARK         ,                     Record the time.
30   6  LOP     QUEUE        CPUQ                  Wait for the CPU.
31   7          SEIZE        CPU                   Access the CPU.
32   8          DEPART       CPUQ                  Leave the waiting line.
33   9          TEST GE      P1,X$SLICE,LAST       Is this last pass thru CPU?
34  10          ASSIGN       1-,X$SLICE            if not, reduce remaining time,
35  11          ADVANCE      X$SWAP                receive CPU service,
36  12          RELEASE      CPU                   leave the CPU,
37  13          TRANSFER     ,LOP                  get back in line.
38  14  LAST    ADVANCE      V$LSTTME              CPU service for last time.
39  15          RELEASE      CPU                   Leave the CPU.
40  16          ASSIGN       1,0                   Reset CPU time needed to 0.
41  17          TABULATE     RSPTME                Make entry in table.
42  18          SAVEVALUE    CTR+,1                Increment completed job counter.
43  19          TEST GE      X$CTR,X$NOJOBS,STRT   If sufficient number
44  20          TERMINATE    1                     completed, stop.
45
46          START        1
```

Figure 7.12–1. A simulation of a time-shared computer system.

RELATIVE CLOCK 4772796   ABSOLUTE CLOCK 4772796

BLOCK	CURRENT	TOTAL	BLOCK	CURRENT	TOTAL
1	0	32	11	0	37326
2	22	5031	12	0	37326
3	0	5009	13	0	37326
4	0	5009	14	0	5000
5	0	5009	15	0	5000
6	9	42335	16	0	5000
7	0	42326	17	0	5000
8	0	42326	18	0	5000
9	0	42326	19	0	5000
10	0	37326	20	0	1

QUEUE	MAXIMUM CONTENTS	AVERAGE CONTENTS	TOTAL ENTRIES	ZERO ENTRIES	PERCENT ZEROS
CPUQ	17	4.957	42335	37542	88.678

		AVERAGE TIME/TRANS	$AVERAGE TIME/TRANS	TABLE NUMBER	CURRENT CONTENTS
		558.899	4936.570	9	

FACILITY	AVERAGE UTILIZATION	NUMBER ENTRIES	AVERAGE TIME/TRANS	SEIZING TRANS. NO.	PREEMPTING TRANS. NO.
CPU	0.966	42326	108.930		

NON-ZERO FULLWORD SAVEVALUES

SAVEX	VALUE	SAVEX	VALUE	SAVEX	VALUE
AVG	800	CTR	5000	NOJOBS	5000
NOTERM	32	OHEAD	15	SLICE	100
SWAP	115				

TABLE ENTRIES	RSPTME MEAN ARGUMENT	ST. DEV.	SUM OF ARGUMENTS	
5000	5646.794	3161.180	28233972.000	NON-WEIGHTED

UPPER LIMIT	OBS. FREQ.	PER CENT OF TOTAL	CUM. PERCENT	CUM. REMAINDER	MULT. OF MEAN	DEVIATION FROM MEAN
600	136	2.72	2.72	97.28	0.106	-1.596
1200	141	2.82	5.54	94.46	0.213	-1.407
1800	247	4.94	10.48	89.52	0.319	-1.217
2400	293	5.86	16.34	83.66	0.425	-1.027
3000	274	5.48	21.82	78.18	0.531	-0.837
3600	333	6.66	28.48	71.52	0.638	-0.647
4200	368	7.36	35.84	64.16	0.744	-0.458
4800	401	8.02	43.86	56.14	0.850	-0.268
5400	383	7.66	51.52	48.48	0.956	-0.078
6000	366	7.32	58.84	41.16	1.063	0.112
6600	314	6.28	65.12	34.88	1.169	0.302

Figure 7.12–2. The output of the simulation of the time-shared computer system in Figure 7.12–1 (continues).

7200	291	5.82	70.94	29.06	1.275	0.491
7800	272	5.44	76.38	23.62	1.381	0.681
8400	227	4.54	80.92	19.08	1.488	0.871
9000	199	3.98	84.90	15.10	1.594	1.061
9600	159	3.18	88.08	11.92	1.700	1.251
10200	129	2.58	90.66	9.34	1.806	1.440
10800	122	2.44	93.10	6.90	1.913	1.630
11400	91	1.82	94.92	5.08	2.019	1.820
12000	63	1.26	96.18	3.82	2.125	2.010
12600	48	0.96	97.14	2.86	2.231	2.200
13200	46	0.92	98.06	1.94	2.338	2.389
13800	34	0.68	98.74	1.26	2.444	2.579
14400	22	0.44	99.18	0.82	2.550	2.769

AVERAGE VALUE OF THE 41 OVERFLOW ITEMS IS   15232.122

49   END

Figure 7.12–2. The output of the simulation of the time-shared computer system in Figure 7.12–1 (concluded).

hold this value until the entire computation request is satisfied; however, P1 will be decremented (line 34) as 100-ms slices of CPU time are allocated to this request. Consequently, P1 will always represent the CPU time yet to be allocated.

The MARK block (line 29) and the TABLE/TABULATE combination (lines 21 and 41) will gather the response time for each CPU request into the table RSPTME. It is certainly possible to obtain the average response time without using a table, by accumulating the response times and dividing the sum by X$NOJOBS. However, this would not give the modeler any sense of the distribution of response times.

After a transaction has captured the CPU (line 31), the TEST block (line 33) determines (by comparing P1 to X$SLICE) if the residual CPU time needed by the transaction is less than 0.1 second. If so, this will be the transaction's last pass through the CPU. In this case, the transaction moves to location LAST, finishes its use of the CPU (line 38), enters its response time in the RSPTME table (line 41), and increments the savevalue CTR (line 42), the count of CPU requests serviced. If this is not the transaction's last pass through the CPU, the balance of the transaction's CPU time is decremented by 100 ms, and the transaction reenters the queue at location LOP.

At this point, a check is made by the TEST block (line 43) to see if as many as X$NOJOBS have been served. If not, the terminal transaction moves to STRT and restarts its cycle at the "think" state; otherwise, it shuts off the simulation.

The analysis in Section 6.4 led us to believe that 31.98 active terminals can be attached to the time-shared computer if the average response time is to be ≤ 6 seconds. The GPSS program in Figure 7.12–1 simulates this system with 32 terminals (X$NOTERM is initialized to 32) and its output is given in Figure 7.12–2. From the RSPTME table we see that the average response time was 5.65 seconds. From the distribution pattern we can also see that almost 59% of all response times were below 6 seconds, with about 4% being greater than 12 seconds. Thus, while the overall interactive nature of the system is maintained, in some relatively few cases response times get as large as 15 seconds. We should note that when the simulation was run with X$NOTERM = 33, an average response time of 6.26 seconds was obtained.

7.13   The LOOP **Block**

The TEST block provides a mechanism for constructing loops similar to the FORTRAN DO loop or the FOR-NEXT loop in BASIC. A transaction parameter or a savevalue can be used as a loop counter that is incremented or decremented (following initialization) until a desired value is attained and the loop is terminated. The construct given in Figure 7.13–1 uses the savevalue COUNT as the loop counter for 15 iterations through the loop. A similar construct, shown in Figure 7.13–2, uses the first transaction parameter as the loop counter.

The LOOP block simplifies loop construction by combining loop counter decrementation with testing and branching into a single block. The restrictions on the LOOP block are that a transaction parameter must be used as the loop counter, the counter will be decremented at each pass through the loop, and the loop is aborted only when the loop counter becomes 0. The LOOP block has two operands:

Operand A: The parameter number to be used as the loop counter; it cannot be defaulted.

Operand B: Must be the symbolic name of a block location; it cannot be defaulted.

Figure 7.13–3 uses the LOOP block, with the transaction's parameter 1 as the loop counter, to construct a loop that is logically equivalent to that given in Figure 7.13–2.

⋮

```
            SAVEVALUE      COUNT,10
            ⋮
CYCLE       SAVEVALUE      COUNT+,1
            ⋮
            TEST  E        25,X$COUNT,CYCLE
            ⋮
```

Figure 7.13–1. The use of a SAVEVALUE for a loop counter.

⋮

```
            ASSIGN         1,15
            ⋮
CYCLE       ASSIGN         1-,1
            ⋮
            TEST E         0,P1,CYLCE
            ⋮
```

Figure 7.13–2. The use of a transaction parameter for a loop counter.

⋮

```
            ASSIGN         1,15
            ⋮
CYCLE       ...
            ⋮
            LOOP           1,CYCLE
```

Figure 7.13–3. The use of the LOOP block.

7.14 Logic Switches, INITIAL Statements, LOGIC and GATE Blocks

Logic switches are GPSS entities used to signal the occurrence of model events. Logic switches are typically used to synchronize the movement of transactions by postponing the progress of certain transactions through portions of the model until some condition is satisfied. The simulation of a traffic signal to control traffic flow or of a "gone to lunch" sign to hold off transaction movement for a specified period of time is greatly simplified by the use of logic switches. Logic switches are actually memory locations representing Boolean variables. By assuming either a set (on) or reset (off) value, they can simulate two-way switches. There are no built-in GPSS entities that simulate more complicated schemes such as three-way switches; but savevalues, in conjunction with TEST blocks, can be used to simulate more complex multiway switches. GPSS provides a way to initialize the value of logic switches through an INITIAL statement, modify the values of logic switches during execution through the LOGIC block, and block the progress of transactions through the GATE block.

7.14.1 Initialization of Logic Switches

If logic switches are not explicitly initialized, they are placed in the reset position before program execution. Basically, the same scheme that was used to initialize savevalues is also used to initialize logic switches. In the case of logic switches, however, explicit initial values are not supplied, since initialization is necessary only when a logic switch is to be given an initial set value. The INITIAL statement for logic switch initialization consists of the word INITIAL in the operation field and a list of symbolic or numeric logic switch names separated by / in the operand field. For instance, the statement

INITIAL   LS2/LS$DOOR/LS$FLAG

initializes to set values for the second logic switch as well as the logic switches named DOOR and FLAG.

Unlike savevalues, there is no special provision for selectively preventing the resetting of logic switches when a CLEAR statement is used. The CLEAR statement puts all logic switches into the reset position, and the RESET statement does not alter the values of logic switches.

7.14.2   The LOGIC Block

The LOGIC block modifies the values of logic switches during program execution. This block has an operand A and an auxiliary operator X:

Operand A: This operand, which cannot be defaulted, is the numeric or symbolic name of the logic switch to be modified.

Auxiliary Operand X: Must be one of I, R, or S, to indicate inversion, resetting, or setting of the logic switch named by operand A.

When a transaction enters a LOGIC block, with an auxiliary operator of I, the logic switch specified by the A operand is inverted—changed from reset to set or from set to reset. If the auxiliary operator is R, the logic switch is reset regardless of its previous value. If the auxiliary operator is S, the logic switch is set again, regardless of its previous value.

7.14.3   Logic Switches and the GATE Block

This section will consider the use of the GATE block only in connection with logic switches. Other uses of the GATE block will be discussed in Section 7.14.4. The GATE block, when used in conjunction with logic switches, controls the flow of transactions through portions of the model much as a TEST block does. In the case of the TEST block, a test is performed by comparing two SNAs and a "true" or "false" outcome is associated with that test. A test is also performed in connection with a GATE block, but in this case the test checks the condition of a logic switch and produces a "true" or "false" value depending on whether the value of the logic switch agrees with the auxiliary operator of the GATE block. The mnemonics LS and LR are used for the auxiliary operator to indicate testing for the set and reset conditions, respectively. The GATE block has A and B operands and an auxiliary operator:

Operand A: This operand is the numeric or symbolic name of the logic switch; it may not be defaulted.

Operand B: This operand is a model block location. The consequences of using or defaulting it will be discussed below.

Auxiliary Operand X: This operand, LS or LR, is used to produce "true" test results when the logic switch in question is in a set (in the case of LS) or reset (in the case of LR) position.

If the B operand is defaulted, the GATE block functions in **refusal mode**: the test is conducted immediately before a transaction enters the GATE block and if the result is "true," then the "gate is open" and the transaction proceeds normally through the model blocks; if the test result is "false," then the "gate is closed" and the transaction is held up in its current block.

If the B operand is used, the GATE block functions in **alternate exit mode**, in which transactions are always allowed entry into the GATE block. If the test result is "true," a transaction entering the GATE block proceeds sequentially through the model. If the test result is "false," the transaction is directed to the block whose location is specified by the B operand.

A common situation in which logic switches and GATE blocks can be used effectively is when it is desirable to allow transactions to enter a system and be processed until a specified time. Following the deadline, additional transactions are not permitted to enter the system, but those transactions already in the system continue their normal course through the model. A bank or a supermarket operation fits this pattern. At closing time, no new customers are allowed to come in. However, those customers who are already in receive the usual service. Figure 7.14–1 shows a program (with its output) that uses the logic switch DOOR and the GATE block to simulate customer service in a bank with 5 tellers. The logic switch DOOR is initialized to "set" by the INITIAL statement and transactions are allowed to move through the GATE block as long as DOOR retains its "set" value. The LOGIC block in the timer segment of the simulation changes the status of DOOR to "reset," effectively locking the bank door at time 480 (if minutes are used for the simulation time, this represents 8 hours of simulated time). Transactions arriving after time 480 are directed to location GOAWAY and are removed from the simulation. The TEST block allows the simulation to continue until the total counts of blocks 3 and 7 are equal. Since this happens when all transactions in the system have been processed, the simulation stops when all customers have received service.

7.14.4 Other Forms of the GATE block

The GATE block, with different auxiliary operators, can also be used to control access to facilities and storages. The role of operand A is expanded to reflect logical conditions associated with facilities and storages. A complete listing of the choices available for operand A and the corresponding choices for the mnemonics of the auxiliary operator is given in Table 7.14–2. In all cases a "true" or "false" test result is obtained and the progress of a transaction attempting to enter the GATE block is controlled in the manner described in Section 7.14.3.

```
1              SIMULATE
2              TELLER      STORAGE        5                Five tellers in the bank.
3              INITIAL     LS$DOOR                         Set logic switch DOOR to "set."
4
5         1    GENERATE    10,5                            Customers arrive,
6         2 GATE  LS                  DOOR,GOAWAY  Move through DOOR, if it is "set."
7         3    QUEUE       LINE
8         4    ENTER       TELLER,1
9         5    DEPART      LINE
10        6    ADVANCE     45,15
11        7    LEAVE       TELLER,1
12        8 GOAWAY  TERMINATE
13
14        9    GENERATE    480                             At end of 8-hour day,
15        10 LOGIC   R                 DOOR              DOOR is "reset" and the simulation
16        11 TEST    E                 N3,N7             continues until all customers
17        12   TERMINATE   1                             are processed.
18
19             START       1
```

RELATIVE CLOCK 520   ABSOLUTE CLOCK 520

BLOCK	CURRENT	TOTAL		BLOCK	CURRENT	TOTAL
1	1	50		11	0	1
2	0	49		12	0	1
3	0	45				
4	0	45				
5	0	45				
6	0	45				
7	0	45				
8	0	49				
9	1	2				
10	0	1				

QUEUE	MAXIMUM CONTENTS	AVERAGE CONTENTS	TOTAL ENTRIES	ZERO ENTRIES	PERCENT ZEROS
LINE	1	0.100	45	35	77.778

	AVERAGE TIME/TRANS	$AVERAGE TIME/TRANS	TABLE NUMBER	CURRENT CONTENTS	
	1.156	5.200	0		
STORAGE	CAPACITY CONTENTS	AVERAGE ENTRIES	TOTAL TIME/TRANS	AVERAGE UTILIZ.	AVERAGE
TELLER	5	3.942	45	45.556	0.788
	CURRENT STATUS	FRACTION AVAIL.	CURRENT CONTENTS	MAXIMUM CONTENTS	
	A	1.000	0	5	

```
23             END
```

Figure 7.14–1. The use of the GATE block and logic switches.

Table 7.14–2. Operand options for the GATE block.

OPERAND A	AUXILIARY OPERATOR	TEST
Logic Switch Name	LS	Logic Switch is "Set"
	LR	Logic Switch is "Reset"
Facility Name	U	Facility is in use
	NU	Facility is not in use
	I	Facility is interrupted (preempted)
	NI	Facility is not interrupted (not preempted)
	FV[†]	Facility is available
	FNV[†]	Facility is not available
Storage Name	SE	Storage is empty
	SF	Storage is full
	SNE	Storage is not empty
	SNF	Storage is not full
	SV[†]	Storage is available
	SNV[†]	Storage is not available

[†]Options available with GPSS V.

As a simple illustration of the use of the SNF auxiliary operator, consider a situation where a barbershop has three barbers who serve customers as they arrive at the shop. If all three barbers are busy when customers arrive, the customers leave the shop rather than wait for a barber to become available. The program given in Figure 7.14–3 simulates this situation under the assumption of uniform interarrival times in the range of 5 to 15 minutes and uniform service times in the range of 15 to 35 minutes.

```
1       SIMULATE
2
3       BARBER    STORAGE        3                Three barbers.
4
5    1            GENERATE       10,5             Customers arrive.
6    2            GATE  SNF      BARBER,GOAWAY    If a barber is available,
7    3            ENTER          BARBER           they get a barber,
8    4            ADVANCE        25,10            receive a haircut,
9    5            LEAVE          BARBER           and leave.
10   6            TERMINATE      1
11   7  GOAWAY    TERMINATE
12
13      START          50
```

Figure 7.14–3. A barbershop with no waiting.

7.15   The SELECT Block

An assumption was embedded within the program describing the bank operation given in Figure 7.14–1. The treatment of tellers as a storage implied
that the customers in the bank joined a single queue and captured a teller
when one became available. We could also consider a bank with five tellers
as having five queues, one for each teller. When customers enter the bank,
they join the shortest queue (or, in case of ties, one of the shortest queues)
expecting to receive the least waiting time in that queue.

GPSS enables transactions to choose among alternatives through the SE-
LECT block. In the selection process, a set of numerically contiguous model
entities (e.g., facilities 3 through 7) is scanned in increasing numerical order
to find an entity satisfying a given condition. If such an entity is found within
the scanning range, the number of that entity is copied into a transaction
parameter; if no such entity is found, the parameter value is set to 0. The
precise syntax of the SELECT block depends on the mode in which the block
is used. The modes, the choices of the auxiliary operators for each mode,
and the basis for the selection process are summarized in Table 7.15–1.

7.15.1   The Relational Mode

In this mode, the SELECT block has six operands, of which only operand F
can be defaulted. The auxiliary operator in this case must be one of G, GE,
E, L, LE, or NE.

Operand A: A parameter number into which the number of the selected
    entity is placed. This operand cannot be defaulted.

Operand B: The starting point for the scanning process. Operand B cannot
    be defaulted.

Operand C: This operand, which also cannot be defaulted, is the ending
    point for the scanning process. The B and C operands together de-
    termine the range of entity numbers over which the search will be
    conducted.

Operand D: The value with which the SNA specified by the E operand must
    be compared.

Operand E: The family name of the SNA.

Table 7.15–1. The modes of the SELECT block.

MODE	AUXILIARY OPERATOR	SELECTION CRITERION
Relational	G,GE, E,NE, LE,L	Select first entity with the SNA of the E operand that has the specified relation to the SNA of the D operand.
MIN-MAX	MIN, MAX	Select entity with minimum/maximum value of the SNA of the E operand over the scanning range.
LOGICAL	LS, LR, U, NU, SF, SNF SE, SNE, I, NI NI	Select first entry satisfying the condition of the auxiliary operator over the scanning range.

Operand F: This operand gives the location to which the transaction entering the SELECT block will move if the selection fails. If it is defaulted, the transaction moves to the next sequential block.

When a transaction enters a SELECT block in the relational mode, a search is initiated over the range specified by the B and C operands for an SNA value whose family name is given by the E operand and that has the relationship to operand D stipulated by the auxiliary operator. In the case of a successful search, the number of the first entity that satisfies the search is placed in the transaction parameter specified by operand A. If the search fails, the value of the parameter given by operand A is modified to 0 and the transaction moves to the next sequential block or the location stipulated by the F operand, depending on whether or not the F operand was defaulted.

The SELECT block

SELECT E   4,2,6,0,F

will cause a scanning of facilities 2 through 6 in search of an idle facility (the F at operand E is the idle/busy SNA for facilities). The value of parameter 4 of the transaction entering this block will become the number of the first idle facility if an idle facility is encountered during the search; otherwise, parameter 4 will be set to 0. The block

SELECT LE   5,4,9,2,Q

will initiate a search of queues 4 through 9 to find a queue whose current content (the Q SNA in operand E) is less than or equal to 2. If such a queue is found, its number will be stored in parameter 5 of the transaction entering the SELECT block; otherwise, parameter 5 is set to 0. The SELECT block

SELECT G   P1,2,5,100,FR

will cause a search of facilities 2 through 5 to find the first facility with fractional utilization (in parts per 1000) greater than 100. In the case of a successful search, the number of the first such facility will be placed in a parameter whose number is in parameter 1 of the transaction entering the SELECT block. If the search fails, the value of the parameter whose number is in parameter 1 will be changed to 0.

7.15.2   The Min-Max Mode

In this mode, operands A, B, C, and E continue to have the same meaning as they did in the relational mode. Operands D and F are necessarily defaulted, and the auxiliary operator is restricted to MIN or MAX. When a transaction enters a SELECT block in Min-Max mode, a scan is conducted over the range specified by the B and C operands to determine the entity number of the SNA whose family name is given by the E operand with a minimum or maximum value stipulated by the auxiliary operator. This entity number is then placed in the parameter specified by the A operand. Note that since there is always a minimum or a maximum value, a search cannot fail in Min-Max mode.

The SELECT block

SELECT MIN   5,1,P3,,Q

will scan queues 1 through the queue number given by parameter 3 for the queue with the minimum content (if there are several such queues, the one with the smallest number will be chosen). This queue number will then be placed in parameter 5 of the transaction entering the SELECT block.

7.15.3   The Logical Mode

In this mode, the auxiliary operator makes explicit the condition to be checked; therefore, the E operand is not needed. Also, as in the Min-Max mode, comparisons will not be made, making operand D unnecessary. Operands A, B, and C will continue to have the same meaning as they had in the other two modes.

When a transaction enters a SELECT block in the logical mode, a search is made over the range indicated by the B and C operands for the first entity (a logic switch, a facility, or a storage) with the logical condition described by the auxiliary operator. If the search is successful, the number of this entity is then placed in the parameter specified by the A operand. If the search fails, a 0 is placed in that parameter and the transaction moves either to the location given by the F operand (if the F operand is not defaulted) or to the next sequential block.

The SELECT block

$$\text{SELECT NU} \quad 3,5,8,,,\text{WAIT}$$

will initiate a scan of facilities 5 through 8 to find the first facility that is not in use (the NU condition). If such a facility is found, its number will be saved in parameter 3. If all facilities in the range 5 through 8 are in use, the transaction will move to block location WAIT with its parameter 3 set to 0. The meanings of the auxiliary operators were given in Table 7.14–2.

7.16 Example: A Multiple-Checkout Supermarket Operation

At busy times when most people do their grocery shopping, a medium-size supermarket operates 6 cash registers. The manager of the supermarket notices that on some occasions, potential shoppers enter the store, see long queues at the cash registers, and leave without making any purchases. To better understand the behavior of customers, the manager surveys the people who enter the store (whether they buy anything or not) and determines that

1. Most of those who leave without buying anything intended to purchase relatively few items.

2. Those customers who intend to purchase 12 or fewer items leave the store without buying anything if they see at least 3 people waiting (in addition to the customer being checked out) at each checkout line when they enter the store.

3. Those who intended to purchase more than 12 items are more tolerant of long lines; they do not change their shopping plans unless the checkout lines have 9 or more people.

A study of cash-register records, customer purchasing patterns, and the checkout operation reveal that

1. During the peak hours of operation, the interarrival times of customers follow the exponential distribution with a mean of 22 seconds.

2. The number of items a shopper purchases is normally distributed with mean 16 and standard deviation 4.

3. On the average, customers spend 1 minute per item doing their shopping.

4. The average purchase price per item is $1.75.

5. With their new computer-assisted checkout system, it takes 4 seconds for a clerk to enter the price of a purchased item into the cash register; an additional 75 seconds is needed to take money (or a check) and make change, regardless of the number of items purchased.

The manager wants to reduce the number of customers who leave immediately upon arrival, primarily to increase the number of people who habitually shop at the store. In fact, the manager is willing to sacrifice some modest reduction in sales during peak hours if as a consequence the regular customer base can be expanded. Due to physical limitations, the obvious solution of adding more checkout stations during peak hours is not feasible. The simulation shown in Figure 7.16–1, with its output given in Figure 7.16–2, investigates the option of converting of one of the 6 checkout stations into an express station available only to customers with 12 or fewer items.

The program in Figure 7.16–1 simulates the operation of the supermarket at peak times with 1 express and 5 regular checkout stations. The time unit used is seconds, and the monetary unit is cents. The INITIAL statement (line 3) initializes savevalues NMCHK (the total number of checkout stations), NMXPR (the number of express checkout stations), and XPRPO (the number of express checkout stations plus 1). The XPRO savevalue is used (lines 25 and 36) when we scan the nonexpress checkout lines.

The variable NOITEMS generates the number of items that a customer will purchase, and the variables PRCH, PRCHTME, and PAYTME calculate the sale total (in cents), the time required to shop, and the time required to check out, respectively. Upon a customer's arrival, through the GENERATE block at line 19, parameter 1 of the customer transaction is set to the number of items that the customer intends to purchase. Hence, the second through the fourth parameters represent the approximate sale value, the time to shop, and the time to check out.

The TEST block at line 24 separates those who will buy 12 or fewer items from those who will buy a larger number. Those with larger purchases look

```
 1                      SIMULATE
 2          RMULT       1234567,1234567,1234567
 3          INITIAL     X$NMCHK,6/X$NMXPR,1/X$XPRPO,2
 3
 4          EXPON       FUNCTION      RN3,C24      Exp. dist. with mean 1
 5          0,0/.1,.104/.2,.222/.3,.355/.4,.509/.5,.69/.6,.915/.7,1.2
 6          .75,1.38/.8,1.6/.84,1.83/.88,2.12/.9,2.3/.92,2.52/.94,2.81/.95,2.99
 7          .96,3.2/.97,3.5/.98,3.9/.99,4.6/.995,5.3/.998,6.2/.999,7/.9998,8
 8
 9          SNORM       FUNCTION      RN2,C23      Standard Normal Distribution
10          0,-4/.00138,-3/.00621,-2.5/.02275,-2/.06681,-1.5/.11507,-1.2/.15866,-1/.21186,-.8
11          .27425,-.6/.34458,-.4/.42074,-.2/.5,0/.57926,.2/.65542,.4 /.72575,.6
12          .78814,.8/.84134,1/.88493,1.2/.93319,1.5/.97725,2/.99379,2.5/.99862,3/1,4
13
14          NOITEMS     FVARIABLE     4*FN$SNORM+16   No. of items to buy.
15          PRCH        VARIABLE      175*P1          Price of all items.
16          PRCHTME     VARIABLE      60*P1           Time to buy items.
17          PAYTME      VARIABLE      75+4*P1         Time to check out.
18
19    1                GENERATE      22,FN$EXPON,,,,F     Customers arrive.
20    2                ASSIGN        1,V$NOITEMS          P1 ⟵ no. of items.
21    3                ASSIGN        2,V$PRCH             P2 ⟵ price of items.
22    4                ASSIGN        3,V$PRCHTME          P3 ⟵ shopping time.
23    5                ASSIGN        4,V$PAYTME           P4 ⟵ checkout time.
24    6                TEST G        P1,12,XPRSS          If > 12 items, check
25    7                SELECT MIN    5,X$XPRPO,X$NMCHK,,Q  nonexpress lines
26    8                TEST LE       Q*P5,8,BOLT          and leave if all
27    9                TRANSFER      ,ENTER               have > 8.
28
29   10 XPRSS  SELECT MIN    5,1,X$NMCHK,,Q         If <= 12 items, check
30   11        TEST LE       Q*P5,2,BOLT            all lines and leave
31    *                                            if all have > 2.
32   12 ENTER  ADVANCE       P3                     Do shopping.
33   13        TEST LE       P1,12,FINDQ            If <= 12, choose the
34   14        SELECT MIN    5,1,X$NMCHK,,Q         shortest overall queue;
35   15        TRANSFER      ,CHKOUT                proceed to check out.
36   16 FINDQ  SELECT MIN    5,X$XPRPO,X$NMCHK,,Q   > 12, choose short-
37    *                                            est nonexpress queue.
38   17 CHKOUT QUEUE         P5                     Get in the queue.
39   18        SEIZE         P5                     Get checkout station.
40   19        DEPART        P5                     Leave the queue.
41   20        ADVANCE       P4                     Check out.
42   21        RELEASE       P5                     Leave checkout station.
43   22        SAVEVALUE     SALES+,P2              Accumulate sales.
44   23        TERMINATE
45
46   24 BOLT   SAVEVALUE     SALELOSS+,P2           Accumulate lost sales.
47   25        TERMINATE
```

Figure 7.16–1. A simulation of the multiple-checkout supermarket (continues).

```
48   26        GENERATE       3600
49   27        TERMINATE      1
50   START     2,NP
51   RESET
52   INITIAL   X$SALES,0/X$SALELOSS,0
53   START     3
```

Figure 7.16–1. A simulation of the multiple-checkout supermarket (concluded).

RELATIVE CLOCK 10800   ABSOLUTE CLOCK 18000

BLOCK	CURRENT	TOTAL		BLOCK	CURRENT	TOTAL
1	1	476		11	0	99
2	0	476		12	40	451
3	0	476		13	0	459
4	0	476		14	0	76
5	0	476		15	0	76
6	0	476		16	0	383
7	0	377		17	27	459
8	0	377		18	0	452
9	0	377		19	0	452
10	0	99		20	6	452

BLOCK	CURRENT	TOTAL
21	0	452
22	0	452
23	0	452
24	0	25
25	0	25
26	0	2
27	0	3

QUEUE	MAXIMUM CONTENTS	AVERAGE CONTENTS		TOTAL ENTRIES	ZERO ENTRIES	PERCENT ZEROS
1	6	1.448		76	11	14.474
2	7	4.705		82	0	0.000
3	7	4.547		81	0	0.000
4	6	4.322		80	0	0.000
5	6	4.140		80	0	0.000
6	6	3.949		80	0	0.000

Figure 7.16–2. The output of the multiple-checkout supermarket simulation in Figure 7.16–1 (continues).

	AVERAGE TIME/TRANS	TIME/TRANS	$AVERAGE NUMBER	TABLE CONTENTS	CURRENT
	205.763	240.585	0		
	619.707	619.707	6		
	606.210	606.210	6		
	583.450	583.450	5		
	558.838	558.838	5		
	533.125	533.125	5		
FACILITY	AVERAGE UTILIZATION	NUMBER ENTRIES	AVERAGE TIME/TRANS	SEIZING TRANS. NO.	PREEMPTING TRANS. NO.
1	0.812	77	113.922	69	
2	1.000	77	140.260	40	
3	1.000	76	142.105	18	
4	1.000	76	142.105	48	
5	1.000	76	142.105	44	
6	1.000	76	142.105	24	

NON-ZERO FULLWORD SAVEVALUES

SAVEX	VALUE	SAVEX	VALUE	SAVEX	VALUE
NMCHK	6	NMXPR	1	SALELOSS	44450

SAVEX	VALUE	SAVEX	VALUE
SALES	1264025	XPRPO	2

```
60
61          END
```

Figure 7.16–2. The output of the multiple-checkout supermarket simulation in Figure 7.16–1 (concluded).

at the nonexpress queues (line 25) and leave for destination BOLT if the shortest waiting line has more than 8 customers. A similar logic is used at lines 29 and 30 for those purchasing 12 or fewer items; in this case, however, the customer considers all the checkout lines, both express and nonexpress. Queue number 1 is modeled as the express line and queues 2 through 6 are nonexpress lines.

All customers who stay in the store end up at location ENTER, where they spend time shopping. Following this, the express customers (at lines 33 and 34) choose the shortest (or one of the shortest) checkout queues and store that queue number as parameter 5. The nonexpress customers make a similar selection from the nonexpress queues at line 36. Eventually, all customers arrive at CHKOUT, with their parameter 5 representing the queue and facility (checkout station) that they will use. Lines 38 through 44 model the checkout process, and the savevalue SALES accumulates the gross receipts from all customers. Before the customers who depart because of long lines leave the simulation, they increment X$SALELOSS.

The timer segment of the program initiates a 2-hour simulation with no output in order to get past the transient state. This is followed by a 3-hour simulation, with its standard GPSS output as given in Figure 7.16–2. By looking at block counts for blocks 8 and 9, we see that in this simulation, all customers who intended to purchase more than 12 items stayed and did their shopping at the store. The total block count of 25 for blocks 24 and 25 also indicates that 25 customers who intended to purchase relatively few items left the store because of long lines. The express queue had an average length of 1.448, considerably shorter than the other five queues. In spite of this, the average utilization of the express checkout station was 0.812, which is relatively high. The total sales during the 3 hours of peak load simulation were \$12,640.25, and the sales lost due to customer departures were \$444.50.

To understand the true impact of these observations, we need to compare these results to those obtained through a simulation that does not allocate a checkout station for express customers. The program in Figure 7.16–1 is very flexible, and we can easily adjust it to this situation simply by initializing X\$NMXPR to 0 and X\$XPRPO to 1. Actually, through the proper setting of these savevalues, we can simulate any number of express line choices. The output of the simulation with no express lines (not given here) shows 14 customer defections, total sales of \$12,986.75, and lost sales of \$234.50.

In summary, the use of an express checkout station reduces customer defections from 25 to 14, but also reduces total sales by \$12,986.75 − \$12,640.25 = \$346.50. The manager must decide if a sales reduction of \$346.50 during 3-hour peak load spans justifies having 9 more customers doing their shopping at the supermarket.

7.17   User Chains and the LINK and UNLINK Blocks

In addition to system-managed chains such as the CEC and FEC (Section 5.1) and the interrupt chains (Section 7.9), GPSS allows the modeler to define and manipulate other chains of transactions. Such chains, called **user chains**, are linked lists of transactions with the same basic structure as the CEC, FEC, or interrupt chains. However, when a transaction is placed on a user chain, it is removed from the model (i.e., it does not contribute to a block count and it is removed from the CEC and the FEC), and its movement is suspended until it is reactivated by another transaction. The suspension and subsequent reactivation are done through the LINK and UNLINK blocks, respectively.

The LINK block can be used in either the conditional or the unconditional mode. In the simpler unconditional mode, a transaction entering the LINK block is transferred to a user chain regardless of other conditions in the model. In the conditional mode, the path that the transaction entering the LINK block takes is determined by the link indicator (to be discussed shortly) of the user chain. The LINK block has three operands:

Operand A: The symbolic or numeric name of the user chain; it cannot be defaulted.

Operand B: This operand, which also cannot be defaulted, indicates the manner in which the linking is to be done. It must be one of FIFO, LIFO, or Pj. With the FIFO specification, the transaction will be placed at the rear of the chain; with LIFO, it will be placed at the front of the chain; and with Pj, it will be located immediately behind those transactions already on the chain with Pj value smaller than the Pj value of the transaction being inserted on the chain. If identical Pj values are encountered, the incoming transaction is placed behind those already on the chain with the same Pj value.

Operand C: If defaulted, the LINK block functions in the unconditional mode; otherwise, this operand must be a block location. In the latter case, if the transaction is unable to get on the chain (as discussed below), it will be directed to the block indicated by this operand.

The link indicator is an on–off switch associated with the user chain. In the conditional mode, it determines whether the transaction entering the LINK block is to be placed on the user chain specified by the A operand of the block or is to be moved to the location given by the C operand. Initially, all link indicators are off. If the link indicator is on when a transaction enters a conditional-mode LINK block, the transaction will be placed on the user chain without affecting the link indicator. If the link indicator is off when a transaction arrives at a LINK block, the transaction will remain active by continuing its movement through the block specified by the C operand **after turning the link indicator on**. In this way, future transactions are forced onto the user chain until some other mechanism alters the status of the link indicator. We will see shortly how the LINK and UNLINK blocks coordinate the status of their link indicator.

The UNLINK block performs a function complementary to the LINK block by reactivating transactions that have been placed on a user chain. Note that a transaction that is on a user chain cannot reinsert itself into the flow of the simulation, because it is not on either the CEC or the FEC. This must be done by another transaction that enters an UNLINK block. The specific actions produced by the UNLINK block are determined by its six operands:

Operand A: As in the case of the LINK block, this is the numeric or symbolic name of the user chain and cannot be defaulted.

Operand B: This operand, which cannot be defaulted, is the location to which any unlinked transactions are to be directed.

Operand C: Called the **unlink count**, this operand specifies the number of transactions that are to be removed from the user chain and sent to the location indicated by the B operand. A constant or an SNA can be used to indicate a precise count. If the designation ALL is used, all transactions from the user chain will be moved.

Operands D and E: Of the variety of D and E operand combinations (D, E), we will consider only the (default, default) and (BACK, default) combinations. The (default, default) combination removes transactions from the front of the user chain; the (BACK, default) combination removes them from the back of the user chain.

Operand F: If this operand is used and if the unlink operation **fails completely** (i.e., the chain is empty and no transactions are available for unlinking), then the transaction that enters the UNLINK block is directed to the location specified by this operand.

If the user chain is empty when a transaction enters an UNLINK block, then the link indicator of that user chain is turned off. Thus, a LINK block never turns a link indicator off and an UNLINK block never turns a link indicator on. Equivalently, a LINK block can only turn a link indicator on (forcing subsequent transactions to be placed on the user chain), and an UNLINK block can only turn a link indicator off (enabling transactions arriving at the link block to continue through the model). The effect of the LINK and UNLINK blocks on the link indicator may seem confusing at first. It may help to view the interactions of these blocks with the user chain and link indicator structures as a scheme to make sure that at any time at most one transaction can be resident in a portion of the program bracketed by the LINK and UNLINK blocks.

Consider the program structure in Figure 7.17–1. Initially, the link indicator of the user chain WAIT is off. When the first transaction enters the LINK block, it is allowed to continue through the model at location MOVEIN. However, in this process, the link indicator is changed to on, which is how it will stay while the transaction is between locations MOVEIN and DONE. If other transactions arrive at the LINK block during the period, they are placed at the rear of user chain WAIT.

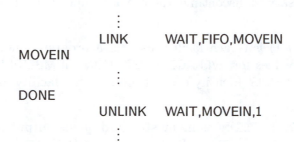

Figure 7.17–1. A user chain that controls entry into a program section.

Let us assume that eventually that first transaction enters the UNLINK block. If the unlinking is successful (i.e., if the WAIT chain is not empty), a transaction is moved from the WAIT chain to the CEC and the transaction in the UNLINK block proceeds normally through the model. Note that in this case, the link indicator is unaffected. In case of an unsuccessful unlinking (i.e., the WAIT chain is empty), in addition to the activities just described, the status of the link indicator is turned off, thus enabling the next transaction that enters the LINK block to move ahead to MOVEIN. Whatever the circumstances, in time, the transaction that did the unlinking will either terminate or stop moving. In either case, the transaction that was unlinked and placed on the CEC will eventually take its turn and move through the simulation blocks.

Through user chains, it is also possible to model queueing disciplines other than FIFO. For example, in many computing systems, jobs that require fewer resources, such as computer time, are processed ahead of those that require more resources. In such situations, the resource needs can be placed in a transaction parameter and transactions can be placed on a user chain according to increasing values of this parameter. Subsequently, when transactions are removed from the front of the chain by an UNLINK block, they will be processed in the desired order.

Problems for Chapter 7

7.1 (Section 7.1) Remove the two ASSIGN blocks from the program in Figure 7.1–2 without changing the logic of the program. Note that this will make it necessary to discontinue the use of P2 and P3 in the definition of the variable ST.

7.2 (Section 7.1) For each facility used in a program, the standard GPSS output gives values for AVERAGE UTILIZATION, NUMBER ENTRIES, and AVERAGE TIME/TRANS. Identify each of these with a facility SNA described in Section 7.1.1.

7.3 (Section 7.1) Identify as many standard queue output items as you can with the SNAs in Section 7.1.2. Do the same for storage and block count outputs and the SNAs in Sections 7.1.3 and 7.1.4, respectively.

7.4 (Section 7.2) Through the use of a savevalue, calculate the total service time used in the simulation in Figure 7.1–2. Is there a way to determine this value from the standard output of the program given in Figure 7.1–2?

7.5 (Section 7.2) Use the savevalue PRI1CTR in the simulation in Figure 7.1–2 to count the number of priority 1 jobs that enter the simulation. Of course, the value of X$PRI1CTR should agree with the SNA N2 (the total number of transactions that have entered block number 2).

7.6 (Section 7.2) Write a GPSS program that will calculate the sum (accurate to 5 significant digits) of the first 100 numbers produced by the RN1 random number generator. Hint: Individual values of RN1 can be obtained through a function that computes $10^5 \times$ RN1.

7.7 (Section 7.3) In the program in Figure 7.3–1, P1 is used at lines 18, 23, and 27. Explain why the use of P1 cannot be eliminated by deleting the ASSIGN block (line 23) and replacing lines 18 and 27, respectively, by

```
CHARGE      FVARIABLE   1500+50*V$ST
            ADVANCE     V$ST.
```

7.8 (Section 7.4) Only the portion of the output of the program in Figure 5.4–1 dealing with queue statistics was given in Figure 5.4–2. Modify the program in Figure 5.4–1 so that its only output is the queue output given in Figure 5.4–2.

7.9 (Section 7.4) Write a program to obtain the first 10 random numbers produced by RN1. You may use the ideas of Problem 7.6 and actually print out $10^5 \times$ RN1 values.

7.10 (Section 7.5) Eliminate the use of transaction parameters from the program in Figure 7.5–1 without affecting the logic of the program.

7.11 (Section 7.6) Using the random number generating algorithm described in Problem 1.1, simulate the queueing model described in that problem. The use of the random number generator was discussed in Section 7.6; you will need to extend its use to obtain service times. The results of a 200 time unit simulation should be identical to those obtained in Problem 1.2.

7.12 (Section 7.6) Implement URN22 (Section 3.5) in GPSS and use it to produce interarrival times in a simple queueing model.

7.13 (Section 7.7) Modify the example in Section 5.6 (Figure 5.6–1) to obtain the true average turnaround time.

7.14 (Section 7.7) Modify the program in Figure 7.1–2 to obtain the average waiting times of jobs in each priority class.

7.15 (Section 7.7) Suppose that a transaction moves from location X to Y and then to Z. We want to determine its transit times between X and Y and between Y and Z. Give a program segment that captures both transit times as transaction parameters without using a MARK block.

7.16 (Section 7.7) Use the MARK block to capture the transit times described in Problem 7.15. Can MARK blocks with defaulted A parameters be used in this situation?

7.17 (Section 7.7) Suppose that a transaction moves to locations U, V, X, and Y, in that order, and we want to determine its transit times between U and X and between V and Y. Give a program segment that uses MARK blocks to capture the two transit times as transaction parameters. Can MARK blocks with defaulted A operands be used in this situation?

7.18 (Section 7.8) Write a GPSS program to produce a frequency table of the first 1000 random numbers produced by RN1. The table should classify the numbers into intervals from 0 to 0.1, 0.1 to 0.2, ..., 0.9 to 1.0.

7.19 (Section 7.8) Use the results of Problem 7.18 to perform a chi-squared test (see Section 3.3.1) with a 0.5 significance level.

7.20 (Section 7.8) Write a GPSS program that tabulates 1000 interarrival times (exponential with mean 100) of transactions. Draw a histogram from the output of the program and compare its shape to that of the exponential probability density function.

7.21 (Section 7.9) The jobs in the simulation in Figure 7.1–2 access the CPU according to their priority, without preempting. If preemption is allowed (i.e., the SEIZE and RELEASE blocks are replaced by PREEMPT and RETURN), what effect would this have on the average residence time of queue number 4? Queue number 1?

7.22 (Section 7.9) A small-town police station operates a single cruiser during the night shift (midnight to 6:00 a.m.). During this time, incidents requiring police attention occur randomly with an average of 1 incident every 30 minutes. The service times for these incidents are normally distributed with a mean of 25 minutes and a standard deviation of 6 minutes. Write a GPSS program that simulates the police service from midnight to 6:00 a.m. under the assumption that half the police calls need to be served with a higher priority than other calls.

7.23 (Section 7.9) In Problem 7.22, assume that half the high-priority calls (a fourth of all calls) are serious enough that service to lower priority calls should be preempted. Modify the simulation in Problem 7.22 to reflect this situation.

7.24 (Section 7.10) Waiting times of those individuals initially classified as noncritical, but later reclassified as critical, are obscured within the NCWAIT table of Figure 7.10–2. Modify the program to produce a third table that classifies the waiting times of these individuals. Hint: P1 can be used as a numeric reference to a table.

7.25 (Section 7.10) In Problem 7.24, make the table classify the turnaround times (time spans between arrivals and departures) instead of waiting times.

7.26 (Section 7.11) Write a program segment in which a transaction will use facility X if the utilization of X is below 90%; otherwise, it will move to alternate location ALT.

7.27 (Section 7.11) Write a program segment in which a transaction will use a facility X if the utilization of X is below 90%; otherwise, it will wait indefinitely until the utilization of X falls below 90%.

7.28 (Section 7.11) We wish to control the entry of a transaction into a program section between locations X and Y so that at most one transaction can be between X and Y. To do this, we use a savevalue SWITCH in the following manner:

```
X       SAVEVALUE   SWITCH,1
   ⋮         ⋮
Y       SAVEVALUE   SWITCH,0
```

Use a TEST block with this structure to impose the desired restriction on the movement of transactions.

7.29 (Section 7.11) Problem 7.28 treats the program segment between X and Y in the same way that facilities are treated. Describe a general mechanism for avoiding the use of facilities in GPSS programs.

7.30 (Section 7.11) A transaction is to join the shorter of two queues that have numeric references 1 and 2. In case the queue lengths are equal, it can join either queue. Show how this can be done.

7.31 (Section 7.11) A transaction is to join the shortest, or any one of the shortest, of 3 queues with numeric references 1, 2, and 3. Show how this can be done. Choosing the shortest of n queues gets more complicated as n gets larger. The SELECT block (Section 7.15; see Problem 7.37) simplifies this type of choice.

7.32 (Section 7.11) Model a queue with a maximum capacity of 3 so that customers who observe 3 people in the queue when they are ready to join it leave the system.

7.33 (Section 7.13) In the telephone switchboard problem in Section 7.11 (Figure 7.11–1), suppose that callers who are successful in their first attempt try to call 5 more times and then give up. Modify the program in Figure 7.11–1 to reflect this situation.

7.34 (Section 7.14) Write a GPSS program segment to simulate the operation of a two-way (red–green) traffic light through a GATE block. The traffic light is to alternate between red and green every minute.

7.35 (Section 7.14) Modify the barbershop simulation in Figure 7.14–3 so that customers arrive over a 6-hour period and the simulation stops when all customers who are in the shop have been served.

7.36 (Section 7.14) Construction on a 10-mile stretch of a 2-lane highway has forced 1-lane traffic on that portion of the highway. To allow safe passage on the 1-lane portion of the road, a flagman is stationed at each end of the 10-mile section. Let us assume that the highway runs east to west. The flagman supervising the entry of cars into the 1-lane section from the east stops the east-to-west flow when there are no more vehicles trying to go east-to-west, gives the flag to the last driver to enter the section, and asks that the flag be delivered to the flagman at the west end. With the arrival of the flag at the west station, west-to-east traffic is opened. When there are no more vehicles trying to go west-to-east, the flagman stops the west-to-east flow and gives the flag to the last driver who enters the 1-lane section. When traffic is initiated in either direction, a vehicle in the kth position in the queue takes $10k$ sec. to move up to the flagman's position and then needs 800 sec. to traverse the section of the highway that is under repair. We wish to study the traffic pattern on this 10-mile stretch of highway when the interarrival times (east-to-west as well as west-to-east) of vehicles are exponentially distributed with a mean of 90 sec.

Write a GPSS program that uses the ideas developed in Section 7.14 (logic switches and GATE blocks) to simulate the traffic flow described above for a 5-hour time period.

7.37 (Section 7.15) Use a SELECT block to have a transaction join the shortest queue described in Problem 7.31.

7.38 (Section 7.15) Suppose that a simulation uses 7 storages. Have a transaction select the storage with the largest remaining capacity and store the storage number in its 4th parameter.

7.39 (Section 7.15) Have a transaction select the facility with the least utilization from among facilities in the range 3 to 7. The number of the selected facility is to be stored in the 3rd parameter of the transaction.

7.40 (Section 7.16) In the program in Figure 7.16–1, the selection of which queue to join depends on the lengths of the queues. Suppose we want to make the selection depend on the total number of items purchased by all customers in each queue. This can be done by having Xk represent the total number of items purchased by the customers in the k-th queue. After a customer selects a checkout counter j, Xj is incremented by P1 (the number of items purchased), and as a customer leaves the j-th queue, Xj is decremented by P1. Modify the program in Figure 7.16–1 to base the queue selection on the total number of items purchased.

7.41 (Section 7.17) When customers arrive at a facility, they capture it if it is idle; otherwise, they wait for the facility. When a customer is done with the facility, the most recent arrival is allowed to capture the facility next (i.e., it is a LIFO system). With a user chain, model this type of server.

Chapter 8

Case Study of a Simulation: Design, Analysis, Programming

In this chapter, we give a detailed discussion of a simulation solution for a realistic optimization problem. The problem we have chosen involves a transportation system, an area of much current interest, and one where analytic solutions are often not available for the most interesting measures of performance such as cost. To focus on the simulation's design, analysis, and programming (rather than on the details of the problem), we have simplified the real-world problem. It should be clear to the reader, however, how to add more complex cost measures, as well as more real-world complexities, such as time delays, breakdowns, and the like; some additional complexities are considered in the code of Section 8.4.

After describing the problem in Section 8.1, in Section 8.2 we detail the statistical design and how the simulation runs to be made are determined. In Section 8.3, we perform the statistical analysis of the simulation's output. The GPSS implementation of a variation of the model is given in Section 8.4. The end result is a capstone to the material of the previous chapters, showing in detail the full impact of the interaction of statistics and the computer implementation of a simulation in a modern setting. GPSS/PC implementations of the models given in Sections 8.1 and 8.4 are included on the disk accompanying this book. Because of the memory requirements of these models the full version of GPSS/PC is needed for their execution.

8.1 Description of the Transportation System Problem

In this problem, trucks arrive randomly throughout a 24-hour day, 7 days a week, at a terminal where they are to be unloaded. Historical data indicates that the interarrival times follow an exponential distribution with a mean of 140 minutes. Past records also show that cargo weights are normally distributed with a mean of 80,000 pounds, and a standard deviation of 7500 pounds. The number of pounds per hour that a crew can unload varies, being a function of the type of cargo. The probability of having each type of cargo, and the unloading rate for each, are given in Table 8.1–1.

Table 8.1–1. Probabilities and unloading rates by type of cargo.

Type of Load	Probability	Unloading Rate (lb/hr)
A	.40	8000
B	.35	7000
C	.25	5000

Each crew consists of a forklift operator (paid \$12/hr) and two laborers (each paid \$8/hr). A crew is paid for an entire day regardless of how many hours it works. In addition, needed labor that the company's own crews cannot provide is purchased from outside contractors, at rates 50% higher than the company pays its own crews. Such labor is available in any quantity needed, but the ability to use it may be limited by the number of forklifts the company has available. Company policy is to unload trucks on the day following their arrival, regardless of any extra labor costs incurred. Note that although the labor pool is unlimited, due to the presence of outside contractors, the pool of forklifts is not. Thus, it is possible that the company may not meet its goal of unloading trucks on the day following their arrival. While this is not considered in the present study, we could allow for an additional restriction, such as meeting the 24-hour unloading goal with probability at least .99. This would involve two optimization criteria, the new one being used to eliminate forklift-crew combinations that could not meet this restriction and the cost criterion being used on combinations remaining to select an optimal one.

Data suggests that forklifts break down at the start of each working day (independently of each other) with a probability of .10 and repairs take two full days. The simulation for this simplified model is given in Figure 8.1–2. The \$75 per day cost of a forklift is not considered in our initial simplified model and analysis; however, it is part of the program given in Section 8.4. We do not wish to have more forklifts than needed to minimize crew cost, since there is a forklift cost, though it is small compared to labor costs.

Our objective is to **determine** the "Number of Crews" (C) and "Number of Forklifts" (F) that are optimal for minimizing the total unloading cost.

```
1             SIMULATE
2             RMULT      12345,12345,12345,12345
3
4              ARRIVE     FUNCTION     RN4,C24     EXPONENTIAL FUNCTION
5             0,0/.1,.104/.2,.222/.3,.355/.4,.509/.5,.69/.6,.915/.7,1.2/.75, 1.38
6             .8,1.6/.84,1.83/.88,2.12/.9,2.3/.92,2.52/.94,2.81/.95,2.99/.96, 3.2
7             .97,3.5/.98,3.9/.99,4.6/.995,5.3/.998,6.2/.999,7/.9998,8
8
9              UNLD       FUNCTION     RN2,D3      UNLOADING FUNCTION
10            .25,5000/.6,7000/1,8000
11
12             NRDIS      FUNCTION     RN3,C25     STANDARD NORMAL
13            0,-5/.00003,-4/.00135,-3/.00621,-2.5/.02275,-2/.06681, -1.5/.11507,-1.2
14            .15866,-1/.21186,-.8/.274225,-.6/.34458,-.4/.42074, -.2/.5,0/.57926,.2
15            .65542,.4/.72575,.6/.78814,.8/.84134,1/.88493,1.2/.93319, 1.5/.97725,2
16            .99379,2.5/.99865,3/.99997,4/1,5
17
18            WEIGHT     FVARIABLE    7500*FN$NRDIS+80000      Weight of cargo.
19            UNRATE     FVARIABLE    (V$WEIGHT*6)/(FN$UNLD)   Time to unload truck.
20            NRMPAY     FVARIABLE    224*X$CREWNUM            8-hour pay per crew.
21            OVERPAY    FVARIABLE    (P7-48)*7                Overtime pay per crew.
22            TOTPAY     FVARIABLE    V$OVERPAY+V$NRMPAY       Total pay per crew.
23
24      1             LOGIC R      SELECTOR         Resets gate SELECTOR.
25      2             LOGIC R      ONEDAY           Resets gate ONEDAY.
26      3             LOGIC R      OTHERDAY         Resets gate OTHERDAY.
27
28      4             GENERATE     14,FN$ARRIVE,,50,,,F   A truck arrives with cargo.
29      5             SAVEVALUE    LASTARR,C1
30      6             QUEUE        CREWQ            Put it in line.
31      7             GATE LR      SELECTOR,DAY2    Send trucks to one gate or
32      *                                          another, depending on day.
33      8             ASSIGN       1,V$UNRATE       Time to unload this truck.
34      9             GATE LS      ONEDAY           Stack up trucks to be
35      *                                          unloaded on the odd days.
36      10            TRANSFER     ,PICKONE         Unload cargo from ONEDAY.
37      11  DAY2      ASSIGN       1,V$UNRATE       Time to unload this truck.
38      12            GATE LR      OTHERDAY         Stack up trucks to be
39      *                                          unloaded on the even days.
40      13  PICKONE   SELECT NU    3,1,X$CREWNUM    Select the available crew.
41      14            TEST E       P3,0,FREE        If there are none,
42      15            LINK         BUSY,FIFO        put truck in waiting line.
43      16  FREE      SELECT L     4,P3,P3,48,X     If there is a crew available,
44      *                                          check for overtime.
45      17            TEST E       P4,0,CHOSEN      If no overtime, put it to work.
46      18            TEST NE      P3,X$CREWNUM,PCKFRE  If this is the last crew,
47      *                                          get the first available crew.
48      19            ASSIGN       2,P3+1           If this is not the last crew,
49      *                                          find another one to use.
50      20            SELECT NU    3,P2,X$CREWNUM   Pick first crew from the rest.
```

Figure 8.1–2. GPSS simulation for a simplified
transportation model (continues).

```
51  21            TEST NE     P3,0,PCKFRE           If none, crews are busy
52      *                                           or have worked overtime.
53  22            TRANSFER    ,FREE                 If there is a free crew, loop.
54  23  PCKFRE    SELECT NU   3,1,X$CREWNUM         All crews are either busy or
55      *                                           have worked overtime.
56  24  CHOSEN    SEIZE       P3                    Put crew to work on truck.
57  25            ENTER       LIFT                  Give the crew forklift-unload.
58  26            DEPART      CREWQ                 Truck leaves waiting line.
59  27            ADVANCE     P1                    Unload the truck.
60  28            SAVEVALUE   P3+,P1                Record time to unload truck.
61  29            LEAVE       LIFT                  Return forklift to storage.
62  30            RELEASE     P3                    Crew ready for another truck.
63  31            UNLINK      BUSY,PICKONE,1        Another truck comes.
64  32            TERMINATE
65
66  33            GENERATE    144,,288,,,,F         This section tallies day's pay.
67  34            SELECT G    8,1,X$CREWNUM,0,X
68  35            TEST NE     P8,0,FINSIM
69  36            ASSIGN      2,1                   Start with the first crew.
70  37  OVRTME    SELECT G    1,P2,X$CREWNUM,48,X   Any crew worked overtime?
71  38            TEST NE     P1,0,DONE             If no, pay regular wages.
72  39            ASSIGN      5,48                  If yes, start
73  40  MOROVR    ASSIGN      5+,1                  at the eighth hour and in-
74  41            SELECT G    6,P1,P1,P5,X          crement by each 10 min. past
75  42            TEST E      P6,0,MOROVR           8 hrs. until overtime is found.
76  43            ASSIGN      7,P5-1                Set work time in P5
77  44            SAVEVALUE   COST+,V$OVERPAY       and calculate pay for crew.
78  45            ASSIGN      2,P1+1                Do same for other crews
79  46            TEST G      P2,X$CREWNUM,OVRTME   by looping around.
80
81  47  DONE      SAVEVALUE   NPAY+,V$NRMPAY        Add reg. wages to X$NPAY
82  48            SAVEVALUE   COST+,V$NRMPAY        and total wages in X$COST.
83  49            ASSIGN      1,0                   Use P1 as a loop counter
84  50            TRANSFER    ,CLRLP                to clear work times.
85  51  FINSIM    SAVEVALUE   ENDSIM+,1
86  52  CLRLP     ASSIGN      1+,1                  Move to the next crew tally.
87  53            SAVEVALUE   P1,0                  Clear the crew tally.
88  54            TEST E      P1,X$CREWNUM,CLRLP    Loop until done.
89  55            TERMINATE
90
91  56            GENERATE    144                   At the beginning of each day,
92  57            LOGIC I     SELECTOR              regulate the flow of trucks
93  58            LOGIC I     ONEDAY                to the day following their
94  59            LOGIC I     OTHERDAY              arrival for unloading.
95  60            TEST NE     X$ENDSIM,1,ENDIT
96  61            TRANSFER    ,NOENDIT
97  62  ENDIT     TERMINATE   1
98  63  NOENDIT   TERMINATE
99
100
```

Figure 8.1–2. GPSS simulation for a simplified
transportation model (continued).

101	64		GENERATE	144,,,,,,F	At the beginning of each day,
102	65		SAVEVALUE	FORKCOUNT,R$LIFT	find out the number of
103	66		TEST NE	X$FORKCOUNT,0,NOBRK	working forklifts.
104	67		ASSIGN	1,0	If there are any, break them
105	68	FORKLP	TRANSFER	.9,,OKAY	with a probability of 10%.
106	69		ENTER	LIFT	The broken ones are removed.
107	70		ASSIGN	1+,1	If more could potentially be
108	71	OKAY	SAVEVALUE	FORKCOUNT-,1	broken, loop back.
109	72		TEST E	X$FORKCOUNT,0,FORKLP	
110	73		ADVANCE	288	Time lapse for repairs.
111	74		LEAVE	LIFT,P1	Return forklift for use.
112	75	NOBRK	TERMINATE		
113					
114			INITIAL	X1-X20,0	
115			INITIAL	X$ENDSIM,0	
116			INITIAL	X$NPAY,0	
117			INITIAL	X$FORKCOUNT,0	Initialize FORKCOUNT to 0.
118			INITIAL	X$COST,0	Initialize COST to 0.
119			INITIAL	X$CREWNUM,15	Initialize CREWNUM to 15 (C=15).
120			INITIAL	X$FORKLIFT,8	Initialize FORKLIFT to 8 (F=8).
121					
122		LIFT	STORAGE	8	Put 8 forklifts in storage.
123			START	1	

Figure 8.1–2. GPSS simulation for a simplified
transportation model (concluded).

Section 8.2 considers in detail what (F, C) should be simulated, and for how long. Section 8.3 then considers how to analyze the resulting output to determine the optimal (F, C) combination.

8.2   Statistical Design

8.2.1   Statistically Designed Experimentation

Since there are two factors in the experiment—the number of forklifts F and the number of crews C—a central composite design (see Section 6.4) will be used. This mandates $2^2 + 2 \times 2 + 1 = 9$ experiments, and in each one we report the total cost of 5 days of simulated operation with specified numbers (F, C). The statistical analysis of the data given in Table 8.2–1 suggests a need for more experiments, and a supplemental central composite design with 9 additional experiments is run, with the results shown in Table

Table 8.2–1. Total cost from the 9 initial 5-day simulations.

No. of Crews	No. of Forklifts	Total Cost ($)
1	8	21,273
5	3	20,608
5	13	22,050
15	1	91,098
15	8	23,898
15	15	23,898
20	3	47,418
20	8	29,498
20	13	29,498

8.2–2. A plot of the (F, C) combinations of the original and supplemental data, at which simulations were made, is given in Figure 8.2–3. Note that the amount of simulation needed to reach statistically valid conclusions is $9 + 9 = 18$ experiments (each on 5 days of operation, or 90 simulated days).

Table 8.2–2. Total cost from the 9 additional 5-day simulations.[†]

No. of Crews	No. of Forklifts	Total Cost ($)
1	10	21,385
5	7	21,819
5	13	21,819
10	5	20,993
10	10	21,497
10	15	21,497
15	7	24,010
15	13	24,010
20	10	29,610

[†]The same seed (but different from the one used in the previous run of Table 8.2–1) was used for all simulations.

8.2.2   Naive Experimentation

To illustrate the costs of naive experimentation, we also ran a naive experiment, with a grid on $1, 2, 3, 4, 6, 9, 12, 15$ forklifts with $1, 3, 5, 7, 10, 15, 20$ crews. The results of the 56 experiments, each obtained through a 10-day simulation, are summarized in Table 8.2–4.

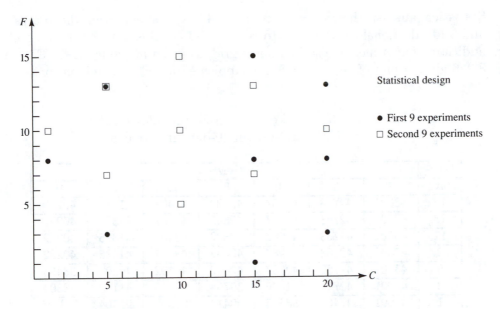

Figure 8.2–3. 18 (F,C) combinations, statistical design (90 simulated days).

Table 8.2–4. Total cost ($) from the initial 56 naive-design 10-day simulations.[†]

Fork-lifts	Crews						
	1	3	5	7	10	15	20
15	43295 #1	43057 #2	42581 #3	42854 #4	42371 #5	48426 #6	59626 #7
12	43295 #8	43057 #9	42581 #10	42854 #11	42371 #12	48426 #13	59626 #14
9	43295 #15	43057 #16	42581 #17	42854 #18	42371 #19	48426 #20	59626 #21
6	43295 #22	43057 #23	42581 #24	41482 #25	40656 #26	48426 #27	59626 #28
4	43295 #29	43057 #30	42077 #31	40012 #32	43946 #33	58506 #34	73066 #35
3	43295 #36	43057 #37	40040 #38	41482 #39	52906 #40	71946 #41	90986 #42
2	43295 #43	40887 #44	42826 #45	54026 #46	70826 #47	98826 #48	126826 #49
1	43295 #50	47754 #51	69706 #52	91658 #53	124586 #54	179466 #55	234346 #56

[†]Experimental reference numbers are noted in each cell.

Statistical analysis showed a need to delete 16 experiments from the analysis and add additional experiments to reach valid conclusions. There were 49 additional experiments, also in a naive grid, as given in Table 8.2–5. Figure 8.2–6 gives a plot of the original and supplemental (F, C) combinations.

Table 8.2–5. Total cost (\$) from the additional 49 naive-design 10-day simulations.[†]

Fork-lifts	Crews						
	1	3	4	5	6	7	9
15	43001	42763	42987	42882	43407	41335	40222
13	43001	42763	42987	42882	43407	41335	40222
11	43001	42763	42987	42882	43407	41335	40222
8	43001	42763	42987	42882	43407	41335	40222
5	43001	42763	42987	42882	41692	40502	38465
3	43001	42763	41426	39764	38066	41188	48804
1	43001	47460	58436	69412	80388	91364	113316

[†]The same seed (but different from the one used in the previous run that produced the data in Table 8.2–4) was used for all simulations.

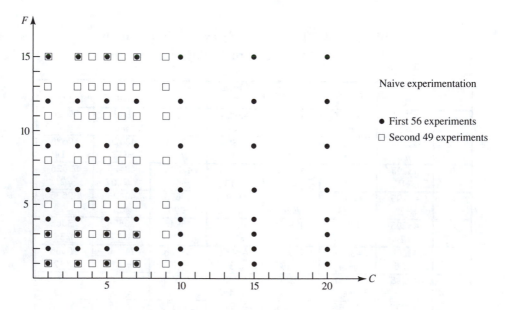

Figure 8.2–6. 105 (F, C) combinations, naive design (1050 simulated days).

Note that the amount of simulation needed to reach statistically valid conclusions is $(56+49)\times10 = 1050$ simulated days with the naive experiment. Statistically designed experimentation reached valid conclusions with only 90 simulated days, or only 8.6% of the effort the naive method needed. One might wonder: With the speed of modern computers, is this saving needed? The answer is: Perhaps not for this illustrative problem. But, for the real-world problems that need to be solved, one will have more factors than F and C. Thus, regardless of computer speed, the naive method will make a large class of problems infeasible while statistically designed experimentation will lead to solutions. Historically, the complexity of problems one wants to solve has increased with increasing computer speed. Hence, our assertion of the need for experimental design has been true for some time and it will continue to be true.

8.3 Statistical Analysis

We noted in Section 8.2 that one can use either statistical design of the simulation experiment or, at much greater cost, naive design. The statistical analysis of the data proceeds similarly in either case. We will now cover the details of the data analysis in the efficient statistical design method, and give final recommendations on choice of (F, C). So that we do not obscure the thrust of how this analysis proceeds, we will omit many of the supporting tables and plots produced and examined in the full analysis, but typical ones will be given. In Section 8.3.2, we give the full output and statistical programming code in SAS (Statistical Analysis System) for the analysis.

8.3.1 Analysis of Data

At the outset, we have the data of Table 8.3–1 (reproduced from Table 8.2–1 but with the addition of code numbers for ease of reference). We first use stepwise regression of Cost on F, C, F^2, C^2, FC, which results in fitting up to a full quadratic model, as well as maximum R^2 improvement regression, and find the model

$$\text{Cost} = 49183 - 9357F + 2099C + 586F^2 + 17C^2 - 205FC \qquad (8.3.1)$$

with $R^2 = 0.773$ and $C_p = 6.00$. (For some details on statistical programming, see Dudewicz, Chen, and Taneja (1989), especially Chapters 16–18.)

A model without the C^2 term could achieve $R^2 = 0.772$, and a better $C_p = 4.01$, but we will not reduce the model at this time. To check if any observations are "outliers," the predictions and residuals (differences between actual and predicted values), as well as Studentized residuals (residuals divided by estimated standard error at the design point), are computed as summarized in Table 8.3–2.

Table 8.3–1. Results of 9 initial 5-day simulations, with code numbers added for ease of reference.

Code No.	C	F	Cost ($)
1	1	8	21273
2	5	3	20608
3	5	13	22050
4	15	1	91098
5	15	8	23898
6	15	15	23898
7	20	3	47418
8	20	8	29498
9	20	13	29498

Table 8.3–2. Residuals and Studentized residuals for model (8.3.1).

Code No.	Actual Cost ($)	Prediction	Residual	Std. Residual
1	21273	12341	8932	.949
2	20608	34241	−13633	−1.469
3	22050	24257	−2207	−.238
4	91098	72710	19198	1.794
5	23898	22611	1287	.171
6	23898	29996	−6098	−.570
7	47418	62969	−15551	−1.472
8	29498	27918	1580	.119
9	29498	22196	7302	.691

Although some residuals in Table 8.3–2 are high, the standard errors are large enough that we would not, at any reasonable statistical significance, delete our observations. For example, for observation number 4 we have $C = 15$ and $F = 1$ so that the prediction is $49183 - (9357)(1) + (2099)(15) +$

$(586)(1^2) + (17)(15^2) - (205)(1)(15) = 72,710$. (You may find $72,647$ if you plug into equation (8.3.1), but that is due to the coefficients given there being rounded off.) However, the standard error is $10,700$ for the residual at this design point, resulting in a Studentized residual of $(91908 - 72710)/10700 = 1.794$. For reference, we give model (8.3.1) with more decimal places in the coefficients:

$$\text{Cost} = 49183.355 - 9357.199F + 2099.592C + 586.564F^2$$
$$+ 17.254680C^2 - 205.257FC. \tag{8.3.2}$$

We might wonder if we are now in a position to use model (8.3.2) to state the optimal (F, C) combination. A plot of function (8.3.2) for F between 1 and 15 and C between 1 and 20 (the range where the model was developed) would show the predicted ranges of minimal cost. We can also obtain a contour plot (on a figure not given, but similar to Figure 8.3–5) of the estimated error of the predictions at each of the (F, C) points in the predicted cost plot, to assess the adequacy of the predictions coming from model (8.3.2). In technical detail, we do this by taking the MSE of the full quadratic model we fitted to find (8.3.2), called the model's "error mean square" on many computer outputs (here MSE $= 315,043,070$), and find its square root SQRT(MSE) $= 17749.453$. Then build the matrix

$$M = \begin{pmatrix} 1 & F_1 & C_1 & FSQ_1 & CSQ_1 & FC_1 \\ 1 & F_2 & C_2 & FSQ_2 & CSQ_2 & FC_2 \\ \vdots & & & & & \\ 1 & F_9 & C_9 & FSQ_9 & CSQ_9 & FC_9 \end{pmatrix} \tag{8.3.3}$$

consisting of the values of F, C, F^2, C^2, and $F \times C$ (the variables in the full quadratic model) at each of the 9 design points that are left in the analysis. We then compute

$$V = (M' * M)^{-1}, \tag{8.3.4}$$

and at any point (C, F) we have the vector of model variables

$$P = (1 \ \ F \ \ C \ \ FSQ \ \ CSQ \ \ FC); \tag{8.3.5}$$

at that point the standard error of prediction is estimated to be

$$W = \sqrt{MSE}\,\sqrt{P * V * P'}. \tag{8.3.6}$$

Since the two standard deviation limits of error of prediction are $\pm 2W$, we contour plot (not given; see Figure 8.3–5 for an example of a similar plot)

$2W$ for a range of F and of C in the range of the experimentation. This plot shows that in the range of the predicted minimum, we have a cost of about \$20,000 and a $2W$ of about $15,000$. Since this means that we cannot distinguish minimum cost from the points nearby with costs of $30,000$ or more, **this plot shows that more experimentation is needed in the range of the predicted minimum cost in order to have confidence in the final (F, C) prediction.**

Therefore, we add a supplemental central composite design, with the 9 additional points, with code numbers 10 through 18 for ease of reference. The simulation results for these 9 combinations are given in Table 8.3–3. Regression on this data set of 18 experiments now yields the model

$$\text{Cost} = 51463.75 + 891.91C - 8356.51F$$
$$+ 509.0375F^2 + 64.40903C^2 - 184.25FC \qquad (8.3.7)$$

with $R^2 = 0.75$.

Table 8.3–3. Results of 9 additional 5-day simulations.

Code No.	C	F	Cost ($)
10	1	10	21385
11	5	7	21819
12	5	13	21819
13	10	5	20993
14	10	10	21497
15	10	15	21497
16	15	7	24010
17	15	13	24010
18	20	10	29610

Again computing the predictions, residuals, and Studentized residuals, we find the largest Studentized residual is that of the experiment with code number 4 (Studentized residual of 3.39, with actual cost of \$91,098 and predicted cost of \$68,723.2). Note that the actual cost is as in Table 8.3–2, but the Studentized residual and predicted cost *differ* from those in Table 8.3–2 because we recomputed these using the data of *both* Table 8.3–2 and Table 8.3–3. **Because this residual is in a region of high cost (which is not of interest, because our goal is to minimize cost) and is hard to fit with a quadratic model**—because the cost is increasing exponentially in this region of $(C, F) = (15, 1)$—**we delete experiment number 4 and**

reanalyze. Note that there were two other experiments with fairly high Studentized residuals, number 2 with actual 20,608 and predicted 34,281, and number 7 with actual of 47,418 and predicted of 63,522. We keep both: number 2 because it is in a region of low cost and number 7 because the model may change markedly when the "worst" outlier is eliminated. Thus, we prefer to sequentially eliminate one point at a time and then reanalyze.

With experiment number 4 eliminated, our quadratic model fitted to 17 data points has $R^2 = 0.929$ and SQRT(MSE)= 2098.576, so our predictions are much more accurate in the restricted range (with a high-cost region gone). We find one high Studentized residual, that of experiment number 7 with actual 47418, predicted of 44884, and Studentized residual of 3.17. Because this residual is in a high-cost region (with $(C, F) = (20, 3)$), we eliminate it and rerun. (Note that no other residuals were markedly high in this run.)

With experiments numbered 4 and 7 eliminated, we analyze the remaining 16 data points and find (with $R^2 = 0.973$ and SQRT(MSE)= 638.1977) the model

$$\text{Cost} = 21380.38 - 451.501C + 95.69437F$$
$$+ 44.79007C^2 + 3.310469F^2 - 11.3685CF. \qquad (8.3.8)$$

No Studentized residuals are high for this model. The cost prediction contour plot[1] of Figure 8.3–4 shows a minimal cost of about \$20,500 in a region around C between 4 and 7, with F between 1 and 7 (note that the points deleted were not near this region). The $2W$ contour plot in Figure 8.3–5 shows a prediction error no larger than 1800 when we restrict ourselves to that part of the region with C between 4 and 7, and F between 4 and 7. **We conclude that the minimal cost is nearly attained at $(C, F) = (5, 5)$ and, from (8.3.8), that the cost minimum is \$20,519.65 per 5 days,** with an error of at most \$1800 in this prediction. If greater accuracy is desired, additional experiments can be performed in this region.

8.3.2 Statistical Programming in SAS for Data Analysis

In this subsection, we detail the code needed to perform analyses such as those discussed in Section 8.3.1, with the associated tables and plots pro-

---

[1]In Figures 8.3–4 and 8.3–5, as well as Figures 8.3–10 and 8.3–12 later in this section, we have used SAS-generated contour plots that can be printed on all output devices. If a more sophisticated graphics system is available (e.g., SASGRAPH with an appropriate graphics printer), clearer plots that are easier to interpret can be obtained. However, the same interpretations are available from all renditions of the contour plots.

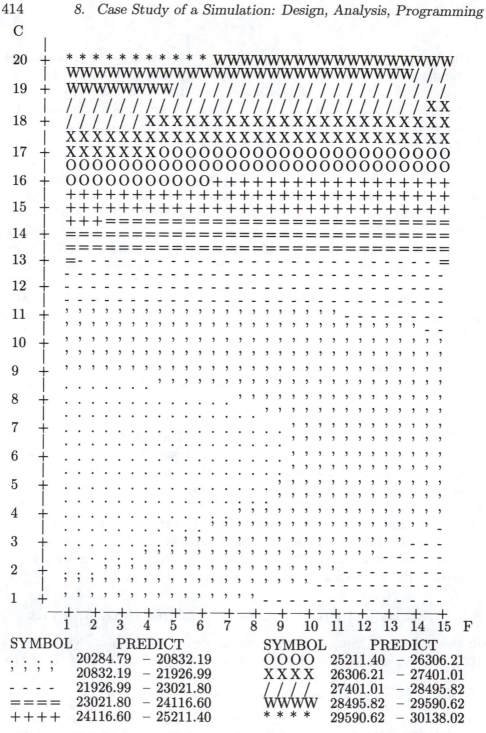

Figure 8.3–4. Contour plot of the predicted cost with model (8.3.8).

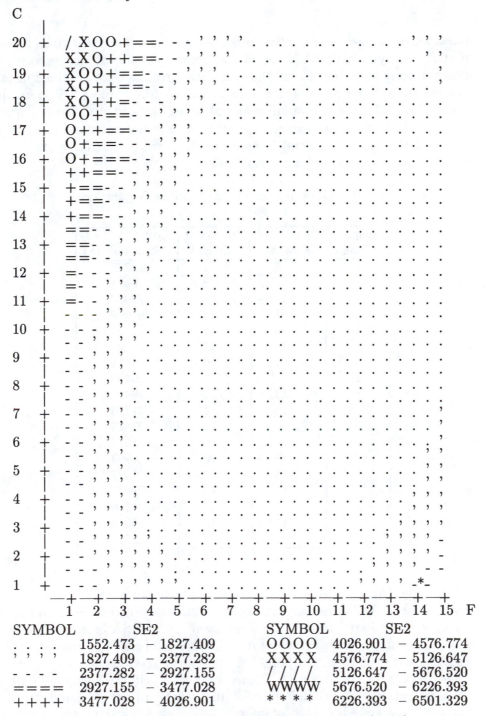

Figure 8.3–5. Contour plot of $2W$ with model (8.3.8).

duced and examined to make the decisions discussed. The code given is that used for analysis with SAS, one of the premier statistical software packages currently available. Recall that simulation results were obtained in two batches (sampling time $= 1$ and 2 in Table 8.3–6). Table 8.3–6 gives the cost (Y) of 5 days of operation of an unloading facility, with various numbers of forklifts (X_1) and crews (X_2) for all 18 simulation runs. The goal is to model the cost (in dollars) Y as a function of X_1 and X_2, with a view toward designing the minimum-cost facility.

Table 8.3–6. Data on cost (Y) at forklifts X_1 and crews X_2.

Sampling Time	Number of Forklifts	Number of Crews	Cost of 5 Days (Simulated) ($)
1	8	1	21273
1	3	5	20608
1	13	5	22050
1	1	15	91098
1	8	15	23898
1	15	15	23898
1	3	20	47418
1	8	20	29498
1	13	20	29498
2	10	1	21385
2	7	5	21819
2	13	5	21819
2	5	10	20993
2	10	10	21497
2	15	10	21497
2	7	15	24010
2	13	15	24010
2	10	20	29610

Since the SAS code and the subsequent output are quite extensive, they are given at the end of this section in Figures 8.3–8 through 8.3–11. In Figure 8.3–8, we have Part I of the code from an interactive SAS session that sets up the data for analysis in the form of Table 8.3–6 (which is printed out on one page with the title "Output A" on the output page).

The Part II SAS code given in Figure 8.3–9 results in the 5 pages of output labeled "Output B" given in Figure 8.3–10. (Another page of output was also produced, but since it contained information more effectively summarized in the plots of Figure 8.3–10, it is omitted.)

From the output of Figure 8.3–10, we see that the result of fitting a full quadratic model to the full data set is the model

$$
\begin{aligned}
Y = \quad & 51463.75 - 8356.51X_1 + 891.91X_2 \\
& + 509.04X_1^2 + 64.41X_2^2 - 184.25X_1X_2,
\end{aligned}
$$

which is the model of (8.3.7). Since $R^2 = 0.75$ and $s = 10,039.67$, this model explains 75% of the variability of the cost Y, but the standard deviation of prediction is \$10,039.67, which is quite large.

Still in Figure 8.3–10, we see that there are three outliers in the data set: observations numbered 2, 4, and 7. Their residuals (actual – predicted) are $-13,674$, $22,375$, and $-16,104$, respectively. Examining the data, we find that observations 4 and 7 have very large values. A plot of the design points in Figure 8.3–7 shows that observations 4 and 7 were at the upper left of the design space.

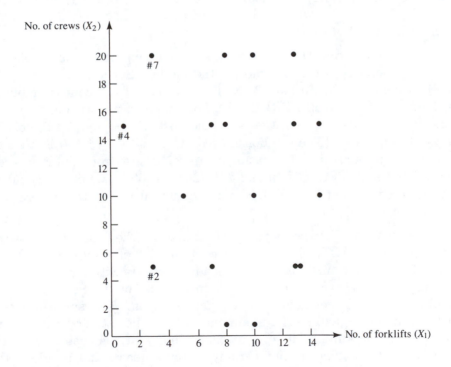

Figure 8.3–7. 18 points where experiments were run. (Outliers 4 and 7 were deleted and outlier 2 was kept.)

We conclude that the failure to model them well is due to the fact that the cost surface becomes very large faster than any quadratic model in this zone. Because we are seeking minimum cost, we therefore delete these points (not because they are bad data—they are not—but because to model them well would require a more complicated, perhaps exponential, model, and since these points represent a high-cost region, we have no interest in them).

We now eliminate the points numbered 4 and 7 and rerun the same analysis with $18 - 2 = 16$ data points. The SAS program for this, Part III, is given in Figure 8.3–11, while the 5 pages of resulting output are given in Figure 8.3–12 labeled "Output C." (Again, an additional page of output was produced, but is not given for reasons similar to those discussed earlier.)

In the output of Figure 8.3–12 (the last figure of this section), we see that $R^2 = 0.97$ (i.e., we explain 97% of the variability of the data) and $s = 638.20$. Since none of the residuals seem extremely large, we will use the model of Figure 8.3–12:

$$Y = 21380.38 + 95.69X_1 - 451.50X_2 + 3.31X_1^2 + 44.79X_2^2 - 11.37X_1X_2,$$

which is the model of (8.3.8).

Uses for Prediction. Using a model for prediction can be accomplished easily by generating a contour plot of the model. Such a plot for model (8.3.8) is shown in Figure 8.3–12 (it is the second of the contour plots in that figure, the one labeled PREDICT). For example, we see that if we use 8 forklifts and 10 crews, then the estimated cost will be about $21,000$ for 5 days of operation. (The symbol plotted at $X_1 = 8$ $X_2 = 10$ is a single right-side quote mark '. This means a predicted cost from the regression equation of between $20,832.19$ and $21,926.99$.) While we could calculate the exact prediction at any (X_1, X_2) combination, it is easy to interpret the contour plot visually.

Uses for Process Control and Optimization. The contour plot of the predictions (PREDICT, in the second contour plot of Figure 8.3–12) can be used for easy optimization of the unloading process with respect to cost. For example, we see that the minimum cost occurs in the range of 1 to 8 forklifts and 3 to 8 crews, in a bullet-shaped region. The surface of cost is a bowl in this region, so the minimum cost will be attained with about 6 crews and 4 forklifts. There is much flexibility in the system with regard to forklifts (setting this number anywhere between 1 and 10 affects the cost only mildly), but crews must be kept below 10, lest the cost increase markedly. These predictions for optimal operation should be tempered by "extrapolation control," discussed next.

Extrapolation and Its Control. It is clear that if we use the model to make predictions far from where we had data, our predictions may be bad. There are two reasons for this. First, although we have pinned down the surface of cost well in the region of the experiments, it may wobble a lot (due to estimation of the true coefficients) far from that zone. Second, the model may change its shape as we move into other zones. For example, at 1 forklift and 15 crews, the prediction of cost shown in the contour plot (Figure 8.3–12, Output C, PREDICT) is about $25,000$. However, recall that we had an observation at that point, which was a cost of $91,098$, and we excluded it from the analysis because its residual was large **and** it was a region of no interest in terms of low-cost operation. Thus, points far from the bulk of the data are not points where we should decide to run the future process, at least not without confirmatory experiments after the present analysis.

One way to control the extrapolation is to use contour plots of the predicted error of estimates; the error increases as we move away from the points where experiments were run. This method is especially valuable in cases where plots such as the design set are not possible because these are more than two dimensional. In Figure 8.3–12 (Output C), the contour plot labeled "SE2" gives the error of prediction as a function of X_1 and X_2 settings. We see that over the zone marked with symbols ' and . this error will be at most 2164, while outside this zone predictions may become several times as variable. Restricting ourselves to the zone marked . may be reasonable, in which case we may decide to run with 6 forklifts and 6 crews since this configuration has minimum cost over this zone.

Simulated Annealing Optimization; ARSTI Optimization. Methods for optimizing systems using simulation are not static, but are constantly developing. For example, an exciting new method called "simulated annealing" (see Johnson (1988) for a complete annotated bibliography, computer code, etc.) burst upon the scene in the 1980s. This method attempts to optimize the function $f(x_1, \ldots, x_k)$ by sequential choice of new $(x_1, \ldots, x_k)$ vectors. Many methods attempt the same thing, and simulated annealing (which dates from 1983) was perhaps just the "new kid on the block" at that time. However, the simulated annealing method seems very adept at avoiding a common flaw of optimization methods, namely the tendency to become trapped in local optima (versus a global optimum). For this reason, there are well over 300 papers on the method and interest in it is still high. Although the methodology is not yet complete (for example, there seems to be no way to form a confidence statement on optimality yet), readers with an interest in systems with very large k may find it worth exploring. What are called "genetic algorithms" have emerged as a strong competitor (e.g., see Parsons and Johnson (1997), where a review is given along with an application to DNA sequence assembly).

Even more recently, in the 1990s, Edissonov (1994) presented a new optimization method called Adaptive Random Search with Translating Intervals (or ARSTI) and provided FORTRAN code for its use. This method, which seeks optima of a multi-parameter objective function, has been compared with many other random search methods on established test functions and has been found to be best for 3 or more parameters (and close to best for 2). Details can be found in Edissonov (1994).

Vehicle Routing; Location Modeling. Two additional areas of high interest where simulation plays a large role are vehicle routing (sometimes with "time windows" within which delivery or pickup must occur), and location modeling (of bank branches, fire stations, service centers, hospitals, depots, etc.). While details are beyond the scope of this book, excellent coverage and references are provided by Golden and Eiselt (1992) for location modeling, and by Golden (1993) for vehicle routing (for detailed test problems, etc., see Golden and Assad (1986)).

```
1            options ls=76    no date NONUMBER;
2
3            data d5day1;
4            input numcrew numfork cost @@;
5            run = 1;
6            cards;
NOTE: SAS WENT TO A NEW LINE WHEN INPUT STATEMENT
            REACHED PAST THE END OF A LINE.
NOTE: DATA SET WORK.D5DAY1 HAS 9 OBSERVATIONS AND 4 VARIABLES. 976
            OBS/TRK.
NOTE: THE DATA STATEMENT USED 0.18 SECONDS AND 68K.
10           ;
11
12           data d5day2;
13           input numcrew numfork cost @@;
14           run = 2;
15           cards;
NOTE: SAS WENT TO A NEW LINE WHEN INPUT STATEMENT
            REACHED PAST THE END OF A LINE.
NOTE: DATA SET WORK.D5DAY2 HAS 9 OBSERVATIONS AND 4 VARIABLES. 976
            OBS/TRK.
NOTE: THE DATA STATEMENT USED 0.14 SECONDS AND 68K.
19           ;
20
21           data d5day;
22           set d5day1 d5day2;
23           crew 2 = numcrew*numcrew;fork2 = numfork*numfork;
24           crewfork = numcrew*numfork;;
25           label
26           numcrew = Number of Crews
27           numfork = Number of Forklifts
28           cost = Cost of Operation (Simulated)
29           crew2 = Number of Crews Squared
30           fork 2 = Number of Forklifts Squared
31           crewfork = Number of Crews times Forklifts;
NOTE; DATA SET WORK.D5DAY HAS 18 OBSERVATIONS AND 7 VARIABLES. 586
            OBS/TRK.
NOTE; THE DATA STATEMENT USED 0.22 SECONDS AND 68K.
32           run;
33
34           proc print label split=* uniform;
35           var numfork numcrew cost; id run;
36           label numfork=Number of * Forklifts
37                numcrew=Number * of Crews
38                    cost=Cost of 5 Days * (Simulated)
39                        run=Sampling* Time;
40           title Output A;
```

Figure 8.3–8. SAS program, Part I, yielding Table 8.3–6 as output.

```
NOTE: THE PROCEDURE PRINT USED 0.48 SECONDS AND 184K
         AND PRINTED PAGE 1.
41       run;
42
43       proc reg data=d5day;
44       model cost = numcrew numfork fork2 crew2 crewfork / p r influence ;
45       title Output B;
NOTE: THE PROCEDURE REG USED 0.78 SECONDS AND 160K
         AND PRINTED PAGES 2 TO 4.
46       run;
47
48       data d5dayex;set d5day end=eof;
49       output;
50       if eof then do ;
51       do numfork = 1 to 15 by .5 ;
52       do numcrew = 1 to 20 by .5 ;
53       crew2=numcrew*numcrew;fork2=numfork*numfork;crewfork=numcrew*numfork;
54       cost = .;
55       output;
56       end;
57       end;
58       end;
59
NOTE: DATA SET WORK.D5DAYEX HAS 1149 OBSERVATIONS AND 7 VARIABLES. 586
         OBS/TRK
NOTE: THE DATA STATEMENT USED 0.38 SECONDS AND 68K.
60       proc reg data=d5dayex noprint;
61       model cost = numcrew numfork fork2 crew2 crewfork / noprint ;
62       output out=plot p=predict l95=l95 u95=u95 r=residual ;
NOTE: DATA SET WORK.PLOT HAS 1149 OBSERVATIONS AND 11 VARIABLES. 382
         OBS/TRK
NOTE: THE PROCEDURE REG USED 2.35 SECONDS AND 164K
         AND PRINTED PAGE 5.
63       run;
64
65       data plot;set plot;se2=u95-predict;
NOTE: DATA SET WORK.PLOT HAS 1149 OBSERVATIONS AND 12 VARIABLES. 350
         OBS/TRK
NOTE: THE DATA STATEMENT USED 0.51 SECONDS AND 68K.
66       run;
67
68
69       proc plot data=plot;
70       plot numcrew*numfork=se2/vaxis=1 to 20 by 1 haxis=1 to 15 by 1 contour=10;
71       plot numcrew*numfork=predict / vaxis=1 to 20 by 1 haxis=1 to 15 by 1 contour=10;
NOTE: THE PROCEDURE PLOT USED 1.39 SECONDS AND 192K
         AND PRINTED PAGES 6 TO 7.
72       run;
```

Figure 8.3–9. SAS program, Part II, yielding Figure 8.3–10 as output.

Output B

DEP VARIABLE: COST   Cost of Operation (Simulated)

SOURCE	DF	SUM OF SQUARES	MEAN SQUARE	F VALUE	PROB>F
MODEL	5	3602751081	720550216	7.149	0.0026
ERROR	12	1209539389	100794949		
C TOTAL	17	4812290471			

ROOT MSE	10039.669	R-SQUARE	0.7487	
DEP MEAN	28659.944	ADJ R-SQ	0.6439	
C.V.	35.03031			

VARIABLE	DF	PARAMETER ESTIMATE	STANDARD ERROR	T FOR HO: PARAMETER= 0	PROB > |T|
INTERCEP	1	51463.748	19966.436	2.578	0.0242
NUMCREW	1	891.910	2008.452	0.444	0.6649
NUMFORK	1	−8356.512	3162.739	−2.642	0.0215
FORK2	1	509.037	152.801	3.331	0.0060
CREW2	1	64.409026	73.678105	0.874	0.3992
CREWFORK	1	−184.250	111.055	−1.659	0.1230

OBS	ACTUAL	PREDICT VALUE	STD ERR PREDICT	RESIDUAL	STD ERR RESIDUAL	STUDENT RESIDUAL	-2-1-0 1 2
1	21273	16672	6199	4601	7897	0.583	| |* |
2	20608	34282	7596	−13674	6565	−2.083	|**** | |
3	22050	22950	5191	−889.942	8593	−0.105	| | |****** |
4	91098	68723	7572	22375	6592	3.394	| |****** |
5	23898	22951	4783	947.294	8827	0.107	| | |
6	23898	27064	6100	−3166	7974	−0.397	| | |
7	47418	63522	7488	−16104	6688	−2.408	|**** | |
8	29498	31312	5243	−1814	8562	−0.212	| | |
9	29498	24553	6777	4945	7407	0.668	| |* |
10	21385	17916	6205	3469	7893	0.439	| | |
11	21819	17532	4393	4287	9027	0.475	| | |
12	21819	22950	5191	−1131	8593	−0.132	| | |
13	20993	28555	4923	−7562	8750	−0.864	| * | |
14	21497	15737	4937	5760	8742	0.659	| |* |
15	21497	28372	5884	−6875	8135	−0.845	| * | |
16	24010	26435	4742	−2425	8849	−0.274	| | |
17	24010	20798	4139	3212	9147	0.351	| | |
18	29610	25554	5214	4056	8580	0.473	| | |

Figure 8.3–10. Output of regression on full data set of Table 8.3–6, using the SAS program of Figure 8.3–9 (continues).

Output B

OBS	COOK'S D
1	0.035
2	0.968
3	0.001
4	2.533
5	0.001
6	0.015
7	1.211
8	0.003
9	0.062
10	0.020
11	0.009
12	0.001
13	0.039
14	0.023
15	0.062
16	0.004
17	0.004
18	0.014

SUM OF RESIDUALS   3.79259E–10
SUM OF SQUARED RESIDUALS   1209539389

OBS	RESIDUAL	RSTUDENT	HAT DIAG H	COV RATIO	DFFITS	DFBETAS INTERCEP	DFBETAS NUMCREW
1	4600.63	0.5658	0.3813	2.2934	0.4442	0.2112	−0.3027
2	−13673.6	−2.4956	0.5724	0.2671	−2.8875	−2.4886	1.3600
3	−899.942	−0.1003	0.2674	2.2881	−0.0606	0.0199	−0.0148
4	22374.8	16.2371	0.5688	0.0000	18.6498	6.0259	2.3158
5	947.294	0.1028	0.2270	2.1678	0.0557	−0.0304	0.0349
6	−3165.88	−0.3827	0.3692	2.4683	−0.2928	−0.0964	0.0401
7	−16104.3	−3.2060	0.5562	0.0724	−3.5903	0.2954	−0.1344
8	−1813.83	−0.2032	0.2727	2.2659	−0.1244	0.0298	0.0090
9	4944.82	0.6514	0.4557	2.4675	0.5960	0.2673	−0.3557
10	3468.8	0.4242	0.3819	2.4741	0.3335	0.0568	−0.1616
11	4286.98	0.4590	0.1915	1.8604	0.2234	0.0446	−0.0100
12	−1130.94	-0.1261	0.2674	2.2808	−0.0762	0.0250	−0.0186
13	−7561.62	−0.8544	0.2405	1.5092	−0.4808	0.0034	−0.2113
14	5759.64	0.6425	0.2418	1.7822	0.3629	−0.1962	0.2469
15	−6874.97	−0.8344	0.3435	1.7765	−0.6035	0.0375	−0.1586
16	−2425.41	−0.2632	0.2231	2.0894	−0.1411	0.0738	−0.0912
17	3211.68	0.3379	0.1699	1.9086	0.1529	−0.0111	0.0318
18	4055.86	0.4569	0.2697	2.0618	0.2776	−0.0070	−0.0790

Figure 8.3–10. Output of regression on full data set of Table 8.3–6, using the SAS program of Figure 8.3–9 (continued).

Output B

OBS	DFBETAS NUMFORK	DFBETAS FORK2	DFBETAS CREW2	DFBETAS CREWFORK
1	−0.0306	−0.0184	0.2593	0.0883
2	2.0393	−1.0941	−0.1889	−2.0402
3	−0.0181	−0.0036	0.0020	0.0371
4	−9.9465	11.2840	−0.8473	−3.0541
5	0.0348	−0.0405	−0.0341	−0.0074
6	0.1532	−0.1649	0.0082	−0.1122
7	0.0241	−0.7942	−1.3607	2.0385
8	−0.0485	0.0507	−0.0412	0.0190
9	−0.2286	0.1508	0.2952	0.3392
10	0.0495	−0.0161	0.1957	−0.0683
11	0.0323	−0.0924	−0.0548	0.0696
12	−0.0227	−0.0045	0.0025	0.0467
13	−0.0443	0.1588	0.3254	−0.0892
14	0.2209	−0.2379	−0.2749	−0.0518
15	0.1210	−0.2863	0.1462	0.1147
16	−0.0831	0.0965	0.0858	0.0257
17	−0.0029	−0.0025	−0.0555	0.0461
18	0.0530	−0.0805	0.1096	0.0489

Figure 8.3–10. Output of regression on full data set of Table 8.3–6, using the SAS program of Figure 8.3–9 (continued).

C   Output B: Contour plot of NUMCREW*NUMFORK

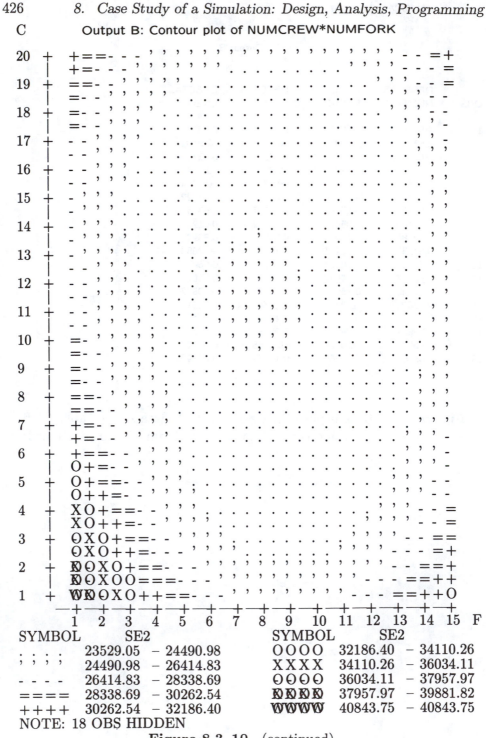

SYMBOL	SE2		SYMBOL	SE2	
. . . .	23529.05	– 24490.98	O O O O	32186.40	– 34110.26
’ ’ ’ ’	24490.98	– 26414.83	X X X X	34110.26	– 36034.11
– – – –	26414.83	– 28338.69	⊖ ⊖ ⊖ ⊖	36034.11	– 37957.97
= = = =	28338.69	– 30262.54	𝕏 𝕏 𝕏 𝕏	37957.97	– 39881.82
+ + + +	30262.54	– 32186.40	𝕎 𝕎 𝕎 𝕎	40843.75	– 40843.75

NOTE: 18 OBS HIDDEN

Figure 8.3–10. (continued).

C   Output B: Contour plot of NUMCREW*NUMFORK

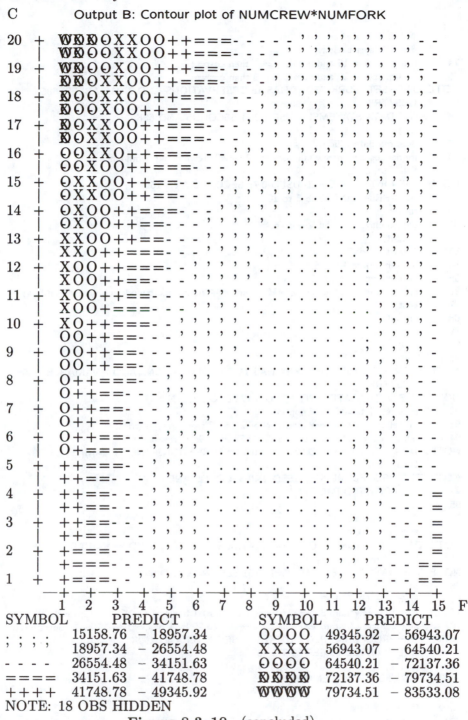

SYMBOL	PREDICT	SYMBOL	PREDICT
. . . .	15158.76 – 18957.34	O O O O	49345.92 – 56943.07
' ' ' '	18957.34 – 26554.48	X X X X	56943.07 – 64540.21
- - - -	26554.48 – 34151.63	⊖ ⊖ ⊖ ⊖	64540.21 – 72137.36
= = = =	34151.63 – 41748.78	⊠ ⊠ ⊠ ⊠	72137.36 – 79734.51
+ + + +	41748.78 – 49345.92	⊞ ⊞ ⊞ ⊞	79734.51 – 83533.08

NOTE: 18 OBS HIDDEN

Figure 8.3–10. (concluded).

```
73       data d5day;set d5day;
74       if numcrew = 15 and numfork = 1 then cost=.;
75       if numcrew = 20 and numfork = 3 then cost=.;
```
NOTE: DATA SET WORK.D5DAY HAS 18 OBSERVATIONS AND 7 VARIABLES. 586
        OBS/TRK.
NOTE: THE DATA STATEMENT USED 0.21 SECONDS AND 68K.
```
76       run;
77       proc reg data=d5day;
78       model cost=numcrew numfork fork2 crew2 crewfork / p r influence ;
79       title Output C;
```
NOTE: THE PROCEDURE REG USED 0.75 SECONDS AND 160K
        AND PRINTED PAGES 8 to 10.
```
80       run;
81       data d5dayex;set d5day end=eof;
82       output;
83       if eof then do ;
84       do numfork = 1 to 15 by .5 ;
85       do numcrew = 1 to 20 by .5 ;
86       crew2=numcrew*numcrew;fork2=numfork+*numfork;crewfork=numcrew*numfork;
87       cost = .;
88       output;
89       end;
90       end;
91       end;
```
NOTE: DATA SET WORK.D5DAYEX HAS 1149 OBSERVATIONS AND 7 VARIABLES. 586
        OBS/TRK
NOTE: THE DATA STATEMENT USED 0.40 SECONDS AND 68K.
```
92       proc reg data=d5dayex noprint;
93       model cost = numcrew numfork fork2 crew2 crewfork / noprint;
94       output out=plot2 p=predict 195=195 u95=u95 r=residual ;
```
NOTE: DATA SET WORK.PLOT2 HAS 1149 OBSERVATIONS AND 11 VARIABLES. 382
        OBS/TRK
NOTE: THE PROCEDURE REG USED 2.40 SECONDS AND 164K
        AND PRINTED PAGE 11.
```
95       run;
96       data plot2;set plot2;
97       se2=u95-predict;
```
NOTE: DATA SET WORK.PLOT2 HAS 1149 OBSERVATIONS AND 12 VARIABLES. 350
        OBS/TRK
NOTE: THE DATA STATEMENT USED 0.51 SECONDS AND 68K.
```
98       run;
99       proc plot data=plot2;
100      plot numcrew*numfork=se2/vaxis=1 to 20 by 1 haxis=1 to 15 by 1 contour=10;
101      plot numcrew*numfork=predict / vaxis=1 to 20 by 1 haxis=1 to 15 by 1 contour=10;
```
NOTE: THE PROCEDURE PLOT USED 1.40 SECONDS AND 192K
        AND PRINTED PAGES 12 TO 13.
```
102      run;
```
NOTE: SAS USED 192K MEMORY.

Figure 8.3–11. SAS program, Part III, yielding Figure 8.3–12 as output.

Output C

DEP VARIABLE: COST   Cost of Operation (Simulated)

SOURCE	DF	SUM OF SQUARES	MEAN SQUARE	F VALUE	PROB>F
MODEL	5	145791549	29158310	71.590	0.0001
ERROR	10	4072964	407296		
C TOTAL	15	149864512			

ROOT MSE	638.198	R-SQUARE	0.9728
DEP MEAN	23585.188	ADJ R-SQ	0.9592
C.V.	2.705926		

VARIABLE	DF	PARAMETER ESTIMATE	STANDARD ERROR	T FOR H0: PARAMETER= 0	PROB > |T|
INTERCEP	1	21380.376	1433.959	14.910	0.0001
NUMCREW	1	−451.501	133.778	−3.375	0.0071
NUMFORK	1	95.694373	296.500	0.323	0.7535
FORK2	1	3.310469	18.268994	0.181	0.8598
CREW2	1	44.790070	5.638366	7.944	0.0001
CREWFORK	1	−11.368501	12.127381	−0.937	0.3706

OBS	ACTUAL	PREDICT VALUE	STD ERR PREDICT	RESIDUAL	STD ERR RESIDUAL	STUDENT RESIDUAL	-2-1-0 1 2
1	21273	21860	417.022	−587.142	483.104	−1.215	| ** | |
2	20608	20389	559.545	219.026	306.929	0.714	| |* |
3	22050	21307	347.776	742.833	535.115	1.388	| |** |
4	.	24614	1578	.	.	.	
5	23898	24299	305.079	−400.835	560.556	−0.715	| * | |
6	23898	24308	396.161	−409.988	500.352	−0.819	| * | |
7	.	29901	1552	.	.	.	
8	29498	29425	466.542	73.147	435.471	0.168	| | |
9	29498	29114	445.118	383.926	457.348	0.839	| |* |
10	21385	22148	402.696	−762.971	495.108	−1.541	| *** | |
11	21819	20677	338.215	1142	541.209	2.110	| |**** |
12	21819	21307	347.776	511.833	535.115	0.956	| |* |
13	20993	21337	348.727	−344.184	534.496	−0.644	| * | |
14	21497	21496	400.553	1.484	496.844	0.003	| | |
15	21497	21819	403.108	−322.371	494.773	−0.652	| * | |
16	24010	24324	320.781	−314.011	551.721	−0.569	| * | |
17	24010	24272	319.800	−262.268	552.290	−0.475	| | |
18	29610	29281	350.431	329.321	533.381	0.617	| |* |

Figure 8.3–12. Output of regression on 16 data points, using the SAS program of Figure 8.3–11 (continues).

Output C

OBS	COOK'S D
1	0.183
2	0.282
3	0.136
4	.
5	0.025
6	0.070
7	.
8	0.005
9	0.111
10	0.262
11	0.290
12	0.064
13	0.029
14	0.000
15	0.047
16	0.018
17	0.013
18	0.027

SUM OF RESIDUALS $-1.81899E-12$
SUM OF SQUARED RESIDUALS   4072964

OBS	RESIDUAL	RSTUDENT	HAT DIAG H	COV RATIO	DFFITS	DFBETAS INTERCEP	DFBETAS NUMCREW
1	−587.142	−1.2489	0.4270	1.2586	−1.0781	−0.2680	0.7673
2	219.026	0.6949	0.7687	5.9455	1.2669	1.1285	−0.3202
3	742.833	1.4657	0.2970	0.7409	0.9526	−0.1552	0.2936
4	.	.	.	.	.	.	.
5	−400.835	−0.6964	0.2285	1.7801	−0.3790	0.1960	−0.2190
6	−409.988	−0.8048	0.3853	2.0174	−0.6372	−0.1867	0.0904
7	.	.	.	.	.	.	.
8	73.1471	0.1596	0.5344	3.9735	0.1710	0.0100	0.0219
9	383.926	0.8260	0.4865	2.3634	0.8039	0.2155	−0.5026
10	−762.971	−1.6742	0.3981	0.6146	−1.3617	−0.1204	0.6469
11	1142.2	2.6885	0.2808	0.0761	1.6801	−0.1010	−0.3243
12	11.833	0.9520	0.2970	1.5050	0.6187	−0.1008	0.1907
13	344.184	−0.6240	0.2986	2.0807	−0.4071	−0.0811	−0.2032
14	.48434	0.0028	0.3939	3.1047	0.0023	−0.0014	0.0007
15	322.371	−0.6317	0.3990	2.4133	−0.5146	−0.0630	−0.1751
16	314.011	−0.5489	0.2526	2.0664	−0.3191	0.0975	−0.2161
17	−262.268	−0.4557	0.2511	2.1912	−0.2639	0.0708	−0.0004
18	329.321	0.5972	0.3015	2.1336	0.3924	−0.0113	−0.0911

Figure 8.3–12. Output of regression on 16 data points, using the SAS program of Figure 8.3–11 (continued).

Output C

OBS	DFBETAS NUMFORK	DFBETAS FORK2	DFBETAS CREW2	DFBETAS CREWFORK
1	−0.2116	0.3186	−0.3388	−0.3765
2	−0.9681	0.7005	0.2484	0.0866
3	−0.0097	0.2662	0.1405	−0.5666
4	.	.	.	.
5	-0.1792	0.1657	0.1972	0.0164
6	0.2116	−0.1594	0.0656	−0.1962
7	.	.	.	.
8	−0.0357	0.0606	0.1002	−0.1057
9	−0.0540	−0.0631	0.2363	0.4006
10	-0.2313	0.1144	−0.6770	0.1510
11	.7518	−1.0628	−0.8004	1.0220
12	-0.0063	0.1729	0.0913	−0.3680
13	.1052	−0.0762	0.1452	0.0711
14	.0017	−0.0018	−0.0019	0.0010
15	.2047	−0.2765	0.0227	0.1816
16	-0.0470	0.0204	0.0926	0.1204
17	-0.0972	0.1238	0.1465	−0.1600
18	-0.0348	-0.0273	0.1730	−0.0202

Figure 8.3–12. Output of regression on 16 data points, using
the SAS program of Figure 8.3–11 (continued).

C   Output C: Contour plot of NUMCREW*NUMFORK

```
20 +  WWOXOO+==- - ' ' ' '   . . . . . . . . . .     ' ' ' '
   |  WWOXO++=- - ' ' ' '   . . . . . . . . .    . ' ' ' '
19 +  WOXOO+==- - ' ' ' '   . . . . . . . . .      ' ' '
   |  WOXO++=- - ' ' ' ' '  . . . . . . . . .      ' ' '
18 +  OXXO+==- - ' ' ' '    . . . . . . . . .        ' '
   |  OXO++=- - ' ' ' ' '   . . . . . . . . .        ' '
17 +  OXO+==- - ' ' ' '     . . . . . . . . .
   |  XXO+==- - ' ' ' '     . . . . . . . . .
16 +  XO++=- - ' ' ' '      . . . . . . . .
   |  XO+==- - ' ' ' '      . . . . . . . .
15 +  OO+==- ' ' ' '        . . . . . . . .
   |  O++=- - ' ' ' '       . . . . . . . .
14 +  O+==- - ' ' ' '       . . . . . . . .
   |  O+==- ' ' ' '         . . . . . . . .
13 +  ++=- - ' ' ' '        . . . . . . . .
   |  +==- - ' ' ' '        . . . . . . . .
12 +  +==- - ' ' ' '        . . . . . . . .
   |  +=- - ' ' ' '         . . . . . . . .
11 +  +=- - ' ' ' '         . . . . . . . .
   |  ==- - ' ' ' '         . . . . . . . .
10 +  ==- - ' ' ' '         . . . . . . . .
   |  ==- ' ' ' '           . . . . . . . .
 9 +  =- - ' ' ' '          . . . . . . . .
   |  =- - ' ' ' '          . . . . . . . .
 8 +  =- - ' ' ' '          . . . . . . . .         '
   |  =- - ' ' ' '          . . . . . . . .         '
 7 +  =- - ' ' ' '          . . . . . . . .         '
   |  =- - ' ' ' '          . . . . . . . .       ; ' '
 6 +  =- - ' ' ' '          . . . . . . . .       ' ' '
   |  =- - ' ' ' '          . . . . . . . .       ' ' '
 5 +  =- - ' ' ' '          . . . . . . . .     ' ' ' '
   |  =- - ' ' ' '          . . . . . . . .     ' ' ' '
 4 +  =- - ' ' ' '          . . . . . . . .     ' ' ' '  -
   |  =- - ' ' ' ' ;        . . . . . . . .     ' ' ' '  -
 3 +  =- - ' ' ' ' '        . . . . . . . .     ' ' ' '  - -
   |  =- - - ' ' ' ' ;      . . . . . . . .     ' ' ' '  - -
 2 +  =' ' ' ' ' ' ' ' '    . . . . . . . .   ; ' ' ' '  - - =
   |  ==- - ' ' ' ' ' ; ;   . . . . . . .   ; ; ' ' ' '  - - =
 1 +  ==- - - ' ' ' ' ' ; ; . . . . . .   ; ' ' ' '  - - ==
   -+-+-+-+-+-+-+-+-+-+-+-+-+-+-+
    1 2 3 4 5 6 7 8 9 10 11 12 13 14 15 F
```

SYMBOL	SE2	SYMBOL	SE2
. . . .	1552.472 – 1756.540	O O O O	3389.082 – 3797.217
' ' ' '	1756.540 – 2164.675	X X X X	3797.217 – 4205.353
- - - -	2164.675 – 2572.811	θ θ θ θ	4205.353 – 4613.489
= = = =	2572.811 – 2980.946	⊠ ⊠ ⊠ ⊠	4613.489 – 5021.624
+ + + +	2980.946 – 3389.082	W W W W	5021.624 – 5225.692

NOTE: 18 OBS HIDDEN

Figure 8.3–12. (continued).

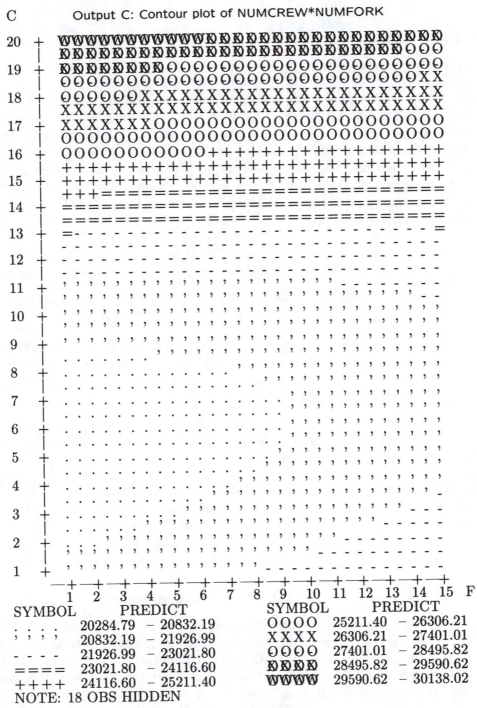

Figure 8.3–12. (concluded).

8.4   The GPSS Program

Consider the transportation system problem introduced in Section 8.1 and solved in Sections 8.2 and 8.3. Suppose that, through new initiatives, the company is considering expanding its business. As the company contemplates the new higher frequency of truck arrivals, it finds it necessary to consider managing work crews itself rather than having an outside contractor supply crews. The following are the specific points of departure from the original conditions of the problem described in Section 8.1, with that prior status shown in brackets.

1. The cost of forklift operation, $75 per forklift per day, is included in the modified version. [No such cost was included in previous sections.]

2. The pay premium for overtime work is now 100%. [The cost of overtime labor, supplied by an outside firm, was at a 50% premium.]

3. As in the earlier version, trucks arrive with normally distributed cargo weights ($\mu = 80,000$ lb and $\sigma = 7500$ lb). The interarrival times of the trucks follow an exponential distribution with a mean of 45 minutes. [The mean had been 140 minutes.]

4. There are, as before, three types of cargo. However, there are new unloading rates, as given in Table 8.4–1. [Figures on unloading rates had been as in Figure 8.1–1.]

In this section, we develop a GPSS program for this modification of the problem. The basic structure of the program is the same as that in Figure 8.1–2, so except for minor adjustments the program is the same as that used to simulate the original problem. As in Section 8.3, the program and its output are given at the end of this section, as Figures 8.4–5 and 8.4–6.

8.4.1   Functions, Variables, and Savevalues

The GPSS program for this problem (Figure 8.4–5) uses the functions EXPON (exponential distribution with $\mu = 1$), SNORM (normal distribution with $\mu = 0$ and $\sigma = 1$), and UNLOAD, which computes the unloading rate (in pounds per minute) for arriving trucks. Note that the rates of 300 lb/min., 400 lb/min., and 450 lb/min. used in the definition of UNLOAD (line 17) are equivalent to the rates of 18,000 lb/hr, 24,000 lb/hr, and 27,000 lb/hr, respectively, given in Table 8.4–1.

Table 8.4–1. Cargo types and unloading rates.

Type of Load	Probability	Unloading Rate (lb/hr)
A	.40	27,000
B	.35	24,000
C	.25	18,000

In the segment of the program that simulates the arrival and unloading of trucks (lines 30 through 48), trucks are modeled as transactions. The variable WEIGHT (line 20) is used to produce an observation from $N(80,000, 7500^2)$ as the weight of an arriving truck. Assuming that the second parameter of a transaction represents the weight to be unloaded, the variable UNLDTME (line 21) computes the time it takes to unload the truck. Since each crew receives $28 per hour ($12 for the forklift operator and $8 each for two laborers), the regular 8-hour wages for a crew would be $8 \times 28 = $224 per day. The variable NORMPAY (line 22) computes the regular wages for X$CREWNUM crews for one day. If the seventh parameter of a transaction represents the number of minutes worked by some crew, and if P7 $\geq$ 480 minutes, (P7–480)/60 would be the hours of overtime worked by the crew. Hence, that crew's overtime compensation, 56 $\times$ (P7–480)/60, is calculated by the variable OVERPAY. With X$FORKLIFT forklifts in the simulation, the cost of forklift operation will be 5 $\times$ 75 $\times$ X$FORKLIFT for a 5-day simulation. The variable FORKEXP (line 24) determines this value. During the simulation, a truck transaction will have P5 represent the time that a crew was assigned to unload it. The variable WAITTME, invoked immediately following the acquisition of a forklift, calculates the amount of time the crew waited for a forklift to become available.

The program uses 30 savevalues, the names and descriptions of which are given in Table 8.4–2.

8.4.2   The Program Logic

The first segment of the program in Figure 8.4–5 (lines 30 through 48) simulates the arrival and unloading of trucks. Upon their arrival through the GENERATE block (line 30), parameters 2 and 3 of the transactions are assigned values representing the cargo weights and their unloading times, respectively. Since all arrivals during a day are to be unloaded on the next day, all transactions are temporarily removed from the simulation and placed on

Table 8.4–2. The savevalues used by the program in Figure 8.4–5.

Savevalue Name	Description
FORKLIFT	Number of forklifts during the simulation
FORKCOUNT	Number of forklifts available (i.e., unbroken)
CREWNUM	Number of crews available
WAGES	Total wages paid
NPAY	Normal wages (wages exclusive of overtime)
COST	Total expenses (wages and forklift expenses)
DAYCOUNT	Day number during the multiple-day simulation
FLAG	Number of times the cargo from one day
	was not completely unloaded the next day
TOTWRKTM	Accumulation of the total minutes worked
	by all crews during the simulation
TOTWT	Accumulation of the weight unloaded by
	all crews during the simulation
X1–X20	Number of minutes worked by crews 1,2,. . .,20

the user-defined chain WAITQ by the LINK block at line 33.

For the moment, let us assume that somehow, at the start of the next day, the transactions from WAITQ are activated and sent to location UNLD (line 35). This is actually done at line 90 in another segment of the program. At the start of the day following their arrival, the reactivated trucks at location UNLD choose the crew or one of the crews that has logged the least working time for that day. The number of the crew that is selected is saved in P4. Following the successful capture of this crew (line 38), the transaction attempts to obtain a forklift. The time that the crew will spend unloading the truck is added to that crew's working time at line 37. If the crew has to wait for a forklift to become available, the waiting time, V$WAITTME is also added to the working time of the crew at line 41. After both a crew and a forklift have been acquired, the unloading is done and the crew and forklift are made available to other trucks (lines 43 and 44). At lines 45 and 46, the total cargo weight unloaded and the total working time of the crews are tallied. These numbers do not have a direct bearing on the simulation but are of general interest as we study the entire operation.

The second segment of the program (lines 52 through 64) simulates the breakdown of forklifts. One minute before the start of each day, beginning with the second day, a transaction "records" X$FORKCOUNT, the number of forklifts that are in use. If this number is positive, the loop (lines 57 through 61) cycles through the available forklifts and, with probability .1, removes them from the simulation for a 2-day period.

The third segment of the program performs the functions of a payroll officer. At the end of each day, starting with the second day of operation, the "payroll officer" checks (line 71) to see if a crew has logged any overtime (i.e., if the savevalue associated with that crew exceeds 480). If the crew has worked overtime, its total work time in minutes is assigned to P7 (line 72) and the wages paid to date are incremented by V$OVERPAY (line 73). Next, the time worked by the crew is cleared for the next day's operation, at line 74. Lines 76 and 77 increment the overall normal (nonovertime) payment, X$NPAY, and overall wages, X$WAGES, by the regular pay for the crew.

The last segment of the program (lines 84 through 95) functions as a "supervisor." At the start of each day, beginning with the second day, the supervisor checks the day within the simulation time span (line 86). If there is a transaction in the ADVANCE block at location WORK (line 42) at this time, then the previous day's cargo is still being unloaded. By incrementing the savevalue FLAG, the supervisor keeps track of the number of days when the previous day's work was not completed.

The supervisor now checks (line 89) to see if this is the sixth simulated day (equivalently fifth unloading day). Recall that on the first day, there were only truck arrivals, and unloading did not begin until the second day. If this is not the last simulation day, all trucks from the chain WAITQ are taken and sent to UNLD at line 35 to be unloaded. If this is the last day, however, then the cost of the 5-day simulation is incremented by the forklift expenses and the accumulated wages paid during that period.

8.4.3 Program Output

The use of the PRINT block at line 94 is mainly for convenience. The results of the simulation that most interest us are the savevalues. We can suppress the voluminous output that this simulation would produce by using

                    START   ,NP

statements and relying on the output produced by the PRINT block.

Figure 8.4–6 shows the output associated with simulations with 8 crews and 7, 8, and 9 forklifts. The use of the RMULT statement enables us to produce the same pattern of truck arrivals with exactly the same cargo weights in all three simulations. Since each simulation uses a different number of forklifts, the forklift breakdown pattern will vary. Table 8.4–3 summarizes the 5-day operating costs for a variety of crew/forklift combinations.

By giving the data in Tables 8.4–3 and 8.4–4, we intend (see problems 8.3 and 8.4) that the reader should now choose a **design** and then **analyze** the data points chosen in that design to optimize the system. As we saw in Sections 8.2 and 8.3, only the naive will use all $17 \times 11 = 187$ data points as good modeling, and optimization will be obtained by statistical design and analysis (at a marked savings in simulation time and effort). Of course, the modeling techniques in Section 8.3 (excluding some outliers and the like) will also be needed here.

Table 8.4–3. Total cost from 5-day simulations.

Fork-lifts	Crews 6	7	8	9	10	11	12	13	14	15	16	17	18
22	33595†	32474	31352	30226	29104	28314	27796	27531	27275	27227	27586	28289	29145
21	33220†	32099	30977	29851	28729	27939	27421	27156	26900	26852	27211	27914	28770
20	32845†	31724	30602	29476	28354	27564	27046	26781	26525	26477	26836	27539	28738
19	32470†	31349	30227	29101	27979	27189	26671	26406	26150	26102	26461	27638	28864
18	32095†	30974	29852	28726	27604	26814	26296	26031	25775	25727	26437	27572	28934
17	31720†	30599	29477	28351	27229	26439	25921	25656	25400	25943	26567	27751	29068
16	31345†	30224	29012	27976	26854	26064	25546	25281	25287	25637	26536	27625	28857
15	30970†	29849	28727	27601	26479	25689	25171	25199	25297	26207	27326	28365	29566
14	30595†	29474	28352	27226	26104	25314	25128	25360	26197	27213	28370	29445	30605
13	30220†	29099	27977	26851	25729	24939	25003	25979	27075	28224	29298	30432	31498
12	29845†	28724	27602	26467	25354	24911	25597	26988	28054	29273	30365	31471	32702
11	29470†	28349	27227	26101	25508	26277	27454	28570	30291	31850	33037	34118	35283
10	29095†	27974	26852	26275	26974	28765	30016	31534	33264	34868	36239	37452	38551
9	28720†	27599	26477	26841	28702	30587	32272	33737	35654	37500	38186	40057	41103
8	28345†	27224	27864	30301	32474	34881	36574	38483	40634	42588	44144	45185	46244
7	27970†	28415	31467	34267	36948	39727	41734	44190	46790†	48673†	50239	51282	52418
6	28539†	32228†	36159†	39273†	42608†	45803†	48290†	51136†	54032†	56755†	58382†	59683†	60978†

†Not all trucks were unloaded on the day following their arrival.

Table 8.4-4.   Total cost from 5-day simulations with different seeds.†

Fork-lifts	Crews												
	6	7	8	9	10	11	12	13	14	15	16	17	18
22	31374‡	30253	29127	28007	27075	26516	26140	25949	26047	26439	27003	27827	28837
21	30999‡	29878	28752	27632	26700	26141	25765	25574	25672	26064	26628	27452	28533
20	30624‡	29503	28377	27257	26325	25766	25390	25199	25297	25689	26335	27156	28345
19	30249‡	29128	28002	26882	25950	25391	25015	24824	24922	25388	26029	26891	28130
18	29874‡	28753	27627	26507	25575	25016	24640	24449	24673	25015	25859	26690	28153
17	29499‡	28378	27252	26132	25200	24641	24265	24074	24298	24938	25939	27029	28166
16	29124‡	28003	26877	25757	24825	24266	23890	23699	24118	24938	26284	27327	28439
15	28749‡	27628	26502	25382	24450	23891	23515	23715	24382	25568	26804	27844	28890
14	28374‡	27253	26127	25007	24075	23516	23655	24184	25085	26234	27530	28589	29585
13	27999‡	26878	25752	24632	23700	23141	23849	25117	26385	27296	28548	29837	30854
12	27624‡	26503	25377	24257	23325	23734	25056	26453	27712	28771	29947	31290	32512
11	27249‡	26128	25002	23882	24140	25519	27017	28702	30027	31208	32561	33825	35076
10	26874‡	25753	24627	24886	26311	27929	29569	31430	32897	34237	35686	37034	38339
9	26499‡	25378	24980	26933	28975	30923	32595	34568	36151	37714	39153	40412	41495
8	26124‡	25839	28497	31003	33286	35174	37579	39885	41992	43567	44673	46412	47637
7	25749‡	29018‡	32165‡	35418‡	37910‡	40521‡	42924‡	45526‡	47835‡	49668‡	51039‡	52750‡	54118‡
6	30007‡	34187‡	38073‡	41896‡	45408‡	48708‡	51578‡	55149‡	57904‡	60106‡	61709‡	63004‡	64649‡

†The same seed (but different from the one used in the previous run of Table 8.4-3) was used for all simulations.

‡Not all trucks were unloaded on the day following their arrival.

```
1
2          SIMULATE
3
4          RMULT       12345,12345,12345,12345
5
6          EXPON     FUNCTION     RN4,C24              Exponential time function
7          0,0/.1,.104/.2,.222/.3,.355/.4,.509/.5,.69/.6,.915/.7,1.2/.75, 1.38
8          .8,1.6/.84,1.83/.88,2.12/.9,2.3/.92,2.52/.94,2.81/.95,2.99/.96, 3.2
9          .97,3.5/.98,3.9/.99,4.6/.995,5.3/.998,6.2/.999,7/.9998,8
10
11         NRDIS     FUNCTION     RN3,C25              Standard normal distribution
12         0,-5/.00003,-4/.00135,-3/.00621,-2.5/.02275,-2/.06681, -1.5/.11507,-1.2
13         .15866,-1/.21186,-.8/.274225,-.6/.34458,-.4/.42074,-.2/.5, 0/.57926,.2
14         .65542,.4/.72575,.6/.78814,.8/.84134,1/.88493,1.2/.93319, 1.5/.97725,2
15         .99379,2.5/.99865,3/.99997,4/1,5
16
17         UNLOAD    FUNCTION     RN2,D3               Unloading function
18         .25,300/.6,400/1,450
19
20         WEIGHT    FVARIABLE    7500*FN$NRDIS+80000  Weight of truck's cargo.
21         UNLDTME   FVARIABLE    P2/(FN$UNLOAD)       Time to unload a truck.
22         NORMPAY   FVARIABLE    224*X$CREWNUM        Regular 8-hour pay.
23         OVERPAY   FVARIABLE    56*(P7-480)/60       Overtime pay per crew.
24         FORKEXP   FVARIABLE    X$FORKLIFT*375       Cost for forklifts.
25         WAITTME   FVARIABLE    C1-P5                Wait time for a forklift.
26
27            *
28            ******* PROGRAM SEGMENT FOR TRUCK ARRIVALS AND UNLOADING
29            *
30    1          GENERATE     45,FN$EXPON,,,,5,F      Truck arrivals.
31    2          ASSIGN       2,V$WEIGHT              P2 ← truck weight.
32    3          ASSIGN       3,V$UNLDTME            P3 ← unloading time.
33    4          LINK         WAITQ,FIFO             Wait for next day.
34            *                                      Next day.
35    5   UNLD   SELECT MIN   4,1,X$CREWNUM,,X       P4 ← crew no.
36            *                                      with least work+wait time.
37    6          SAVEVALUE    P4+,P3                 Inc. X$(P4) by work time.
38    7          SEIZE        P4                     Get crew.
39    8          ASSIGN       5,C1                   P5 ← time of crew assign.
40    9          ENTER        LIFT                   Get a forklift.
41    10         SAVEVALUE    P4+,V$WAITTME          Inc. x$(P4) by wait time.
42    11  WORK   ADVANCE      P3                     Unload the truck.
43    12         LEAVE        LIFT                   Return the forklift.
44    13         RELEASE      P4                     Release the crew.
45    14         SAVEVALUE    TOTWT+,P2              Accum. load weights.
46    15         SAVEVALUE    TOTWRKTM+,P3           Accum. unloading times.
47    16         TERMINATE
48
```

Figure 8.4–5. GPSS simulation of a transportation system (continues).

```
49                *
50                ******* PROGRAM SEGMENT FOR BREAKING FORKLIFTS
51                *
52   17                  GENERATE    1440,,1439,,,1              Forklift breakers at start
53                *                                             of days 2, ... 5.
54   18                  SAVEVALUE   FORKCOUNT,R$LIFT           Number of forklifts left.
55   19                  TEST NE     X$FORKCOUNT,0,NONE         If none, can't break any.
56   20                  ASSIGN      1,0                        But if there are...
57   21         FORKLP   RANSFER     .90,,OKAY                  10% chance of breaking.
58   22                  ENTER       LIFT                       Break one.
59   23                  ASSIGN      1+,1                       If more should be
60   24         OKAY     SAVEVALUE   FORKCOUNT-,1               considered, loop
61   25                  TEST E      X$FORKCOUNT,0,FORKLP       back and try again.
62   26                  ADVANCE     2880                       Keep broken lifts 2 days
63   27                  LEAVE       LIFT,P1                    then return to the pool.
64   28         NONE     TERMINATE
65
66                *
67                ******* PROGRAM SEGMENT FOR CALCULATING WAGES
68                *
69   29                  GENERATE    1440,,2880,,2              To get end of day wages.
70   30                  ASSIGN      2,X$CREWNUM                P2 is the loop counter.
71   31         OVRTME   TEST G      X*P2,480,JUMP              X*P2=savevalue that P2
72   32                  ASSIGN      7,X*P2                     points to; P7=work time.
73   33                  SAVEVALUE   WAGES+,V$OVERPAY           Account for overtime.
74   34         JUMP     SAVEVALUE   P2,0                       Clear P2-th savevalue.
75   35                  LOOP        2,OVRTME                   Loop to get other crews.
76   36                  SAVEVALUE   NPAY+,V$NORMPAY            Compute normal pay.
77   37                  SAVEVALUE   WAGES+,V$NORMPAY           WAGES=normal pay
78   38                  TERMINATE                             + overtime pay.
79
80                *
81                ******* PROGRAM SEGMENT FOR MANAGING THE MOVEMENT OF
82                ******* TRUCKS AND CALCULATING OVERALL WAGES AND COSTS
83                *
84   39                  GENERATE    1440,,1440                 Start unloading at
85                *                                             start of days 2, ... 5.
86   40                  SAVEVALUE   DAYCOUNT+,1                Keep track of days.
87   41                  TEST NE     W$WORK,0,CHKDAY            If at the end unloading
88   42                  SAVEVALUE   FLAG+,1                    continues, inc. FLAG.
89   43         CHKDAY   TEST NE     X$DAYCOUNT,6,LAST          If not yet 5 days,
90   44                  UNLINK      WAITQ,UNLD,ALL             unload another batch.
91   45                  TRANSFER    ,FIN
92   46         LAST     SAVEVALUE   COST+,X$WAGES
93   47                  SAVEVALUE   COST+,V$FORKEXP
94   48                  PRINT       ,,X                        Get dump of savevalues.
95   49         FIN      TERMINATE   1
96
```

Figure 8.4–5. GPSS simulation of a transportation system (concluded).

```
 97              INITIAL     X1-X20,0
 98              INITIAL     X$FLAG,0/X$DAYCOUNT,0/X$NPAY,0/X$COST,0
 99              INITIAL     X$FORKCOUNT,0/X$WAGES,0/X$TOTWT,0
100              INITIAL     X$TOTWRKTM,0/X$FORKLIFT,7/X$CREWNUM,8
101     LIFT     STORAGE     7
102              START       1,NP
103              RESET
104              START       10,NP
```

PRINT BLOCK 48 AT CLOCK TIME 15840
NON-ZERO FULLWORD SAVEVALUES

SAVEX	VALUE	SAVEX	VALUE	SAVEX	VALUE
COST	67854	CREWNUM	8	DAYCOUNT	11
FORKLIFT	7	NPAY	17920	TOTWRKTM	67011
WAGES	62604				

SAVEX	VALUE
FLAG	3
TOTWT	26364094

```
105
106              RMULT       12345,12345,12345,12345
107              CLEAR
108              INITIAL     X$FLAG,0/X$DAYCOUNT,0/X$NPAY,0/X$COST,0
109              INITIAL     X$FORKCOUNT,0/X$WAGES,0/X$TOTWT,0
110              INITIAL     X$TOTWRKTM,0/X$FORKLIFT,8/X$CREWNUM,8
111     LIFT     STORAGE     8
112              START       1,NP
113              RESET
114              START       10,NP
```

PRINT BLOCK 48 AT CLOCK TIME 15840
NON-ZERO FULLWORD SAVEVALUES

SAVEX	VALUE	SAVEX	VALUE	SAVEX	VALUE
COST	61050	CREWNUM	8	DAYCOUNT	11
NPAY	17920	TOTWRKTM	68111	TOTWT	26733344

SAVEX	VALUE
FORKLIFT	8
WAGES	55050

```
115
116              RMULT       12345,12345,12345,12345
117              CLEAR
118              INITIAL     X$FLAG,0/X$DAYCOUNT,0/X$NPAY,0/X$COST,0
119              INITIAL     X$FORKCOUNT,0/X$WAGES,0/X$TOTWT,0
120              INITIAL     X$TOTWRKTM,0/X$FORKLIFT,9/X$CREWNUM,8
121     LIFT     STORAGE     9
122              START       1,NP
```

Figure 8.4–6. Output of the simulation in Figure 8.4–5 (continues).

```
123             RESET
124             START       10,NP
```

PRINT BLOCK 48 AT CLOCK TIME 15840
NON-ZERO FULLWORD SAVEVALUES

SAVEX	VALUE	SAVEX	VALUE	SAVEX	VALUE
COST	56526	CREWNUM	8	DAYCOUNT	11
NPAY	17920	TOTWRKTM	68111	TOTWT	26733344

SAVEX	VALUE
FORKLIFT	9
WAGES	49776

```
125
126             END
```

Figure 8.4–6. Output of the simulation in Figure 8.4–5 (concluded).

Problems for Chapter 8

8.1 (Section ()8.2) In Section 8.2, it was noted that the designs in Tables 8.2–1 and 8.2–2 are each justified by the central composite design of Section 6.4.

  a. Are the designs of Tables 8.2–1 and 8.2–2 basic, or modified, central composite designs? (If the latter, note what the modifications are and what justification there is for them.)

  b. Can you find an unmodified CCD that might be used, to avoid the possible danger of an ad hoc design?

8.2 (Section 8.4) Pretend that you are a naive simulationist who has just burned up 100 hours of computer time simulating a complex system that has cost depending on two factors, called Crews and Forklifts, resulting in the output in Table 8.4–3. Give a complete analysis of this data in regard to a study that seeks to find the (C, F) at which minimal cost is attained. Be sure to discuss

  a. Residuals and their use.

  b. The need for a second stage of simulation.

c. The (C, F) you recommend.

d. How much possible error there is in the predicted cost at the (C, F) you recommended in part c.

8.3 (Section 8.4) You are a statistical simulationist who knows how to simulate a desired system, perhaps in GPSS, and how to use statistical tools to find how long to run and how to analyze the resulting data. At this stage in the study of this text, we assume that you do not need to pretend. Your naive counterpart in Problem 8.2 used 100 hours of computer time to generate 221 data points, which he or she then analyzed to find a cost minimum. Use

- A central composite design

- Statistical analysis complete enough to include the Part a through d considerations in Problem 8.2

- Two stages of experimentation (using such data from Table 8.4–4 as needed in the second stage called for by your statistical analysis of your CCD design's first stage data from Table 8.4–3)

to answer the following question: At $100/221 = 0.452$ hours per data point, what percent savings over the naive simulationist can you attain, with no loss whatever in the quality of the final results?

Appendix A:
Using GPSS/PC[†]

With the advent of what were, at the time, powerful microcomputers in the 1980s, the use of microcomputers as platforms for simulation began to receive serious attention (see Karian (1985a), (1985b), and (1987) and Karian and Dudewicz (1986)). This trend has continued (see the Introduction to Chapter 1 for a discussion on computer speed) and now a personal computer, with suitable software, is likely to have the power to execute simulations one could only dream of being able to perform just a few years ago. We see this trend continuing and therefore emphasize the desirability of having suitable computer simulation languages in one's software arsenal. GPSS/PC is an interactive adaptation of GPSS to the IBM-PC family of computers. The student version of the software, contained on the disk that accompanies this book, is a limited version of GPSS/PC intended for educational use. The files on this disk will allow you to run many of the programs contained in the text. The student version can accommodate models that have up to 100 blocks. In addition, the commercial version of GPSS/PC has the following features that the student version does not have.

- Access to FORTRAN while a simulation is running and the ability to pass values to and from a FORTRAN program.

- The ability to suspend a simulation, return to DOS to use any set of commands, and continue with the simulation where it was left off.

[†] Adapted from "Introduction to Simulation Modeling Using GPSS/PC[tm]" and from "Debugging and Design of Models Using GPSS/PC[tm]" (copyright ©1994 by Minuteman Software, P. O. Box 131, Holly Springs, NC 27540; tel: 800-223-1430; e-mail: minutemn@mindspring.com; web site: www.mindspring/~minutemn/home/htm).

- The option of having the software keep a Session Journal of all inter-active queries and debugging activities

- The ability to produce 2-dimensional animation.

Details on GPSS/PC, its use, documentation (see Minuteman (1992)), tutorial manual, pricing, etc., are available from Minuteman Software.

A.1 Running a GPSS/PC Program

To install the student version of GPSS/PC on your computer, create a directory or a subdirectory with the name GPSSPC and then copy all the files from the attached floppy disk to this directory.

There are two ways in which you can create a file containing the code for a model: you can create the file through the use of any text editor or you can enter the program lines during a GPSS/PC session. In the latter case, you invoke GPSS/PC by typing the DOS command

GPSSPC

entering the code for the program, and saving the program in a file.

To run an existing program (e.g., one of the examples supplied on the disk), **issue the following command at the DOS** prompt:

GPSSPC @FILE NAME<CR>

The <CR> symbol indicates that you should press the Carriage Return or Enter key. For the FILE NAME, substitute the name of the file that you wish to run (e.g., GPSS/PC @BARBER.GPS<CR>). GPSS/PC will proceed to load the program, checking the code for errors as it loads. If the loaded program already has a START statement, then execution will start immediately after loading; otherwise, you should type in the appropriate START statement (e.g., START<SP> 1, where <SP> designates a space) to execute the program.

To **interrupt a running simulation**, enter <ESC> <CR> (the ESC key followed by a carriage return) to **continue the execution of an interrupted simulation**, enter CONT<CR>. **To end your session**, interrupt an executing program, if necessary, and then enter END<CR>.

A.2 Interacting with GPSS/PC

GPSS/PC facilitates interaction between the user and the modeling environment through a number of windows (the Data, Blocks, Storages, Facilities, Matrices, and Tables Windows). You can access any window, during or after the execution of the model, by pressing the <ALT> key and then pressing the alphabetic key that is the first letter of the name of the window.

The Data Window. After a model has been loaded, you will automatically be in this window. You can display the entire program in this window by entering DIS<CR> on the command line at the bottom of the screen. You can also display a selected sequence of lines by entering

$$DIS<SP>LN1,LN2<CR>$$

where <SP> designates pressing the space bar and LN1 and LN2 are the starting and ending line numbers, respectively, to be used for the display. You will notice that during this process, GPSS/PC will prompt you once you press the space bar (for operand A, the first line number) and again after you enter a comma (for operand B, the second line number).

The Blocks Window. This window gives a symbolic representation of the written code. In this window, you will see the blocks flash as transactions move through the model. If you have a color monitor, you will also see a color change depending on the number of transactions that are resident in a block.

The Storage, Facilities, Matrices, and Tables Windows. These windows will display information on model entities associated with the name of the window. Obviously, if the model does not use tables, the Tables Window will not give any information about the executing model.

When accessing the various windows, you will see a menu at the bottom of the screen. You can enter the commands in the window manually, or you can use the keyboard arrows to move the # sign over the menu items and in the blocks screen to select a specific block with which you want to interact. Once the # is over the menu item or block, press the INS key (0 on the numeric keypad) and the menu item or block will be selected. If you are selecting a block, do this first and then select the action on the menu. For instance, select a block and then move to the menu line and select EDIT.

This will cause the line of code related to that block to be displayed so that you may make changes to it.

A.3 GPSS/PC Output

The GPSSREPT.EXE file, included on the disk, produces reports that can be directed to character-type printers. Unformatted reports are written on the user's default disk drive by GPSS/PC through the REPORT statement. The GPSSREPT.EXE program formats reports into easily readable DOS files. To format a report produced during a GPSS/PC run, execute the GPSSREPT program and follow its simple instructions. In the commercial version of GPSS/PC, the report writer can be run from inside GPSS/PC.

A.4 The GPSS/PC Models on the Disk

Many of the GPSS models that were used as examples in this book are on the disk. The name of the file specifies the program that the file represents. The naming convention that is used is CSSFFAB.GPS where C (one digit) represents the chapter number, SS (one or two digits) represents the section number, FF (one or two digits) represents the figure number, and AB indicates the version of GPSS/PC under which the model will run. The AB specification will be A if the model requires the full commercial version of GPSS/PC for its execution, it will be AB if the model executes under both the commercial and student versions of GPSS/PC, and it will be B if the model has been somewhat modified to enable its execution under the student version.

Table A.4–1 summarizes the program files on the disk and their connection to the examples used in the book.

With the naming convention described above, 2101AB.GPS is the model of the program given in Figure 2.10–1 and it will run under either the limited or full version of GPSS/PC. The presence of distinct A and B renditions of a program (e.g., 5113A.GPS) and 5113B.GPS means that the model presented in the book exceeded the capacity of the limited version. However, with some

Table A.4–1. Disk files for examples in this book.

FILE NAME	FIGURE NUMBER IN BOOK	WINDOWS THAT CAN BE VIEWED
2101AB.GPS	2.10–1	D, B, F, S
2103AB.GPS	2.10–3	D, B, F, S
2105AB.GPS	2.10–5	D, B, F, S
561AB.GPS	5.6–1	D, B, F, S
5113A.GPS	5.11–3	D, B, F
5113B.GPS	5.11–3	D, B, F
712A.GPS	7.1–2	D, B, F
712B.GPS	7.1–2	D, B, F
731AB.GPS	7.3–1	D, B, F, S
751AB.GPS	7.5–1	D, B, F, S
7101AB.GPS	7.10–1	D, B, F
7102AB.GPS	7.10–2	D, B, F, T
7121A.GPS	7.12–1	D, B, F, T
7121B.GPS	7.12–1	D, B, F, T
7161A.GPS	7.16–1	D, B, F
7161B.GPS	7.16–1	D, B, F
812A.GPS	8.1–2	D, B, F, S
845A.GPS	8.4–5	D, B, F, S
BARBER.GPS	A.5–1	D, B, F, T

minor modifications (as given in 5113B.GPS), the program runs under the limited version.

A.5 Debugging in GPSS/PC

GPSS/PC's interactivity makes debugging and testing of design alternatives much easier than in older versions of GPSS. In this section, through an annotated session, we describe a few simple debugging techniques that are valuable when dealing with complex models. The session uses the BARBER.GPS program that is on the disk and given in Figure A.5–1.

To start the session, type GPSSPC @BARBER.GPS and once the model has been read in, use the following commands interactively to examine the running of this simple model.

```
;        GPSS/PC Program File A:BARBER.GPS. (V 2, # 39397) 06-25-1990 13:33:05
10       ***********************************************************************
20       *                                                                     *
30       *                        Barber Shop Simulation                       *
40       *                                                                     *
50       ***********************************************************************
51   WAITIME   QTABLE        BARBER,0,100,20          ;Histogram of Waiting times
60             GENERATE      100,50                   ;Create next customer.
70             TEST LE       Q$BARBER,1,FINIS         ;Wait if line 1 or less
80   *                                                    else leave shop
90             SAVEVALUE     CUSTNUM+,1               ;Total customers who stay
100           ASSIGN        CUSTNO,X$CUSTNUM         ;Assign number to customer
110           QUEUE         BARBER                   ;Begin queue time.
120           SEIZE         BARBER                   ;Own or wait for barber.
130           DEPART        BARBER                   ;End queue time.
140           ADVANCE       500,50                   ;Haircut takes a few minutes.
150           RELEASE       BARBER                   ;Haircut done. Give up the barber.
160  FINIS    TERMINATE     1                        ;Customer leaves.
```

Figure A.5–1. The BARBER.GPS program.

1. Type STOP<CR>.

2. Press <ALT>+B (i.e., hold down <ALT> and B at the same time) to move to the Blocks Window and see the block diagram of the model.

3. Type START<SP>100<CR>. You will see the block diagram in the Blocks Window (shown in Figure A.5–2) with a trace message across the top indicating that the very first transaction has stopped on the first block entry in the model. In this case, it happens to be the first GENERATE block, but it could have been another GENERATE block if the model were to have more than one of these blocks. It would all be related to the transaction creation and movement time in the model.

4. Type STEP<SP>1<CR>. The first transaction will now enter the next block.

5. Hold down the <CTRL> key and one of your function keys (keys F1–F10). For demonstration, we will choose <F1>. The last command entered (which was STEP<SP>1<CR>) will now be loaded into that function key. From this point on, by pressing only <F1>, you can step through the model a block entry at a time.

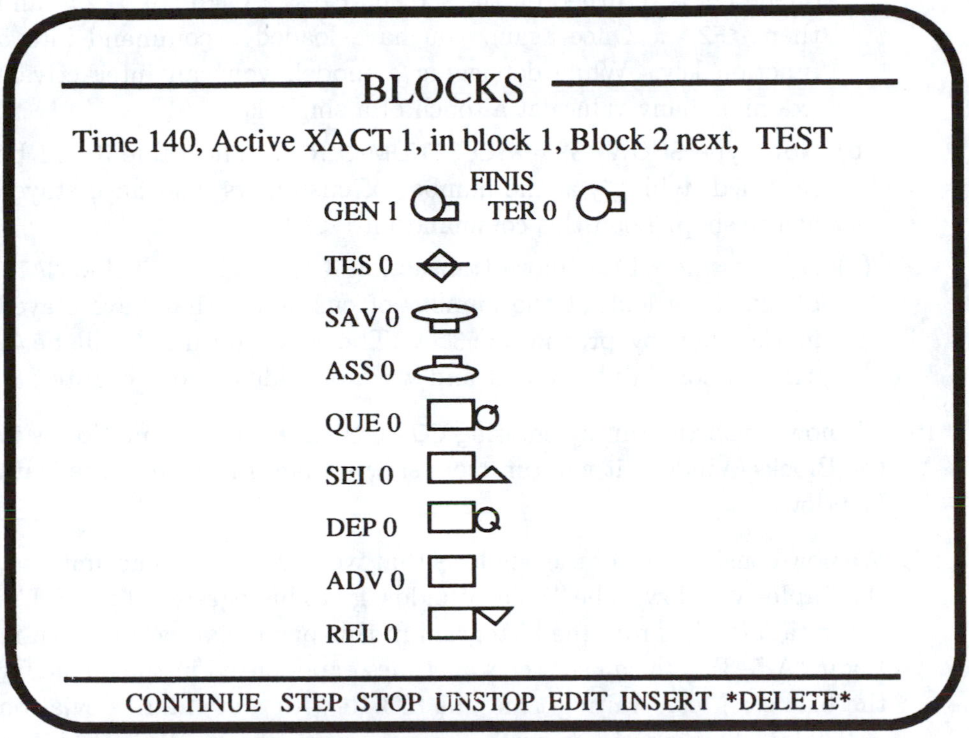

Figure A.5–2. The Blocks Window for the program in Figure A.5–1.

6. Type STOP<SP> ,,OFF before continuing to remove the total model STOP.

7. Type <F1> 18 times—until transaction 4 is ready to enter the GENERATE block.

8. Now step one more time by pressing <F1>. When this transaction enters the TEST block, it will find that the queue for the barber is greater than 1. This customer will then choose not to wait and will leave the shop. You will see it has been scheduled for the TERMINATE block.

9. Before allowing the simulation to progress, we examine some other values related to the simulation.

(a) Type SHOW<SP>Q$BARBER<CR> and see that the size of the queue (2) is returned on the screen. Press <CTRL>+<F2> and then <F2> . Once again, you have loaded a command into a function key. While debugging a model, you can interactively examine many values at a touch of a single key.

(b) Now type SHOW<SP>X$CUSTNUM<CR> . The value 3 will be returned, telling you the number of customers who have stayed at the shop. Load this command into <F3>.

(c) By pressing <F1> move transaction 4 through the TERMINATE block. Now look at the number of customers that have stayed in the shop by pressing <F3> . The value returned will be 3. Transaction 4 did not wait and was not added to our counter.

10. We now finish the run by entering CONT<CR>. If the simulation is in the Blocks Window, it will run more slowly than if it were in the Data Window.

11. We now consider one of the windows that we have not yet encountered, the Tables Window. The Tables Window gives histograms of the SNAs being tabulated. From the histogram in this particular case (shown in Figure A.5–3), you can see the wait times experienced in the queue for the barber. Mean waiting time was 873.10 with a standard deviation of 239.09. This screen, as with all graphics screens in GPSS/PC, can be observed during the run of the simulation. You can often see a specific distribution developing as the simulation runs.

12. If you do not need to do a step-by-step process from the beginning of a model as you did above, but know from running the model that transaction 4 is of particular interest to you, you can put a stop on that transaction. Use the following command to clear out the old transactions and start a new run with transaction 4 as the main point of interest. Type

```
CLEAR<CR>
STOP<SP>4<CR>
<ALT>+B
START<SP>100<CR>
```

The simulation will now run and stop just as transaction 4 becomes the active transaction. You can now use the STEP command or display other values related to this transaction.

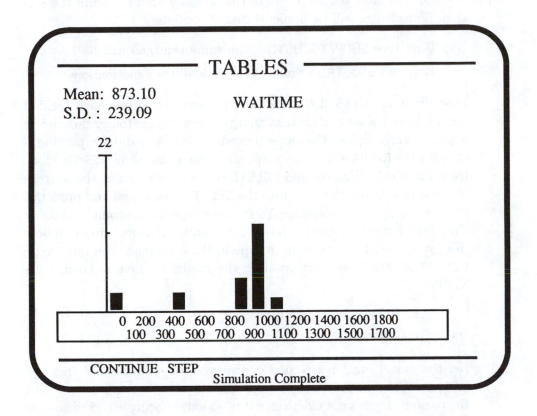

Figure A.5–3. The histogram of WAITIME in the Tables Window.

13. Before going on to our next debugging technique, type

    CLEAR<CR>
    STOP<SP>,,OFF<CR>

to remove the transaction stop that was put on before the last run.

14. Select the RELEASE block (block number 9) in the model, by typing
    STOP<SP>,9 if you do not wish to use any of the above methods. In
    this case you must know the block number. Type START<SP>100<CR>.
    The model will now stop when the first transaction is ready to enter
    the RELEASE block.

15. We will now alter the structure of the model without leaving the session. The change will be implemented immediately.

    (a) First type SHOW P$CUSTNO; the value returned will be 1.

    (b) Hold down <CTRL> and <F4> to load the function key.

    Now eliminate the ASSIGN block (we will assume that you have decided you no longer want each transaction to carry the customer number in a parameter). Follow the same procedure that you did for putting a stop on the RELEASE block, except this time choose the ASSIGN block from the Blocks Window and DELETE it from the menu at the bottom of the screen (move the # sign to the DELETE menu item and press INS or 0 on the keypad). Warning: You cannot type in this command using the block number since the delete command will delete line numbers from your model. If you wish to type in the command, you must type DEL<SP> 100. Now start running the model by first removing the STOP:

    (a) Type STOP<SP>,,OFF<CR>

    (b) Enter CONT<CR>

    Let the model run for a second or two and hit <ESC> and then hit <F4> . Chances are that you will get an error message saying that the parameter does not exist since it is no longer being assigned to new transactions.

16. Add the lines

         170<SP> <SP> START<SP>100<CR> and
         180<SP> <SP> REPORT<SP> BARBER.URP,NOW<CR>

    to the model by simply typing them in at the command line of the Data Window. Typing SAVE<SP> BARBER1.GPS<CR> will save the program in its modified form. Now end your GPSS/PC session by typing END<CR>.

17. Re-enter GPSS/PC with BARBER1.GPS as the active file (i.e., at the DOS prompt enter GPSSPC @BARBER1.GPS). Since this program has a START statement, it will execute immediately following its loading and produce an unformatted report in file BARBER.URP. End your GPSS/PC session.

18. The file BARBER.URP on the default disk drive that you created in the previous step is the unformatted output of the simulation run. Obtain the formatted output by entering

    GPSSREPT<SP> BARBER.URP<SP> BARBER.FRP

    at the DOS prompt. The formatted output file, BARBER.FRP, can now be printed.

Appendix B:
GPSS Block Statement Formats

BLOCK STATEMENT FORMATS

OPERATION	A	B	C	D	E	F	G	H	I
ADVANCE	Mean time [k, SNAj, SNA*SNAj]	Spread [k, SNAj, SNA*SNAj] or Function modifier FNj, [FN*SNAj]							
ALTER {G, GE, L, LE, E, NE, MIN, MAX}	Group no. k, SNAj, SNA*SNAj	Count ALL or k, SNAj, SNA*SNAj	Member attribute to be altered PR or kPx, SNAjPx, SNA*SNAjPx	Value to replace attribute k, SNAj, SNA*SNAj	Matching transaction attribute [PR or kPx, SNAjPx, SNA*SNAjPx]	Matching SNA [k, SNAj, SNA*SNAj]	Alternate exit [k, SNAj, SNA*SNAj]		
ASSEMBLE	No. of transactions to assemble k, SNAj, SNA*SNAj								
ASSIGN	Parameter no. or range k, SNAj, SNA*SNAj [±]	SNA value to be assigned k, SNAj, SNA*SNAj	No. of function modifier [k, SNAj, SNA*SNAj]	Parameter type Px				Note: The parameter type operand may optionally be coded as the C operand if a function modifier is not specified.	
BUFFER									
CHANGE	"From" block no. k, SNAj, SNA*SNAj	"To" block no. k, SNAj, SNA*SNAj							
COUNT {G, GE; L, LE; E, NE; I, NI; SNE, SE; SNF, SF; LR, LS}	Parameter in which to place count kPx, SNAjPx, SNA*SNAjPx	Lower limit k, SNAj, SNA*SNAj	Upper limit k, SNAj, SNA*SNAj	Comparison value if conditional operator is specified [k, SNAj, SNA*SNAj]	Mnemonic of SNA to be counted {Any SNA except MX, MH, MB, ML}				

[ ] Indicates optional operand

{ } Indicates that one of the items within the braces must be selected

BLOCK STATEMENT FORMATS

OPERATION	A	B	C	D	E	F	G	H	I
DEPART	Queue no. k, SNAj, SNA*SNAj	No. of units [k, SNAj, SNA*SNAj]							
ENTER	Storage no. k, SNAj, SNA*SNAj	No. of units [k, SNAj, SNA*SNAj]							
EXAMINE	Group no. k, SNAj, SNA*SNAj	Numeric value—numeric mode [k,SNAj, SNA*SNAj]	Alternate exit k, SNAj, SNA*SNAj						
EXECUTE	Block no. k, SNAj, SNA*SNAj								
FAVAIL	Facility no. or range k, SNAj, SNA*SNAj								
FUNAVAIL	Facility no. or range k, SNAj, SNA*SNAj	Remove or continue option [RE CO]	Alternate block no. [k,SNAj, SNA*SNAj]	Parameter no. [kPx, SNAjPx, SNA*SNAjPx]	Remove or continue option [RE CO]	Alternate block no. [k, SNAj, SNA*SNAj]	Remove or continue option [RE CO]	Alternate block no. [k, SNAj, SNA*SNAj]	
		Options for controlling transactions (spans B, C, D)			Options for preempted transactions (spans E, F)		Options for delayed transactions (spans G, H)		

[ ] Indicates optional operand

{ } Indicates that one of the items within the braces must be selected

BLOCK STATEMENT FORMATS

OPERATION	A	B	C	D	E	F	G	H	I
GATE {LS LR}	Logic switch no. k, SNAj, SNA*SNAj	Next block if condition is false [k, SNAj, SNA*SNAj]							
GATE {NI I NU U FV FNV}	Facility no. k, SNAj, SNA*SNAj	Next block if condition is false [k, SNAj, SNA*SNAj]							
GATE {SE SF SNE SNF SV SNV}	Storage no. k, SNAj, SNA*SNAj	Next block if condition is false [k, SNAj, SNA*SNAj]							
GATE {M NM}	Match block no. k, SNAj, SNA*SNAj	Next block if condition is false [k, SNAj, SNA*SNAj]							
GATHER	No. of transactions to be gathered k,SNAj, SNA*SNAj								

[ ] Indicates optional operand

{ } Indicates that one of the items within the braces must be selected

BLOCK STATEMENT FORMATS

OPERATION	A	B	C	D	E	F	G	H	I
GENERATE	Mean time [k, SNAi, SNA*SNAi]	Spread [k, SNAi, SNA*SNAi] or Function modifier FNj, FN*SNAi	Initialization interval [k, SNAi, SNA*SNAi]	Creation limit [k, SNAi, SNA*SNAi]	Priority level [k, SNAi, SNA*SNAi]	Fullword, halfword, byte & floating point parameters in any sequence [kPx, SNAjPx, SNA*SNAjPx]	[kPx, SNAjPx, SNA*SNAjPx]	[kPx, SNAjPx, SNA*SNAjPx]	[kPx, SNAjPx, SNA*SNAjPx]
						Note: Operands A–I may be a constant, FNj, Vi, Xj, XFj, XBj, XHj, RNj, C1, AC1, or Nj. Likewise, elements of functions or variables specified are restricted to these SNAs.			
HELP HELPA HELPB HELPC HELPAPL1 HELPBPL1 HELPCPL1	Help routine name. When using HELPB or HELPBPL1, the B-G operands reference either fullword or floating-point savevalues. An XL (floating-point) or XF(fullword) suffix should be used with each of these operands.	B-G operands SNA values to be passed to help routine [k, SNAi, SNA*SNAi]							
INDEX	Parameter no. kPx, SNAjPx, SNA*SNAjPx	Increment k, SNAi, SNA*SNAi							
JOIN	Group no. k, SNAi, SNA*SNAi	Numeric value-numeric mode [k,SNAi, SNA*SNAi]							
LEAVE	Storage no. k, SNAi, SNA*SNAi	No. of units [k, SNAi, SNA*SNAi]							

[ ] Indicates optional operand

{ } Indicates that one of the items within the braces must be selected

BLOCK STATEMENT FORMATS

OPERATION	A	B	C	D	E	F	G	H	I
LINK	User chain no. k, SNAj, SNA*SNAj	Ordering of chain LIFO, FIFO or parameter number kPx, SNAjPx, SNA*SNAjPx	Alternate block exit [k, SNAj, SNA*SNAj]						
LOGIC {S R I}	Logic switch no. k, SNAj, SNA*SNAj								
LOOP	Parameter no. kPx SNAjPx, SNA*SNAjPx	Next block if Pxj ≠ 0 k, SNAj, SNA*SNAj							
MARK	Parameter no. [kPx, SNAjPx, SNA*SNAjPx]								
MATCH	Conjugate MATCH block no. k, SNAj, SNA*SNAj								
MSAVEVALUE	Matrix no. or range k, SNAj, SNA*SNAj [±]	Row no. or range k, SNAj, SNA*SNAj	Column no. or range k, SNAj, SNA*SNAj	SNA value to be saved k, SNAj, SNA* SNAj	Msavevalue type H, MH, MX, MB, ML				

[ ] Indicates optional operand

{ } Indicates that one of the items within the braces must be selected

BLOCK STATEMENT FORMATS

OPERATION	A	B	C	D	E	F	G	H	I
PREEMPT	Facility no. k, SNAj, SNA*SNAj	Priority option [PR]	Block no. for preempted transaction [k, SNAj, SNA*SNAj]	Parameter no. of preempted transaction [kPx, SNAjPx, SNA*SNAjPx]	Remove option [RE]				
PRINT	Lower limit [k, SNAj, SNA*SNAj]	Upper limit [k, SNAj, SNA*SNAj]	Entity mnemonic	Paging indicator [Any alphameric character]					
PRIORITY	Priority no. k, SNAj, SNA*SNAj	Buffer option [BUFFER]							
QUEUE	Queue no. k, SNAj, SNA*SNAj	No. of units [k, SNAj, SNA*SNAj]							
RELEASE	Facility no. k, SNAj, SNA*SNAj								

[ ] Indicates optional operand

{ } Indicates that one of the items within braces must be selected

BLOCK STATEMENT FORMATS

OPERATION	A	B	C	D	E	F	G	H	I
REMOVE {G, GE, L, LE, E, NE, MIN, MAX}	Group no. k, SNAj, SNA*SNAj	Count—no. of members to be removed—transaction mode [k, SNAj, SNA*SNAj, ALL]	Numeric value to be removed—numeric mode [k,SNAj, SNA*SNAj]	Transaction attribute for comparison—transaction mode {PR, or parameter no. kPx, SNAjPx, SNA*SNAjPx}	Comparison SNA [k, SNAj, SNA*SNAj]	Alternate exit [k, SNAj, SNA*SNAj]			
RETURN	Facility no. k, SNAj, SNA*SNAj								
SAVAIL	Storage no. or range k, SNAj, SNA*SNAj								
SAVEVALUE	Savevalue no. or range k, SNAj, SNA*SNAj [±]	SNA value to be saved k, SNAj, SNA*SNAj	Savevalue type X, XF, H, XH, XB, XL						
SCAN {G, GE, L, LE, E, NE, MIN, MAX}	Group no. k, SNAj, SNA*SNAj	Transaction attribute for comparison PR or parameter no. kPx, SNAjPx, SNA*SNAjPx	Comparison value for B operand k, SNAj, SNA*SNAj	Attribute to be obtained if match is made {PR or parameter no. kPx, SNAjPx, SNA*SNAjPx}	Parameter no. in which to place D operand value {kPx, SNAjPx, SNA*SNAjPx}	Alternate exit [k, SNAj, SNA*SNAj]			

[ ] Indicates optional operand

{ } Indicates that one of the items within the braces must be selected

BLOCK STATEMENT FORMATS

OPERATION	A	B	C	D	E	F	G	H	I
SEIZE	Facility no. k, SNAj, SNA*SNAj								
SELECT G,GE L,LE E,NE U,NU I,NI SE,SNE SF,SNF LR,LS MIN,MAX	Parameter in which to place entity number kPx, SNAjPx, SNA*SNAjPx	Lower limit k, SNAj, SNA*SNAj	Upper limit k, SNAj, SNA*SNAj	Comparison value if conditional operator is specified [k,SNAj, SNA*SNAj]	SNA mnemonic to be examined if conditional operator is specified [Any SNA except MX,MH, MB,ML]	Alternate exit [k, SNAj, SNA*SNAj]			
SPLIT	No. of copies k, SNAj, SNA*SNAj	Next block for copies k, SNAj, SNA*SNAj	Parameter for serial numbering [kPx, SNAjPx, SNA*SNAjPx]	No. of fullword, halfword, byte, & floating-point parameters in any sequence [k Px, SNAjPx, SNA*SNAjPx]	[kPx, SNAjPx, SNA*SNAjPx]	[kPx, SNAjPx, SNA*SNAjPx]	[kPx, SNAjPx, SNA*SNAjPx]		
				Any SNA except MX, MH, MB, ML					
SUNAVAIL	Storage no. or range k, SNAj, SNA*SNAj								
TABULATE	Table no. k, SNAj, SNA*SNAj	Weighting factor [k, SNAj, SNA*SNAj]							
TERMINATE	Termination count [k, SNAj, SNA*SNAj]								

[ ] Indicates optional operand

{ } Indicates that one of the items within the braces must be selected

BLOCK STATEMENT FORMATS

OPERATION	A	B	C	D	E	F	G	H	I
TEST {E, NE, GE, LE, G, L}	First SNA k, SNAj, SNA*SNAj	Second SNA k, SNAj, SNA*SNAj	Next block if relation is false [k, SNAj, SNA*SNAj]						
TRACE									
TRANSFER	Selection mode With ALL selection mode a block name or number is the only valid operand	Next block A [k, SNAj, SNA*SNAj]	Next block B [k, SNAj, SNA*SNAj]	Indexing factor [k]					
UNLINK [G, GE, L, LE, E, NE]	User chain no. k, SNAj, SNA*SNAj	Next block for the unlinked transaction (s) k, SNAj, SNA*SNAj	Transaction unlink count ALL or k, SNAj, SNA*SNAj	[Parameter no. kPx, SNAjPx, SNA*SNAjPx, or BACK, or BVj, BV*SNAj]	Match argument [k, SNAj, SNA*SNAj]	Next block B [k, SNAj, SNA*SNAj]			
UNTRACE									
WRITE	Jobtape no. {JOBTA1, JOBTA2, JOBTA3}								

[ ] Indicates optional operand

{ } Indicates that one of the items within the braces must be selected

Appendix C:
The Normal Distribution

A random variable X is said to have the normal distribution $N(\mu, \sigma^2)$ if (for some $-\infty < \mu < +\infty$, $\sigma^2 > 0$)

$$f_X(x) = \frac{1}{\sqrt{2\pi}\,\sigma}\, e^{-\frac{1}{2}(\frac{x-\mu}{\sigma})^2} \qquad -\infty < x < +\infty.$$

The table on the following two pages gives, for various values of y (when $\mu = 0$ and $\sigma^2 = 1$), the probability $P[X \le y] = \int_{-\infty}^{y} f_X(x)\, dx$. A table entry is either a number, a, or a number, a, and an exponent, b. If an exponent is not present, then a is the value of the integral to 4 significant digits. If an exponent, b, is present, then it signifies the number of 0s (in case $y < 0$) or the number of 9s (in case $y > 0$) that must be inserted immediately following the decimal point of a. For example, if $y = -3.26$, $a = .5771$ and $b = 3$; hence the value of the integral is .0005771. if $y = 4.52$, then $a = .6908$ and $b = 5$; hence the value of the integral is .999996908.

The graph below shows normal distribution densities with $\mu = 0$ and various selections of σ.

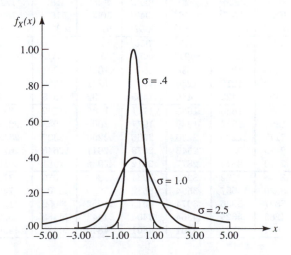

The Normal Distribution (continues)

y	9	8	7	6	5	4	3	2	1	0
-4.90	$.3019^6$	$.3179^6$	$.3347^6$	$.3524^6$	$.3710^6$	$.3906^6$	$.4111^6$	$.4327^6$	$.4554^6$	$.4792^6$
-4.80	$.5042^6$	$.5304^6$	$.5580^6$	$.5869^6$	$.6173^6$	$.6492^6$	$.6826^6$	$.7178^6$	$.7546^6$	$.7933^6$
-4.70	$.8339^6$	$.8765^6$	$.9211^6$	$.9679^6$	$.1017^5$	$.1069^5$	$.1123^5$	$.1179^5$	$.1239^5$	$.1301^5$
-4.60	$.1366^5$	$.1434^5$	$.1506^5$	$.1581^5$	$.1660^5$	$.1742^5$	$.1828^5$	$.1919^5$	$.2013^5$	$.2112^5$
-4.50	$.2216^5$	$.2325^5$	$.2439^5$	$.2558^5$	$.2682^5$	$.2813^5$	$.2949^5$	$.3092^5$	$.3241^5$	$.3398^5$
-4.40	$.3561^5$	$.3732^5$	$.3911^5$	$.4098^5$	$.4294^5$	$.4498^5$	$.4712^5$	$.4935^5$	$.5169^5$	$.5413^5$
-4.30	$.5668^5$	$.5934^5$	$.6212^5$	$.6503^5$	$.6807^5$	$.7124^5$	$.7455^5$	$.7801^5$	$.8163^5$	$.8540^5$
-4.20	$.8934^5$	$.9345^5$	$.9774^5$	$.1022^4$	$.1069^4$	$.1118^4$	$.1168^4$	$.1222^4$	$.1277^4$	$.1335^4$
-4.10	$.1395^4$	$.1458^4$	$.1523^4$	$.1591^4$	$.1662^4$	$.1737^4$	$.1814^4$	$.1894^4$	$.1978^4$	$.2066^4$
-4.00	$.2157^4$	$.2252^4$	$.2351^4$	$.2454^4$	$.2561^4$	$.2673^4$	$.2789^4$	$.2910^4$	$.3036^4$	$.3167^4$
-3.90	$.3304^4$	$.3446^4$	$.3594^4$	$.3747^4$	$.3908^4$	$.4074^4$	$.4247^4$	$.4427^4$	$.4615^4$	$.4810^4$
-3.80	$.5012^4$	$.5223^4$	$.5442^4$	$.5669^4$	$.5906^4$	$.6152^4$	$.6407^4$	$.6673^4$	$.6948^4$	$.7235^4$
-3.70	$.7532^4$	$.7841^4$	$.8162^4$	$.8496^4$	$.8842^4$	$.9201^4$	$.9574^4$	$.9961^4$	$.1036^3$	$.1078^3$
-3.60	$.1121^3$	$.1166^3$	$.1213^3$	$.1261^3$	$.1311^3$	$.1363^3$	$.1417^3$	$.1473^3$	$.1531^3$	$.1591^3$
-3.50	$.1653^3$	$.1718^3$	$.1785^3$	$.1854^3$	$.1926^3$	$.2001^3$	$.2078^3$	$.2158^3$	$.2241^3$	$.2326^3$
-3.40	$.2415^3$	$.2507^3$	$.2602^3$	$.2701^3$	$.2803^3$	$.2909^3$	$.3018^3$	$.3131^3$	$.3248^3$	$.3369^3$
-3.30	$.3495^3$	$.3624^3$	$.3758^3$	$.3897^3$	$.4041^3$	$.4189^3$	$.4342^3$	$.4501^3$	$.4665^3$	$.4834^3$
-3.20	$.5009^3$	$.5190^3$	$.5377^3$	$.5571^3$	$.5770^3$	$.5976^3$	$.6190^3$	$.6410^3$	$.6637^3$	$.6871^3$
-3.10	$.7114^3$	$.7364^3$	$.7622^3$	$.7888^3$	$.8164^3$	$.8447^3$	$.8740^3$	$.9043^3$	$.9354^3$	$.9676^3$
-3.00	$.1001^2$	$.1035^2$	$.1070^2$	$.1107^2$	$.1144^2$	$.1183^2$	$.1223^2$	$.1264^2$	$.1306^2$	$.1350^2$
-2.90	$.1395^2$	$.1441^2$	$.1489^2$	$.1538^2$	$.1589^2$	$.1641^2$	$.1695^2$	$.1750^2$	$.1807^2$	$.1866^2$
-2.80	$.1926^2$	$.1988^2$	$.2052^2$	$.2118^2$	$.2186^2$	$.2256^2$	$.2327^2$	$.2401^2$	$.2477^2$	$.2555^2$
-2.70	$.2635^2$	$.2718^2$	$.2803^2$	$.2890^2$	$.2980^2$	$.3072^2$	$.3167^2$	$.3264^2$	$.3364^2$	$.3467^2$
-2.60	$.3573^2$	$.3681^2$	$.3793^2$	$.3907^2$	$.4025^2$	$.4145^2$	$.4269^2$	$.4396^2$	$.4527^2$	$.4661^2$
-2.50	$.4799^2$	$.4940^2$	$.5085^2$	$.5234^2$	$.5386^2$	$.5543^2$	$.5703^2$	$.5868^2$	$.6037^2$	$.6210^2$
-2.40	$.6387^2$	$.6569^2$	$.6756^2$	$.6947^2$	$.7143^2$	$.7344^2$	$.7549^2$	$.7760^2$	$.7976^2$	$.8198^2$
-2.30	$.8424^2$	$.8656^2$	$.8894^2$	$.9137^2$	$.9387^2$	$.9642^2$	$.9903^2$	$.1017^1$	$.1044^1$	$.1072^1$
-2.20	$.1101^1$	$.1130^1$	$.1160^1$	$.1191^1$	$.1222^1$	$.1255^1$	$.1287^1$	$.1321^1$	$.1355^1$	$.1390^1$
-2.10	$.1426^1$	$.1463^1$	$.1500^1$	$.1539^1$	$.1578^1$	$.1618^1$	$.1659^1$	$.1700^1$	$.1743^1$	$.1786^1$
-2.00	$.1831^1$	$.1876^1$	$.1923^1$	$.1970^1$	$.2018^1$	$.2068^1$	$.2118^1$	$.2169^1$	$.2222^1$	$.2275^1$
-1.90	$.2330^1$	$.2385^1$	$.2442^1$	$.2500^1$	$.2559^1$	$.2619^1$	$.2680^1$	$.2743^1$	$.2807^1$	$.2872^1$
-1.80	$.2938^1$	$.3005^1$	$.3074^1$	$.3144^1$	$.3216^1$	$.3288^1$	$.3362^1$	$.3438^1$	$.3515^1$	$.3593^1$
-1.70	$.3673^1$	$.3754^1$	$.3836^1$	$.3920^1$	$.4006^1$	$.4093^1$	$.4182^1$	$.4272^1$	$.4363^1$	$.4457^1$
-1.60	$.4551^1$	$.4648^1$	$.4746^1$	$.4846^1$	$.4947^1$	$.5050^1$	$.5155^1$	$.5262^1$	$.5370^1$	$.5480^1$
-1.50	$.5592^1$	$.5705^1$	$.5821^1$	$.5938^1$	$.6057^1$	$.6178^1$	$.6301^1$	$.6426^1$	$.6552^1$	$.6681^1$
-1.40	$.6811^1$	$.6944^1$	$.7078^1$	$.7215^1$	$.7353^1$	$.7493^1$	$.7636^1$	$.7780^1$	$.7927^1$	$.8076^1$
-1.30	$.8226^1$	$.8379^1$	$.8534^1$	$.8691^1$	$.8851^1$	$.9012^1$	$.9176^1$	$.9342^1$	$.9510^1$	$.9680^1$
-1.20	$.9853^1$	.1003	.1020	.1038	.1056	.1075	.1093	.1112	.1131	.1151
-1.10	.1170	.1190	.1210	.1230	.1251	.1271	.1292	.1314	.1335	.1357
-1.00	.1379	.1401	.1423	.1446	.1469	.1492	.1515	.1539	.1562	.1587
-.90	.1611	.1635	.1660	.1685	.1711	.1736	.1762	.1788	.1814	.1841
-.80	.1867	.1894	.1922	.1949	.1977	.2005	.2033	.2061	.2090	.2119
-.70	.2148	.2177	.2206	.2236	.2266	.2296	.2327	.2358	.2389	.2420
-.60	.2451	.2483	.2514	.2546	.2578	.2611	.2643	.2676	.2709	.2743
-.50	.2776	.2810	.2843	.2877	.2912	.2946	.2981	.3015	.3050	.3085
-.40	.3121	.3156	.3192	.3228	.3264	.3300	.3336	.3372	.3409	.3446
-.30	.3483	.3520	.3557	.3594	.3632	.3669	.3707	.3745	.3783	.3821
-.20	.3859	.3897	.3936	.3974	.4013	.4052	.4090	.4129	.4168	.4207
-.10	.4247	.4286	.4325	.4364	.4404	.4443	.4483	.4522	.4562	.4602
-0.00	.4641	.4681	.4721	.4761	.4801	.4840	.4880	.4920	.4960	.5000

The Normal Distribution (concluded)

y	0	1	2	3	4	5	6	7	8	9
.00	.5000	.5040	.5080	.5120	.5160	.5199	.5239	.5279	.5319	.5359
.10	.5398	.5438	.5478	.5517	.5557	.5596	.5636	.5675	.5714	.5753
.20	.5793	.5832	.5871	.5910	.5948	.5987	.6026	.6064	.6103	.6141
.30	.6179	.6217	.6255	.6293	.6331	.6368	.6406	.6443	.6480	.6517
.40	.6554	.6591	.6628	.6664	.6700	.6736	.6772	.6808	.6844	.6879
.50	.6915	.6950	.6985	.7019	.7054	.7088	.7123	.7157	.7190	.7224
.60	.7257	.7291	.7324	.7357	.7389	.7422	.7454	.7486	.7517	.7549
.70	.7580	.7611	.7642	.7673	.7704	.7734	.7764	.7794	.7823	.7852
.80	.7881	.7910	.7939	.7967	.7995	.8023	.8051	.8078	.8106	.8133
.9	.8159	.8186	.8212	.8238	.8264	.8289	.8315	.8340	.8365	.8389
1.00	.8413	.8438	.8461	.8485	.8508	.8531	.8554	.8577	.8599	.8621
1.10	.8643	.8665	.8686	.8708	.8729	.8749	.8770	.8790	.8810	.8830
1.20	.8849	.8869	.8888	.8907	.8925	.8944	.8962	.8980	.8997	$.0148^1$
1.30	$.0320^1$	$.0490^1$	$.0658^1$	$.0824^1$	$.0988^1$	$.1149^1$	$.1309^1$	$.1466^1$	$.1621^1$	$.1774^1$
1.40	$.1924^1$	$.2073^1$	$.2220^1$	$.2364^1$	$.2507^1$	$.2647^1$	$.2785^1$	$.2922^1$	$.3056^1$	$.3189^1$
1.50	$.3319^1$	$.3448^1$	$.3574^1$	$.3699^1$	$.3822^1$	$.3943^1$	$.4062^1$	$.4179^1$	$.4295^1$	$.4408^1$
1.60	$.4520^1$	$.4630^1$	$.4738^1$	$.4845^1$	$.4950^1$	$.5053^1$	$.5154^1$	$.5254^1$	$.5352^1$	$.5449^1$
1.70	$.5543^1$	$.5637^1$	$.5728^1$	$.5818^1$	$.5907^1$	$.5994^1$	$.6080^1$	$.6164^1$	$.6246^1$	$.6327^1$
1.80	$.6407^1$	$.6485^1$	$.6562^1$	$.6638^1$	$.6712^1$	$.6784^1$	$.6856^1$	$.6926^1$	$.6995^1$	$.7062^1$
1.90	$.7128^1$	$.7193^1$	$.7257^1$	$.7320^1$	$.7381^1$	$.7441^1$	$.7500^1$	$.7558^1$	$.7615^1$	$.7670^1$
2.00	$.7725^1$	$.7778^1$	$.7831^1$	$.7882^1$	$.7932^1$	$.7982^1$	$.8030^1$	$.8077^1$	$.8124^1$	$.8169^1$
2.10	$.8214^1$	$.8257^1$	$.8300^1$	$.8341^1$	$.8382^1$	$.8422^1$	$.8461^1$	$.8500^1$	$.8537^1$	$.8574^1$
2.20	$.8610^1$	$.8645^1$	$.8679^1$	$.8713^1$	$.8745^1$	$.8778^1$	$.8809^1$	$.8840^1$	$.8870^1$	$.8899^1$
2.30	$.8928^1$	$.8956^1$	$.8983^1$	$.0097^2$	$.0358^2$	$.0613^2$	$.0863^2$	$.1106^2$	$.1344^2$	$.1576^2$
2.40	$.1802^2$	$.2024^2$	$.2240^2$	$.2451^2$	$.2656^2$	$.2857^2$	$.3053^2$	$.3244^2$	$.3431^2$	$.3613^2$
2.50	$.3790^2$	$.3963^2$	$.4132^2$	$.4297^2$	$.4457^2$	$.4614^2$	$.4766^2$	$.4915^2$	$.5060^2$	$.5201^2$
2.60	$.5339^2$	$.5473^2$	$.5604^2$	$.5731^2$	$.5855^2$	$.5975^2$	$.6093^2$	$.6207^2$	$.6319^2$	$.6427^2$
2.70	$.6533^2$	$.6636^2$	$.6736^2$	$.6833^2$	$.6928^2$	$.7020^2$	$.7110^2$	$.7197^2$	$.7282^2$	$.7365^2$
2.80	$.7445^2$	$.7523^2$	$.7599^2$	$.7673^2$	$.7744^2$	$.7814^2$	$.7882^2$	$.7948^2$	$.8012^2$	$.8074^2$
2.90	$.8134^2$	$.8193^2$	$.8250^2$	$.8305^2$	$.8359^2$	$.8411^2$	$.8462^2$	$.8511^2$	$.8559^2$	$.8605^2$
3.00	$.8650^2$	$.8694^2$	$.8736^2$	$.8777^2$	$.8817^2$	$.8856^2$	$.8893^2$	$.8930^2$	$.8965^2$	$.8999^2$
3.10	$.0324^3$	$.0646^3$	$.0957^3$	$.1260^3$	$.1553^3$	$.1836^3$	$.2112^3$	$.2378^3$	$.2636^3$	$.2886^3$
3.20	$.3129^3$	$.3363^3$	$.3590^3$	$.3810^3$	$.4024^3$	$.4230^3$	$.4429^3$	$.4623^3$	$.4810^3$	$.4991^3$
3.30	$.5166^3$	$.5335^3$	$.5499^3$	$.5658^3$	$.5811^3$	$.5959^3$	$.6103^3$	$.6242^3$	$.6376^3$	$.6505^3$
3.40	$.6631^3$	$.6752^3$	$.6869^3$	$.6982^3$	$.7091^3$	$.7197^3$	$.7299^3$	$.7398^3$	$.7493^3$	$.7585^3$
3.50	$.7674^3$	$.7759^3$	$.7842^3$	$.7922^3$	$.7999^3$	$.8074^3$	$.8146^3$	$.8215^3$	$.8282^3$	$.8347^3$
3.60	$.8409^3$	$.8469^3$	$.8527^3$	$.8583^3$	$.8637^3$	$.8689^3$	$.8739^3$	$.8787^3$	$.8834^3$	$.8879^3$
3.70	$.8922^3$	$.8964^3$	$.0039^4$	$.0426^4$	$.0799^4$	$.1158^4$	$.1504^4$	$.1838^4$	$.2159^4$	$.2468^4$
3.80	$.2765^4$	$.3052^4$	$.3327^4$	$.3593^4$	$.3848^4$	$.4094^4$	$.4331^4$	$.4558^4$	$.4777^4$	$.4988^4$
3.90	$.5190^4$	$.5385^4$	$.5573^4$	$.5753^4$	$.5926^4$	$.6092^4$	$.6253^4$	$.6406^4$	$.6554^4$	$.6696^4$
4.00	$.6833^4$	$.6964^4$	$.7090^4$	$.7211^4$	$.7327^4$	$.7439^4$	$.7546^4$	$.7649^4$	$.7748^4$	$.7843^4$
4.10	$.7934^4$	$.8022^4$	$.8106^4$	$.8186^4$	$.8263^4$	$.8338^4$	$.8409^4$	$.8477^4$	$.8542^4$	$.8605^4$
4.20	$.8665^4$	$.8723^4$	$.8778^4$	$.8832^4$	$.8882^4$	$.8931^4$	$.8978^4$	$.0226^5$	$.0655^5$	$.1066^5$
4.30	$.1460^5$	$.1837^5$	$.2199^5$	$.2545^5$	$.2876^5$	$.3193^5$	$.3497^5$	$.3788^5$	$.4066^5$	$.4333^5$
4.40	$.4588^5$	$.4832^5$	$.5065^5$	$.5288^5$	$.5502^5$	$.5707^5$	$.5902^5$	$.6089^5$	$.6268^5$	$.6439^5$
4.50	$.6602^5$	$.6759^5$	$.6908^5$	$.7051^5$	$.7187^5$	$.7318^5$	$.7442^5$	$.7561^5$	$.7675^5$	$.7784^5$
4.60	$.7888^5$	$.7987^5$	$.8081^5$	$.8172^5$	$.8258^5$	$.8340^5$	$.8419^5$	$.8494^5$	$.8566^5$	$.8634^5$
4.70	$.8699^5$	$.8761^5$	$.8821^5$	$.8877^5$	$.8931^5$	$.8983^5$	$.0321^6$	$.0789^6$	$.1235^6$	$.1661^6$
4.80	$.2067^6$	$.2454^6$	$.2822^6$	$.3174^6$	$.3508^6$	$.3827^6$	$.4131^6$	$.4420^6$	$.4696^6$	$.4958^6$
4.90	$.5208^6$	$.5446^6$	$.5673^6$	$.5889^6$	$.6094^6$	$.6290^6$	$.6476^6$	$.6653^6$	$.6821^6$	$.6981^6$

Appendix D:
The Student's t-Distribution

A random variable X is said to have the t-distribution with n degrees of freedom if (for some integer $n \geq 1$)

$$f_X(x) = \frac{\Gamma(\frac{n+1}{2})}{\sqrt{n\pi}\,\Gamma(\frac{n}{2})} \frac{1}{(1+\frac{x^2}{n})^{(n+1)/2}} \qquad -\infty < x < +\infty.$$

The following graph shows t-distribution densities with several selections of n, and the $N(0, 1)$ density (labeled $n = \infty$).

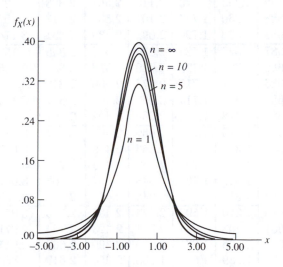

The *t*-Distribution

This table gives values of y such that (for various values of n, γ)

$$P[X \leq y] = \int_{-\infty}^{y} f_X(x)\, dx = \gamma.$$

n	γ									
	.60	.75	.90	.95	.975	.99	.995	.9975	.999	.9995
1	.325	1.000	3.078	6.314	12.71	31.82	63.66	127.3	318.3	636.6
2	.289	.816	1.886	2.920	4.303	6.965	9.925	14.09	22.33	31.60
3	.277	.765	1.638	2.353	3.182	4.541	5.841	7.453	10.21	12.92
4	.271	.741	1.533	2.132	2.776	3.747	4.604	5.598	7.173	8.610
5	.267	.727	1.476	2.015	2.571	3.365	4.032	4.773	5.893	6.869
6	.265	.718	1.440	1.943	2.447	3.143	3.707	4.317	5.208	5.959
7	.263	.711	1.415	1.895	2.365	2.998	3.499	4.029	4.785	5.408
8	.262	.706	1.397	1.860	2.306	2.896	3.355	3.833	4.501	5.041
9	.261	.703	1.383	1.833	2.262	2.821	3.250	3.690	4.297	4.781
10	.260	.700	1.372	1.812	2.228	2.764	3.169	3.581	4.144	4.587
11	.260	.697	1.363	1.796	2.201	2.718	3.106	3.497	4.025	4.437
12	.259	.695	1.356	1.782	2.179	2.681	3.055	3.428	3.930	4.318
13	.259	.694	1.350	1.771	2.160	2.650	3.012	3.372	3.852	4.221
14	.258	.692	1.345	1.761	2.145	2.624	2.977	3.326	3.787	4.140
15	.258	.691	1.341	1.753	2.131	2.602	2.947	3.286	3.733	4.073
16	.258	.690	1.337	1.746	2.120	2.583	2.921	3.252	3.686	4.015
17	.257	.689	1.333	1.740	2.110	2.567	2.898	3.222	3.646	3.965
18	.257	.688	1.330	1.734	2.101	2.552	2.878	3.197	3.610	3.922
19	.257	.688	1.328	1.729	2.093	2.539	2.861	3.174	3.579	3.883
20	.257	.687	1.325	1.725	2.086	2.528	2.845	3.153	3.552	3.850
21	.257	.686	1.323	1.721	2.080	2.518	2.831	3.135	3.527	3.819
22	.256	.686	1.321	1.717	2.074	2.508	2.819	3.119	3.505	3.792
23	.256	.685	1.319	1.714	2.069	2.500	2.807	3.104	3.485	3.768
24	.256	.685	1.318	1.711	2.064	2.492	2.797	3.091	3.467	3.745
25	.256	.684	1.316	1.708	2.060	2.485	2.787	3.078	3.450	3.725
26	.256	.684	1.315	1.706	2.056	2.479	2.779	3.067	3.435	3.707
27	.256	.684	1.314	1.703	2.052	2.473	2.771	3.057	3.421	3.690
28	.256	.683	1.313	1.701	2.048	2.467	2.763	3.047	3.408	3.674
29	.256	.683	1.311	1.699	2.045	2.462	2.756	3.038	3.396	3.659
30	.256	.683	1.310	1.697	2.042	2.457	2.750	3.030	3.385	3.646
40	.255	.681	1.303	1.684	2.021	2.423	2.704	2.971	3.307	3.551
60	.254	.679	1.296	1.671	2.000	2.390	2.660	2.915	3.232	3.460
120	.254	.677	1.289	1.658	1.980	2.358	2.617	2.860	3.160	3.373
∞	.253	.674	1.282	1.645	1.960	2.326	2.576	2.807	3.090	3.291

Appendix E:
The Chi-Square Distribution

A random variable X is said to have the chi-square distribution with n degrees of freedom $\chi_n^2(0)$ if (for some integer $n > 0$)

$$f_X(x) = \begin{cases} \frac{1}{2^{n/2}\Gamma(n/2)} x^{(n/2)-1} e^{-x/2}, & 0 \leq x < \infty \\ 0, & \text{otherwise.} \end{cases}$$

The graph below shows chi-square distribution densities with various selections of n. The table on the following two pages gives values of y such that (for various values of n, γ) $P[X \leq y] = \int_0^y f_X(x)\,dx = \gamma$. When an entry consists of a number, a, and an exponent, b, then the value of y is $a \times 10^{-b}$. For example, when $n = 2$ and $\gamma = 0.01$, we get $a = .201007$ and $b = 1$; hence $y = 0.0201007$.

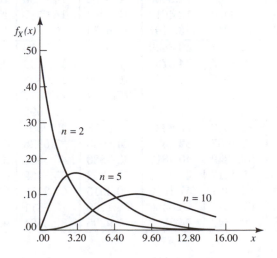

The Chi-Square Distribution (continues)

y such that $P[X \leq y] = \int_0^y f_X(x)\,dx = \gamma$

n	γ						
	.005	.01	.025	.05	.10	.25	.50
1	$.392704^4$	$.157088^3$	$.982069^3$	$.393214^2$	$.157908^1$	.101531	.454936
2	$.100251^1$	$.201007^1$	$.506356^1$	.102587	.210721	.575364	1.38629
3	$.717218^1$	.114832	.215795	.351846	.584374	1.21253	2.36597
4	.206989	.297109	.484419	.710723	1.06362	1.92256	3.35669
5	.411742	.554298	.831212	1.14548	1.61031	2.67460	4.35146
6	.675727	.872090	1.23734	1.63538	2.20413	3.45460	5.34812
7	.989256	1.23904	1.68987	2.16735	2.83311	4.25485	6.34581
8	1.34441	1.64650	2.17973	2.73264	3.48954	5.07064	7.34412
9	1.73493	2.08790	2.70039	3.32511	4.16816	5.89883	8.34283
10	2.15586	2.55821	3.24697	3.94030	4.86518	6.73720	9.34182
11	2.60322	3.05348	3.81575	4.57481	5.57778	7.58414	10.3410
12	3.07382	3.57057	4.40379	5.22603	6.30380	8.43842	11.3403
13	3.56503	4.10692	5.00875	5.89186	7.04150	9.29907	12.3398
14	4.07467	4.66043	5.62873	6.57063	7.78953	10.1653	13.3393
15	4.60092	5.22935	6.26214	7.26094	8.54676	11.0365	14.3389
16	5.14221	5.81221	6.90766	7.96165	9.31224	11.9122	15.3385
17	5.69722	6.40776	7.56419	8.67176	10.0852	12.7919	16.3382
18	6.26480	7.01491	8.23075	9.39046	10.8649	13.6753	17.3379
19	6.84397	7.63273	8.90652	10.1170	11.6509	14.5620	18.3377
20	7.43384	8.26040	9.59078	10.8508	12.4426	15.4518	19.3374
21	8.03365	8.89720	10.2829	11.5913	13.2396	16.3444	20.3372
22	8.64272	9.54249	10.9823	12.3380	14.0415	17.2396	21.3370
23	9.26042	10.1957	11.6886	13.0905	14.8480	18.1373	22.3369
24	9.88623	10.8564	12.4012	13.8484	15.6587	19.0373	23.3367
25	10.5197	11.5240	13.1197	14.6114	16.4734	19.9393	24.3366
26	11.1602	12.1981	13.8439	15.3792	17.2919	20.8434	25.3365
27	11.8076	12.8785	14.5734	16.1514	18.1139	21.7494	26.3363
28	12.4613	13.5647	15.3079	16.9279	18.9392	22.6572	27.3362
29	13.1211	14.2565	16.0471	17.7084	19.7677	23.5666	28.3361
30	13.7867	14.9535	16.7908	18.4927	20.5992	24.4776	29.3360
40	20.7065	22.1643	24.4330	26.5093	29.0505	33.6603	39.3353
50	27.9907	29.7067	32.3574	34.7643	37.6886	42.9421	49.3349
60	35.5345	37.4849	40.4817	43.1880	46.4589	52.2938	59.3347
70	43.2752	45.4417	48.7576	51.7393	55.3289	61.6983	69.3345
80	51.1719	53.5401	57.1532	60.3915	64.2778	71.1445	79.3343
90	59.1963	61.7541	65.6466	69.1260	73.2911	80.6247	89.3342
100	67.3276	70.0649	74.2219	77.9295	82.3581	90.1332	99.3341

The Chi-Square Distribution (concluded)

y such that $P[X \leq y] = \int_0^y f_X(x)\, dx = \gamma$

n	γ						
	.75	.90	.95	.975	.99	.995	.999
1	1.32330	2.70554	3.84146	5.02389	6.63490	7.87944	10.8276
2	2.77259	4.60517	5.99146	7.37776	9.21034	10.5966	13.8155
3	4.10834	6.25139	7.81473	9.34840	11.3449	12.8382	16.2662
4	5.38527	7.77944	9.48773	11.1433	13.2767	14.8603	18.4668
5	6.62568	9.23636	11.0705	12.8325	15.0863	16.7496	20.5150
6	7.84080	10.6446	12.5916	14.4494	16.8119	18.5476	22.4577
7	9.03715	12.0170	14.0671	16.0128	18.4753	20.2777	24.3219
8	10.2189	13.3616	15.5073	17.5345	20.0902	21.9550	26.1245
9	11.3888	14.6837	16.9190	19.0228	21.6660	23.5894	27.8772
10	12.5489	15.9872	18.3070	20.4832	23.2093	25.1882	29.5883
11	13.7007	17.2750	19.6751	21.9200	24.7250	26.7568	31.2641
12	14.8454	18.5493	21.0261	23.3367	26.2170	28.2995	32.9095
13	15.9839	19.8119	22.3620	24.7356	27.6882	29.8195	34.5282
14	17.1169	21.0641	23.6848	26.1189	29.1412	31.3193	36.1233
15	18.2451	22.3071	24.9958	27.4884	30.5779	32.8013	37.6973
16	19.3689	23.5418	26.2962	28.8454	31.9999	34.2672	39.2524
17	20.4887	24.7690	27.5871	30.1910	33.4087	35.7185	40.7902
18	21.6049	25.9894	28.8693	31.5264	34.8053	37.1565	42.3124
19	22.7178	27.2036	30.1435	32.8523	36.1909	38.5823	43.8202
20	23.8277	28.4120	31.4104	34.1696	37.5662	39.9968	45.3147
21	24.9348	29.6151	32.6706	35.4789	38.9322	41.4011	46.7970
22	26.0393	30.8133	33.9244	36.7807	40.2894	42.7957	48.2679
23	27.1413	32.0069	35.1725	38.0756	41.6384	44.1813	49.7282
24	28.2412	33.1962	36.4150	39.3641	42.9798	45.5585	51.1786
25	29.3389	34.3816	37.6525	40.6465	44.3141	46.9279	52.6197
26	30.4346	35.5632	38.8851	41.9232	45.6417	48.2899	54.0520
27	31.5284	36.7412	40.1133	43.1945	46.9629	49.6449	55.4760
28	32.6205	37.9159	41.3371	44.4608	48.2782	50.9934	56.8923
29	33.7109	39.0875	42.5570	45.7223	49.5879	52.3356	58.3012
30	34.7997	40.2560	43.7730	46.9792	50.8922	53.6720	59.7031
40	45.6160	51.8051	55.7585	59.3417	63.6907	66.7660	73.4020
50	56.3336	63.1671	67.5048	71.4202	76.1539	79.4900	86.6608
60	66.9815	74.3970	79.0819	83.2977	88.3794	91.9517	99.6072
70	77.5767	85.5270	90.5312	95.0232	100.425	104.215	112.317
80	88.1303	96.5782	101.879	106.629	112.329	116.321	124.839
90	98.6499	107.565	113.145	118.136	124.116	128.299	137.208
100	109.141	118.498	124.342	129.561	135.807	140.169	149.449

Appendix F:
The Chi-Square Distribution with 99 Degrees of Freedom

This appendix gives y such that

$$P[X \leq y] = \gamma$$

for $\gamma = .01(.01).99$ where X has the χ^2 distribution with 99 degrees of freedom. These are used in the CSCS test of Section 3.3.8 that combines individual chi-square test results on random number generators. The chi-square percentage points given in the table on the following page have been reprinted with permission from *The Handbook of Random Number Generation and Testing* by Edward J. Dudewicz and Thomas G. Ralley, copyright ©1981 by American Sciences Press, Inc., 20 Cross Road, Syracuse, New York 13224–2104. (An additional significant digit was added to four of the table entries.)

Chi-Square Distribution with 99 Degrees of Freedom

γ	y	γ	y	γ	y
0.01	69.2299	0.34	92.6659	0.67	104.630
0.02	72.2880	0.35	93.0320	0.68	105.036
0.03	74.2754	0.36	93.3953	0.69	105.449
0.04	75.7949	0.37	93.7560	0.70	105.868
0.05	77.0463	0.38	94.1145	0.71	106.296
0.06	78.1226	0.39	94.4710	0.72	106.731
0.07	79.0746	0.40	94.8259	0.73	107.175
0.08	79.9336	0.41	95.1793	0.74	107.629
0.09	80.7204	0.42	95.5315	0.75	108.093
0.10	81.4493	0.43	95.8828	0.76	108.569
0.11	82.1306	0.44	96.2333	0.77	109.057
0.12	82.7724	0.45	96.5834	0.78	109.558
0.13	83.3805	0.46	96.9332	0.79	110.074
0.14	83.9599	0.47	97.2829	0.80	110.607
0.15	84.5143	0.48	97.6329	0.81	111.157
0.16	85.0469	0.49	97.9832	0.82	111.728
0.17	85.5603	0.50	98.3341	0.83	112.321
0.18	86.0566	0.51	98.6859	0.84	112.939
0.19	86.5377	0.52	99.0388	0.85	113.585
0.20	87.0052	0.53	99.3929	0.86	114.263
0.21	87.4605	0.54	99.7485	0.87	114.978
0.22	87.9048	0.55	100.106	0.88	115.735
0.23	88.3390	0.56	100.465	0.89	116.542
0.24	88.7642	0.57	100.827	0.90	117.407
0.25	89.1812	0.58	101.191	0.91	118.342
0.26	89.5907	0.59	101.558	0.92	119.364
0.27	89.9934	0.60	101.928	0.93	120.495
0.28	90.3898	0.61	102.301	0.94	121.765
0.29	90.7806	0.62	102.678	0.95	123.225
0.30	91.1663	0.63	103.059	0.96	124.955
0.31	91.5472	0.64	103.444	0.97	127.103
0.32	91.9238	0.65	103.834	0.98	129.996
0.33	92.2966	0.66	104.229	0.99	134.642

Appendix G:
Random Numbers

This appendix gives a collection of high-quality random numbers, 100 from each of the recommended random number generators that passed the testing discussed in Chapter 3. In each case, we also give the initialization or interaction used to generate these numbers. Thus, the tables here have several uses, including

- Providing random numbers for small studies

- Providing information that can be used to debug user-written routines for these high-quality generators (e.g., to speed up the basic routines)

The random numbers and initialization sequences for URNs 03, 12, 13, 14, and 15 have been reprinted with permission from _The Handbook of Random Number Generation and Testing_ by Edward J. Dudewicz and Thomas G. Ralley, copyright ©1981 by American Sciences Press, Inc., 20 Cross Road, Syracuse, New York 13224–2104. The outputs for URNs 03, 12, 13, 14, and 15 were obtained from runs on an IBM 370 system, those for URNs 22, 30 and 35 were obtained from runs on a VAX–11/785 computer, and those for URNS 36, 37 and 41 were obtained by running the programs shown in Figures 3.5–6, 3.5–7, and 3.5–8 (or 3.5–9) on a Sun Microsystems Sparc II workstation.

First 100 Random Numbers from URN03

0.40461671	0.69982326	0.38076901	0.33324194	0.48070574
0.57326061	0.96209621	0.10630631	0.11367542	0.26607591
0.32773596	0.24037737	0.94269621	0.61472696	0.14796633
0.77102375	0.74892873	0.27971303	0.52059972	0.03753376
0.41901445	0.23634988	0.92231232	0.02918905	0.96767312
0.05293137	0.16268539	0.56070369	0.83777303	0.77634692
0.85339433	0.00715560	0.34516907	0.55060714	0.85858345
0.02856642	0.74005908	0.94184327	0.71476948	0.16410995
0.85193664	0.22301644	0.95967376	0.63919353	0.32922477
0.03220677	0.46482444	0.15618962	0.57308775	0.95899922
0.91968334	0.08164358	0.44686884	0.03235793	0.83455026
0.27747023	0.15427011	0.24137986	0.23438716	0.73517567
0.41612756	0.00852907	0.87635487	0.53929967	0.72356665
0.88154095	0.16159558	0.56786507	0.99764025	0.82346386
0.10402644	0.74350578	0.04634136	0.14816207	0.57634336
0.48876947	0.70188856	0.60613978	0.37054616	0.39479268
0.27875596	0.24910325	0.32856989	0.24220198	0.37551349
0.70865047	0.36383653	0.47309804	0.70417929	0.18997937
0.27251816	0.24489319	0.90349859	0.88638568	0.86696678
0.06100678	0.37100679	0.65815705	0.61329663	0.02584821

Initialization Used in the Generation

```
      DIMENSION X(10000)
C
      NBATCH=100
      IX=524287
      JX=654345465
C
      CALL RSTRT(IX,JX,NBATCH)
      CALL URN03(X,IX,JX)
C
      WRITE(6,100)
100   FORMAT(' THESE ARE THE FIRST 100 NUMBERS FROM URN03',//)
      WRITE(6,110) (X(I),I=1,NBATCH)
110   FORMAT(4F11.8)
      STOP
      END
```

First 100 Random Numbers from URN12

0.21085471	0.20177358	0.53907311	0.54959267	0.54857844
0.88280052	0.62516850	0.46515036	0.99301422	0.41408068
0.54328817	0.94749588	0.46223176	0.25762159	0.05807477
0.57931852	0.03778372	0.09054917	0.39867902	0.16550606
0.57652152	0.69622103	0.03073253	0.20628929	0.54897475
0.80065238	0.71738249	0.12819904	0.72929245	0.97924389
0.49438995	0.39783579	0.36174679	0.75888300	0.18428940
0.64345145	0.42879963	0.04020694	0.07461166	0.90934426
0.64573163	0.96537840	0.38516927	0.16806555	0.80683321
0.35602415	0.14940369	0.21543938	0.10815895	0.84654546
0.13479573	0.07339454	0.07184356	0.83773160	0.42733830
0.02560912	0.69698101	0.39568561	0.30507123	0.41096735
0.46206605	0.22106236	0.02567938	0.29982960	0.80399781
0.81361967	0.15315223	0.17811376	0.69864636	0.95708138
0.04850708	0.76419032	0.96798182	0.64931315	0.14957386
0.08358914	0.93876529	0.60281080	0.66484302	0.92651904
0.07780331	0.44050103	0.30637538	0.81319982	0.22876382
0.71185476	0.68677521	0.59200680	0.07555521	0.85431290
0.00522494	0.79542786	0.45726794	0.22502446	0.64758307
0.87883300	0.15967810	0.32789767	0.32554442	0.46705616

Initialization Used in the Generation

```
      DIMENSION X(10000)
      REAL*8 C2,ZZ2
      NBATCH=100
      IX=1
      L=452807053
      C2=0.D0
      ZZ2=2147483648.D0
C
      CALL URN12(IX,X,L,C2,ZZ2,NBATCH)
C
      WRITE(6,100)
100   FORMAT(' THESE ARE THE FIRST 100 NUMBERS FROM URN12',//)
      WRITE(6,110) (X(I),I=1,NBATCH)
110   FORMAT(4F11.8)
      STOP
      END
```

First 100 Random Numbers from URN13

0.79629713	0.84357864	0.04583855	0.14499283	0.21417087
0.46331453	0.83064628	0.65813637	0.52331352	0.91584021
0.29460627	0.20976347	0.38656980	0.67632109	0.51638800
0.26010877	0.21219134	0.44585949	0.64709449	0.70371783
0.50613201	0.17664093	0.77622896	0.88482797	0.67356128
0.06192477	0.96563065	0.92134291	0.45709676	0.92837769
0.13887870	0.27490109	0.19925541	0.46825200	0.51366544
0.50453782	0.11662322	0.74890655	0.26463830	0.14144111
0.77324426	0.43717241	0.10421008	0.62230128	0.94072783
0.17606169	0.46883833	0.21875435	0.03257476	0.55132008
0.99827367	0.90041912	0.30022216	0.08532375	0.99527615
0.06902802	0.97830731	0.69581914	0.55330104	0.10115176
0.56664997	0.05700738	0.03965328	0.63701761	0.43645358
0.30047154	0.62161005	0.55292171	0.94822514	0.49344867
0.52301306	0.55473465	0.26891392	0.33194447	0.69882894
0.33000124	0.07576895	0.48619270	0.61517835	0.19613385
0.33441490	0.03595783	0.82233423	0.18219954	0.33864743
0.43318117	0.00818853	0.48180085	0.19944656	0.42947173
0.21968776	0.21455054	0.97070920	0.40593958	0.93973738
0.43363965	0.59920037	0.54287410	0.70707458	0.31700009

Initialization Used in the Generation

```
      DIMENSION X(10000)
C
      NBATCH=100
      IX=524287
C
      CALL URN13(IX,X,NBATCH)
C
      WRITE(6,100)
100   FORMAT(' THESE ARE THE FIRST 100 NUMBERS FROM URN13',//)
      WRITE(6,110) (X(I),I=1,NBATCH)
100   FORMAT(4F11.8)
      STOP
      END
```

First 100 Random Numbers from URN14

0.86368763	0.54198301	0.08967870	0.26754546	0.12438130
0.03406118	0.67197025	0.40013207	0.04980449	0.37747371
0.34327376	0.37563765	0.41319585	0.83154762	0.50485718
0.07502693	0.93380928	0.29614687	0.21132481	0.85894322
0.63979006	0.71967971	0.25977814	0.23725903	0.27299809
0.84604347	0.81547713	0.40951645	0.11845273	0.11216986
0.58494413	0.95559037	0.12368262	0.93208897	0.92476177
0.33849943	0.49657154	0.05912191	0.70647287	0.56211722
0.09185153	0.33407879	0.27173746	0.89581060	0.21371126
0.44815111	0.76221216	0.80300951	0.88733625	0.14929938
0.50181627	0.80946219	0.45151806	0.55538237	0.99605954
0.60512793	0.91912615	0.10340029	0.33306050	0.14950144
0.52462018	0.00890318	0.81116223	0.77674758	0.96696675
0.28550005	0.80211377	0.92576182	0.87015760	0.29452968
0.13719344	0.23051918	0.53925550	0.22048879	0.33474755
0.95925939	0.22920144	0.41614306	0.22348809	0.50468326
0.43910897	0.45153236	0.85561454	0.79578459	0.20289922
0.57011843	0.52883327	0.80020678	0.17325830	0.22113061
0.74961591	0.71164966	0.97458720	0.66529918	0.86510730
0.28213584	0.93588984	0.01323017	0.67708445	0.72654724

Initialization Used in the Generation
of the Random Numbers from URN14

```
      DIMENSION   X(10000)
      INTEGER TABLE(64)
      DATA TABLE/            453816693,    859273406,    1973808427,   2059993932,
    * 509510129,    583551786,    2052114759,   810618840,    1560248493,   1724450006,
    * 1084172707,   2001645988,   557869481,    1854755266,   1609787697,   1517139120,
    * 222050533,    73145838,     1785734939,   1310401788,   1741957729,   240883034,
    * 1905540151,   126963336,    267106845,    247276550,    1192674707,   19119444,
    * 1373938713,   254375410,    1443045167,   2009808224,   458941525,    1668052766,
    * 806675723,    1844566700,   1226880721,   1868649354,   726922023,    2139021624,
    * 192583565,    1545500470,   737174915,    321051908,    479937161,    1751223842,
    * 1985910815,   887331344,    1163899845,   1077642318,   632497915,    635970652,
    * 1816864577,   133989818,    574549527,    1837418472,   717428989,    106954342,
    * 879430003,    2005340340,   613106937,    435722834,    265606415,    197249728/

      NBATCH=100
      IX=197249728
      JX=0
      CALL URN14(IX,JX,X,NBATCH,TABLE)
      WRITE(6,100)
100   FORMAT(' THESE ARE THE FIRST 100 NUMBERS FROM URN14',//)
      WRITE(6,110) (X(I),I=1,NBATCH)
110   FORMAT(4F11.8)
      STOP
      END
```

First 100 Random Numbers from URN15

0.83243245	0.44883275	0.14152670	0.75742060	0.04374329
0.46891183	0.67829138	0.72205967	0.07497674	0.30082101
0.01165536	0.36970013	0.35753775	0.91369539	0.48843426
0.89160603	0.77659327	0.39165443	0.16264361	0.73787570
0.35909444	0.49800259	0.07921064	0.09138095	0.88178033
0.30294549	0.79977250	0.69617641	0.31564170	0.79223859
0.56809491	0.45091248	0.07373881	0.37644845	0.84628755
0.72845179	0.46826035	0.22481138	0.85281724	0.93770468
0.19187820	0.01281876	0.52352017	0.74677759	0.81517756
0.03342155	0.17075717	0.68975455	0.82499152	0.66803432
0.60911494	0.26906741	0.43294567	0.85229510	0.99799860
0.23066646	0.94310820	0.58117259	0.72482383	0.21493024
0.10082030	0.28265041	0.65343523	0.21524769	0.16260439
0.71186769	0.06235589	0.05132990	0.27307856	0.82883924
0.26692080	0.32140332	0.30057424	0.06281126	0.03914542
0.32041419	0.39753735	0.37260604	0.51003790	0.70446223
0.95995981	0.91693074	0.68546641	0.59086514	0.63067389
0.10024905	0.97777730	0.36563385	0.12746346	0.87648368
0.20914817	0.65771383	0.93472850	0.83619654	0.19865656
0.20671988	0.78280365	0.90928185	0.03690434	0.72599161

Initialization Used in the Generation

```
      DIMENSION X(10000)
C
      NBATCH=100
      IX=524287
C
      CALL URN15(IX,X,NBATCH)
C
      WRITE(6,100)
100   FORMAT(' THESE ARE THE FIRST 100 NUMBERS FROM URN15',//)
      WRITE(6,110) (X(I),I=1,NBATCH)
110   FORMAT(4F11.8)
      STOP
      END
```

First 100 Random Numbers from URN22

0.72924829	0.45097476	0.37858468	0.46684903	0.79917455
0.18918025	0.49316698	0.55244821	0.04606479	0.65183419
0.53619230	0.26759505	0.52306610	0.65443653	0.27860975
0.29742420	0.79526931	0.45999330	0.27989304	0.93458003
0.50922829	0.88979524	0.26957136	0.02739745	0.31697077
0.85509509	0.56293970	0.68491942	0.70137715	0.41905820
0.93282402	0.22312468	0.99980086	0.24836189	0.10927600
0.58674181	0.67280632	0.06381774	0.82907277	0.23102421
0.61489886	0.44939274	0.10689592	0.19786084	0.05412161
0.12900358	0.15006542	0.87059802	0.33739549	0.57045978
0.08870316	0.64194715	0.64847505	0.52592975	0.44320142
0.47935331	0.45539773	0.86830556	0.00087041	0.12220001
0.23429948	0.83284140	0.52364147	0.39522457	0.76647300
0.52343971	0.45869887	0.87251955	0.05554289	0.29296172
0.57391703	0.87649536	0.66163337	0.35693562	0.18997735
0.54617840	0.99940276	0.75003880	0.43103933	0.45922041
0.89682031	0.48472780	0.66584200	0.04199368	0.46156049
0.52139252	0.06110919	0.75214714	0.05188882	0.90940821
0.91913545	0.76799864	0.89891565	0.20846140	0.22142595
0.66913778	0.67709172	0.04952365	0.55026627	0.34261489

Interaction Used in the Generation

$****$ Which test or generator would you like to run? $****$
urn22 100 no 100

SEED=32007779
The initialization for the next random number (following the 100
already generated) is
SEED=1471519871
These are the 100 numbers to be printed for generator URN22

First 100 Random Numbers from URN30

0.19481865	0.73246359	0.60873993	0.32257841	0.10845195
0.18848224	0.06172308	0.71325549	0.63791228	0.04875328
0.67071617	0.39221781	0.37919311	0.53831462	0.88386748
0.40938184	0.79985896	0.43589519	0.01971257	0.13363735
0.67898969	0.39135272	0.29794773	0.18178704	0.36562744
0.13553518	0.68636952	0.30588367	0.45690279	0.36894231
0.95572393	0.11698765	0.43186733	0.26726933	0.03731502
0.46987497	0.36161550	0.43872419	0.55469963	0.74701394
0.03814769	0.10910354	0.57425703	0.87323778	0.88275409
0.40473460	0.66496818	0.30473412	0.28834950	0.35217032
0.14513280	0.23482212	0.99910532	0.10036084	0.37362427
0.83065385	0.41346381	0.23292546	0.58857266	0.82825009
0.01337966	0.55689025	0.16955324	0.13123797	0.03324680
0.57107876	0.56603311	0.04355976	0.22564493	0.73487607
0.55717717	0.65998265	0.52246616	0.23606059	0.66328167
0.84948987	0.29426527	0.00646368	0.80204340	0.42997186
0.18503648	0.52319426	0.93209440	0.05593764	0.88446826
0.51120751	0.42688788	0.92299604	0.66377109	0.45279579
0.05664913	0.98469037	0.27848648	0.71534298	0.14802752
0.95599895	0.86189932	0.99853890	0.96590879	0.90056199

Interaction Used in the Generation

∗∗∗∗ Which test or generator would you like to run? ∗∗∗∗
urn30 100 no 100

IX=524287
The initialization for the next random number (following the 100
already generated) is
IX=1933942168
These are the 100 numbers to be printed for generator URN30

First 100 Random Numbers from URN35[1]

0.01368889	0.04335332	0.10444348	0.48736173	0.25020087
0.74309373	0.25196218	0.21655028	0.41771212	0.80356592
0.04313701	0.58336937	0.87661457	0.75075704	0.57582968
0.36938384	0.77155304	0.81657499	0.35439712	0.43319324
0.24497868	0.50333726	0.49962920	0.83489895	0.89867747
0.84398371	0.04421853	0.24837773	0.27516842	0.16111489
0.19627959	0.59165072	0.54489833	0.61640197	0.21689019
0.54715407	0.71889871	0.28780669	0.93717939	0.31555527
0.72937393	0.82087016	0.84630120	0.17242445	0.36610839
0.17857364	0.30915889	0.27801126	0.61760706	0.76778322
0.64940363	0.25492862	0.18265249	0.02657438	0.98356098
0.27436501	0.69538349	0.27791855	0.70076013	0.43334776
0.90856558	0.02787220	0.75220937	0.96032381	0.33384833
0.91468388	0.64523208	0.74751252	0.19578518	0.87102157
0.65749955	0.98686731	0.14526296	0.13080156	0.40933812
0.14581917	0.07786910	0.49962920	0.93492365	0.89830667
0.26361164	0.67131203	0.62523949	0.57777643	0.41780484
0.46217787	0.86524320	0.64841479	0.23790248	0.74949014
0.25508311	0.00101971	0.79642791	0.26633087	0.05830913
0.32878068	0.75326002	0.23147519	0.50905383	0.70520979

Interaction Used in the Generation

∗ ∗ ∗∗ Which test or generator would you like to run? ∗ ∗ ∗∗
urn35 100 no 100

These are the 100 numbers to be printed for generator URN35

---

[1]The careful reader will notice that these numbers differ in their last few digits from the numbers that should be produced by the algorithm of URN35 (given in (3.5.24)). For example, the first number should be 443/32363=0.01368847. The discrepency is caused by the inherent limitations of floating-point arithmetic on whatever computer may be in use. The greater the floating-point precision of the computer, the closer the numbers will be to the exact values, rather than to those in this table.

First 100 Random Numbers from URN36[2]

0.11769113	0.30095156	0.73271124	0.03693928	0.19552592
0.68124967	0.72671596	0.39872321	0.94929111	0.49730195
0.82807935	0.64092109	0.04364824	0.43464634	0.74686449
0.71121969	0.28619361	0.22659412	0.50911921	0.97936202
0.45536975	0.72396264	0.77673080	0.41894048	0.37984154
0.55806961	0.18827685	0.04260805	0.40214985	0.18605179
0.29118464	0.56726997	0.73445491	0.98257027	0.01051171
0.16081834	0.58918206	0.91696502	0.97465993	0.99697222
0.34238238	0.22992114	0.39333574	0.21978565	0.36262059
0.39870934	0.55806889	0.79438622	0.51320285	0.55506318
0.41060242	0.59550655	0.71055691	0.39475541	0.99946606
0.96037124	0.29657702	0.49293618	0.02004587	0.80750181
0.35774630	0.81429532	0.00846238	0.29800210	0.49884890
0.08211148	0.13456987	0.89088666	0.95408444	0.54525039
0.91438924	0.51901454	0.29339744	0.55717485	0.53728695
0.48719115	0.35683394	0.25602902	0.87821257	0.56900048
0.42473806	0.50985928	0.64030743	0.65393041	0.00660090
0.16603447	0.44819686	0.17135337	0.00637422	0.51892775
0.08727851	0.39292503	0.39706153	0.78492270	0.45931928
0.58996346	0.91239898	0.24880614	0.19967332	0.89737128

Note: x_0 is set to 32767.

---

[2]The numbers shown here differ, usually in the last four significant digits, from those shown in Dudewicz, Karian, and Marshall (1985) where, to perform division by 2^{35} on a microcomputer, x_i was divided by 2^{20}, truncated, and divided a second time by 2^{15}. The numbers shown here result from direct division by 2^{35}.

First 100 Random Numbers from URN37[3]

0.88305812	0.81522504	0.72337977	0.59102849	0.12841475
0.89999474	0.62745540	0.37957139	0.18865783	0.95683121
0.24926954	0.25473737	0.69837178	0.95434138	0.94983179
0.37951508	0.57909783	0.62550599	0.14590735	0.66979236
0.28591600	0.58340220	0.42769285	0.70304762	0.07746468
0.92911376	0.01242956	0.68172874	0.43439762	0.97682869
0.65895423	0.07631952	0.75127851	0.72112438	0.16818736
0.57309351	0.42691296	0.67144031	0.59455894	0.95155550
0.47811807	0.67873817	0.29337179	0.51657038	0.80192943
0.37934950	0.92910230	0.71017954	0.21124568	0.88101089
0.56963874	0.52810479	0.36891052	0.31608663	0.99145700
0.36593978	0.74523226	0.88697882	0.32073393	0.92002889
0.72394083	0.02872508	0.13539806	0.76803362	0.98158233
0.31378368	0.15311793	0.25418655	0.26535105	0.72400619
0.63022785	0.30423267	0.69435901	0.91479765	0.59659117
0.79505836	0.70934297	0.09646424	0.05140108	0.10400775
0.79580899	0.15774840	0.24491889	0.46288664	0.10561445
0.65655152	0.17114981	0.74165390	0.10198758	0.30284407
0.39645621	0.28680041	0.35298447	0.55876326	0.34910843
0.43941539	0.59838366	0.98089823	0.19477306	0.35794041

Note: x_0 is set to 32767.

---

[3]The numbers shown here differ, usually in the last two significant digits, from those shown in Dudewicz, Karian, and Marshall (1985) where, to perform division by 2^{31} on a microcomputer, x_i was multiplied by $.4656613 \times 10^{-9}$ (see Figure 3.5–7).

First 100 Random Numbers from URN41

0.67460162	0.15152637	0.18676730	0.99911913	0.16606654
0.30185115	0.66726522	0.85055565	0.05640866	0.79791909
0.66087965	0.04146227	0.01945484	0.63037293	0.02815871
0.45902274	0.69340918	0.50495090	0.59523033	0.79470998
0.71767453	0.94974704	0.20854943	0.38893471	0.07358414
0.85697244	0.97879382	0.72114767	0.41651274	0.24538072
0.67286257	0.62257449	0.50480539	0.73655240	0.38089954
0.24579348	0.74893101	0.44526388	0.25323342	0.28239765
0.32223641	0.69119824	0.27009427	0.87701379	0.19264518
0.14628606	0.67493564	0.14298250	0.85261481	0.73899597
0.87015966	0.14203439	0.44263388	0.84392562	0.24201153
0.64732891	0.38151020	0.99207529	0.54937865	0.96625354
0.53466928	0.54421737	0.17919800	0.55641217	0.36965135
0.47680630	0.73914459	0.12526315	0.46596403	0.63690489
0.12624222	0.36089806	0.16441625	0.52963396	0.37800670
0.39848434	0.05902798	0.90868376	0.28070040	0.58392542
0.45257897	0.14745668	0.87724569	0.05332366	0.81482850
0.57544205	0.68531368	0.10386257	0.70040688	0.67145523
0.60407332	0.28875609	0.50134813	0.21240417	0.15605478
0.62594361	0.82720216	0.18327817	0.87854960	0.26724132

Appendix H:
Tables for Selection of the Best

This appendix gives tables of the solutions $h_{n_0}(k, P^*)$ which solve equation (6.3.10) and are needed for the goal of selection of the best in Section 6.3. Note that the table for $k = 2$ is that needed for the solution of the Behrens-Fisher problem given in Section 6.2, since for all values of p the quantity needed in equation (6.2.5) is

$$F_{T_1+T_2}^{-1}(p) = h_{n_0}(2, p).$$

The table entries are from Table 4 on pp. 17–23 of "New tables for multiple comparisons with a control (unknown variances)" by E.J. Dudewicz, J.S. Ramberg, and H.J. Chen, *Biometrische Zeitschrift*, Volume 17 (1975), pp. 13–26. Reprinted with the permission of Akademie–Verlag Berlin.

Solution $h_{n_0}(k, P^*)$ of Equation (6.3.10) (continues)

k	P^*	n_0												
		2	3	4	5	6	7	8	9	10	15	20	25	30
2	.75	2.00	1.37	1.21	1.14	1.10	1.07	1.05	1.04	1.03	1.00	0.99	0.98	0.98
	.80	2.75	1.76	1.54	1.44	1.38	1.35	1.32	1.30	1.29	1.25	1.24	1.23	1.22
	.85	3.93	2.27	1.94	1.80	1.72	1.68	1.64	1.62	1.60	1.55	1.53	1.51	1.51
	.90	6.16	3.04	2.50	2.29	2.18	2.11	2.06	2.02	2.00	1.93	1.90	1.88	1.87
	.95	12.63	4.57	3.50	3.11	2.91	2.79	2.71	2.66	2.61	2.50	2.45	2.42	2.41
	.975	25.42	6.54	4.59	3.94	3.63	3.45	3.33	3.24	3.18	3.02	2.95	2.91	2.88
	.99	63.7	10.28	6.31	5.14	4.60	4.30	4.11	3.98	3.89	3.64	3.54	3.48	3.45
3	.75	3.52	2.15	1.86	1.74	1.67	1.63	1.60	1.57	1.56	1.51	1.49	1.48	1.47
	.80	4.59	2.59	2.20	2.04	1.95	1.89	1.85	1.83	1.80	1.75	1.72	1.71	1.70
	.85	6.31	3.17	2.62	2.40	2.28	2.21	2.16	2.13	2.10	2.03	1.99	1.98	1.96
	.90	9.64	4.05	3.22	2.90	2.73	2.63	2.57	2.52	2.48	2.39	2.34	2.32	2.30
	.95	19.40	5.86	4.29	3.75	3.48	3.32	3.21	3.14	3.08	2.94	2.87	2.84	2.81
	.975	38.7	8.25	5.50	4.63	4.22	3.98	3.82	3.72	3.64	3.43	3.35	3.30	3.27
	.99	96.2	12.83	7.44	5.91	5.23	4.86	4.62	4.46	4.34	4.04	3.92	3.85	3.81
4	.75	4.77	2.66	2.25	2.08	1.99	1.93	1.89	1.86	1.84	1.78	1.75	1.74	1.73
	.80	6.16	3.13	2.59	2.38	2.26	2.19	2.14	2.11	2.08	2.01	1.98	1.96	1.95
	.85	8.41	3.77	3.03	2.75	2.60	2.51	2.45	2.41	2.37	2.28	2.24	2.22	2.21
	.90	12.80	4.75	3.66	3.26	3.06	2.93	2.85	2.80	2.75	2.63	2.58	2.55	2.54
	.95	25.76	6.80	4.80	4.14	3.81	3.62	3.50	3.41	3.34	3.17	3.10	3.06	3.03
	.975	51.4	9.53	6.10	5.05	4.57	4.29	4.11	3.99	3.90	3.67	3.57	3.51	3.48
	.99	128	14.79	8.21	6.40	5.62	5.19	4.92	4.74	4.60	4.27	4.13	4.05	4.01
5	.75	5.95	3.05	2.53	2.32	2.21	2.14	2.09	2.06	2.03	1.96	1.93	1.91	1.90
	.80	7.65	3.56	2.89	2.63	2.49	2.40	2.35	2.30	2.27	2.19	2.15	2.13	2.12
	.85	10.43	4.25	3.34	3.00	2.83	2.72	2.65	2.60	2.56	2.46	2.41	2.39	2.37
	.90	15.90	5.32	4.00	3.53	3.29	3.15	3.05	2.99	2.94	2.81	2.75	2.72	2.69
	.95	32.04	7.58	5.20	4.42	4.05	3.84	3.70	3.60	3.53	3.34	3.26	3.21	3.18
	.975	64.1	10.61	6.58	5.37	4.83	4.52	4.32	4.18	4.08	3.83	3.72	3.66	3.62
	.99	160	16.47	8.82	6.78	5.90	5.43	5.13	4.93	4.79	4.43	4.28	4.20	4.14
6	.75	7.10	3.39	2.76	2.52	2.38	2.30	2.25	2.21	2.18	2.10	2.06	2.04	2.03
	.80	9.12	3.93	3.13	2.82	2.66	2.57	2.50	2.45	2.42	2.32	2.28	2.26	2.24
	.85	12.43	4.66	3.60	3.21	3.01	2.89	2.81	2.75	2.70	2.59	2.54	2.51	2.49
	.90	18.96	5.82	4.28	3.74	3.47	3.31	3.21	3.13	3.08	2.93	2.87	2.84	2.81
	.95	38.29	8.26	5.53	4.66	4.25	4.01	3.86	3.75	3.67	3.46	3.38	3.33	3.30
	.975	76.7	11.56	6.97	5.64	5.03	4.69	4.48	4.33	4.22	3.95	3.83	3.77	3.73
	.99	192	17.97	9.34	7.09	6.13	5.62	5.30	5.09	4.93	4.55	4.39	4.30	4.25
7	.75	8.23	3.68	2.96	2.67	2.53	2.44	2.37	2.33	2.30	2.21	2.17	2.14	2.13
	.80	10.58	4.25	3.33	2.99	2.81	2.70	2.63	2.58	2.54	2.43	2.38	2.36	2.34
	.85	14.42	5.03	3.81	3.37	3.15	3.02	2.93	2.87	2.82	2.70	2.64	2.61	2.59
	.90	22.01	6.27	4.51	3.92	3.62	3.45	3.33	3.25	3.19	3.04	2.97	2.93	2.91
	.95	44.53	8.88	5.81	4.85	4.41	4.15	3.98	3.87	3.79	3.57	3.47	3.42	3.39
	.975	89.3	12.43	7.32	5.86	5.20	4.84	4.61	4.45	4.34	4.05	3.93	3.86	3.82
	.99	223	19.34	9.79	7.35	6.32	5.78	5.44	5.21	5.05	4.64	4.48	4.39	4.33

Solution $h_{n_0}(k, P^*)$ of Equation (6.3.10) (continued)

k	P^*	n_0												
		2	3	4	5	6	7	8	9	10	15	20	25	30
8	.75	9.36	3.95	3.13	2.81	2.65	2.55	2.48	2.43	2.40	2.30	2.25	2.23	2.21
	.80	12.02	4.55	3.51	3.13	2.93	2.81	2.73	2.68	2.63	2.52	2.47	2.44	2.42
	.85	16.40	5.37	4.01	3.52	3.28	3.13	3.04	2.97	2.92	2.78	2.72	2.69	2.67
	.90	25.05	6.68	4.72	4.07	3.75	3.56	3.44	3.35	3.29	3.12	3.05	3.01	2.98
	.95	50.76	9.45	6.06	5.03	4.54	4.27	4.09	3.97	3.88	3.65	3.55	3.50	3.46
	.975	101.9	13.24	7.63	6.05	5.35	4.97	4.72	4.56	4.44	4.13	4.00	3.93	3.89
	.99	256	20.62	10.20	7.58	6.49	5.91	5.56	5.32	5.15	4.73	4.55	4.46	4.40
9	.75	10.49	4.20	3.28	2.93	2.75	2.64	2.57	2.52	2.48	2.37	2.33	2.30	2.28
	.80	13.47	4.82	3.67	3.25	3.04	2.91	2.82	2.76	2.72	2.60	2.54	2.51	2.49
	.85	18.37	5.68	4.18	3.65	3.39	3.23	3.13	3.06	3.00	2.86	2.80	2.76	2.74
	.90	28.08	7.06	4.91	4.21	3.86	3.66	3.53	3.44	3.37	3.20	3.12	3.08	3.05
	.95	57.0	9.99	6.29	5.18	4.66	4.37	4.19	4.06	3.96	3.72	3.62	3.56	3.53
	.975	114.5	13.99	7.91	6.22	5.48	5.08	4.82	4.65	4.52	4.20	4.07	4.00	3.95
	.99	287	21.8	10.58	7.79	6.64	6.03	5.66	5.41	5.23	4.80	4.62	4.53	4.46
10	.75	11.60	4.43	3.42	3.04	2.85	2.73	2.65	2.60	2.55	2.44	2.39	2.36	2.35
	.80	14.90	5.08	3.82	3.36	3.13	3.00	2.90	2.84	2.79	2.66	2.60	2.57	2.55
	.85	20.34	5.98	4.33	3.77	3.48	3.32	3.21	3.13	3.08	2.92	2.86	2.82	2.80
	.90	31.12	7.41	5.09	4.33	3.96	3.75	3.62	3.52	3.45	3.26	3.18	3.14	3.11
	.95	63.2	10.49	6.50	5.32	4.77	4.47	4.27	4.14	4.04	3.79	3.68	3.62	3.58
	.975	127.1	14.70	8.17	6.38	5.60	5.17	4.91	4.73	4.60	4.26	4.13	4.05	4.01
	.99	318	22.9	10.92	7.98	6.77	6.14	5.75	5.49	5.31	4.86	4.68	4.58	4.51
11	.75	12.72	4.64	3.54	3.14	2.93	2.81	2.72	2.66	2.62	2.50	2.45	2.42	2.40
	.80	16.34	5.32	3.95	3.46	3.22	3.07	2.98	2.91	2.86	2.72	2.66	2.63	2.61
	.85	22.31	6.25	4.48	3.87	3.57	3.40	3.28	3.20	3.14	2.98	2.91	2.87	2.85
	.90	34.15	7.75	5.25	4.44	4.06	3.83	3.69	3.59	3.51	3.32	3.24	3.19	3.16
	.95	69.4	10.97	6.70	5.44	4.87	4.55	4.35	4.21	4.10	3.84	3.73	3.67	3.63
	.975	139.7	15.38	8.41	6.52	5.71	5.26	4.99	4.80	4.66	4.32	4.18	4.10	4.05
	.99	350	24.0	11.25	8.16	6.90	6.24	5.84	5.57	5.38	4.91	4.73	4.62	4.56
12	.75	13.84	4.85	3.66	3.23	3.01	2.88	2.79	2.73	2.68	2.55	2.50	2.47	2.45
	.80	17.77	5.54	4.07	3.56	3.30	3.14	3.04	2.97	2.92	2.77	2.71	2.68	2.66
	.85	24.27	6.52	4.61	3.97	3.65	3.47	3.35	3.26	3.20	3.03	2.96	2.92	2.90
	.90	37.17	8.07	5.40	4.54	4.14	3.91	3.76	3.65	3.57	3.37	3.28	3.24	3.21
	.95	75.6	11.42	6.88	5.56	4.96	4.63	4.42	4.27	4.16	3.89	3.78	3.72	3.68
	.975	152.3	16.02	8.63	6.66	5.81	5.34	5.06	4.86	4.72	4.37	4.23	4.15	4.10
	.99	382	25.0	11.55	8.32	7.01	6.32	5.91	5.64	5.44	4.96	4.77	4.67	4.60
13	.75	14.95	5.04	3.77	3.31	3.08	2.94	2.85	2.78	2.73	2.60	2.54	2.51	2.49
	.80	19.21	5.76	4.19	3.64	3.37	3.21	3.10	3.03	2.97	2.82	2.76	2.72	2.70
	.85	26.24	6.77	4.74	4.06	4.73	3.53	3.41	3.32	3.25	3.08	3.01	2.97	2.94
	.90	40.20	8.38	5.54	4.64	4.22	3.97	3.82	3.71	3.63	3.42	3.33	3.28	3.25
	.95	81.8	11.86	7.05	5.67	5.05	4.70	4.48	4.33	4.22	3.94	3.82	3.76	3.72
	.975	164.8	16.64	8.85	6.78	5.90	5.42	5.12	4.92	4.78	4.42	4.27	4.19	4.14
	.99	413	26.0	11.84	8.47	7.11	6.41	5.98	5.70	5.50	5.01	4.81	4.71	4.64

Solution $h_{n_0}(k, P^*)$ of Equation (6.3.10) (continued)

k	P^*	n_0												
		2	3	4	5	6	7	8	9	10	15	20	25	30
14	.75	16.06	5.23	3.87	3.39	3.15	3.00	2.90	2.83	2.78	2.65	2.59	2.55	2.53
	.80	20.64	5.97	4.30	3.72	3.44	3.27	3.16	3.08	3.02	2.86	2.80	2.76	2.74
	.85	28.20	7.01	4.86	4.14	3.80	3.60	3.46	3.37	3.30	3.12	3.05	3.01	2.98
	.90	43.23	8.67	5.67	4.73	4.29	4.04	3.87	3.76	3.68	3.46	3.37	3.32	3.29
	.95	88.1	12.27	7.22	5.77	5.12	4.76	4.54	4.38	4.27	3.98	3.86	3.80	3.75
	.975	177.4	17.24	9.05	6.90	5.98	5.49	5.18	4.98	4.83	4.46	4.31	4.22	4.17
	.99	445	27.0	12.11	8.61	7.21	6.48	6.05	5.76	5.55	5.05	4.85	4.74	4.68
15	.75	17.17	5.41	3.97	3.46	3.21	3.05	2.95	2.88	2.83	2.69	2.63	2.59	2.57
	.80	22.07	6.17	4.40	3.80	3.50	3.32	3.21	3.13	3.07	2.91	2.84	2.80	2.77
	.85	30.17	7.24	4.97	4.22	3.86	3.65	3.52	3.42	3.35	3.16	3.09	3.01	3.01
	.90	46.25	8.96	5.80	4.82	4.36	4.09	3.93	3.81	3.72	3.50	3.41	3.35	3.32
	.95	94.3	12.68	7.37	5.87	5.20	4.83	4.59	4.43	4.32	4.02	3.90	3.83	3.79
	.975	190.0	17.81	9.24	7.01	6.06	5.55	5.24	5.03	4.88	4.50	4.34	4.26	4.20
	.99	476	27.9	12.37	8.75	7.30	6.55	6.11	5.81	5.60	5.09	4.88	4.78	4.71
16	.75	18.28	5.58	4.06	3.53	3.26	3.11	3.00	2.93	2.87	2.72	2.66	2.63	2.60
	.80	3.50	6.36	4.50	3.87	3.56	3.37	3.26	3.17	3.11	2.94	2.87	2.83	2.81
	.85	32.13	7.46	5.07	4.29	3.92	3.70	3.56	3.47	3.39	3.20	3.12	3.07	3.05
	.90	49.28	9.23	5.92	4.90	4.42	4.15	3.97	3.85	3.77	3.54	3.44	3.39	3.35
	.95	100.5	13.07	7.52	5.96	5.27	4.88	4.64	4.48	4.36	4.06	3.93	3.86	3.82
	.975	202.6	18.37	9.42	7.11	6.14	5.62	5.29	5.08	4.92	4.53	4.38	4.29	4.24
	.99	508	28.8	12.62	8.88	7.39	6.62	6.16	5.86	5.65	5.12	4.92	4.81	4.74
17	.75	19.40	5.74	4.15	3.59	3.32	3.15	3.05	2.97	2.91	2.76	2.69	2.66	2.63
	.80	24.93	6.55	4.59	3.94	3.61	3.42	3.30	3.21	3.15	2.98	2.90	2.86	2.84
	.85	34.09	7.68	5.17	4.37	3.98	3.75	3.61	3.51	3.43	3.24	3.15	3.11	3.08
	.90	52.30	9.50	6.03	4.97	4.48	4.20	4.02	3.90	3.81	3.57	3.47	3.42	3.38
	.95	106.7	13.45	7.66	6.04	5.33	4.94	4.69	4.52	4.40	4.09	3.96	3.89	3.85
	.975	215.3	18.90	9.60	7.21	6.21	5.67	5.34	5.12	4.96	4.57	4.41	4.32	4.27
	.99	542	29.6	12.86	9.00	7.47	6.68	6.22	5.91	5.69	5.16	4.95	4.84	4.76
18	.75	20.51	5.90	4.23	3.66	3.37	3.20	3.09	3.01	2.95	2.79	2.72	2.69	2.66
	.80	26.36	6.73	4.68	4.00	3.67	3.47	3.34	3.25	3.19	3.01	2.93	2.89	2.87
	.85	36.05	7.89	5.27	4.43	4.03	3.80	3.65	3.55	3.47	3.27	3.18	3.14	3.10
	.90	55.3	9.76	6.14	5.04	4.54	4.25	4.06	3.94	3.84	3.60	3.50	3.45	3.41
	.95	112.9	13.81	7.80	6.13	5.39	4.99	4.73	4.56	4.44	4.12	3.99	3.92	3.87
	.975	227.7	19.43	9.77	7.31	6.28	5.73	5.39	5.16	5.00	4.60	4.43	4.35	4.29
	.99	571	30.5	13.09	9.12	7.55	6.74	6.27	5.95	5.73	5.19	4.98	4.86	4.79
19	.75	21.61	6.06	4.32	3.71	3.42	3.24	3.13	3.04	2.98	2.82	2.75	2.71	2.69
	.80	27.79	6.91	4.77	4.06	3.72	3.51	3.38	3.29	3.22	3.04	2.96	2.92	2.89
	.85	38.01	8.09	5.37	4.50	4.08	3.85	3.69	3.59	3.51	3.30	3.21	3.16	3.13
	.90	58.4	10.01	6.25	5.11	4.59	4.29	4.10	3.97	3.88	3.63	3.53	3.47	3.44
	.95	119.1	14.17	7.93	6.20	5.45	5.04	4.78	4.60	4.47	4.15	4.02	3.95	3.90
	.975	240.2	19.94	9.93	7.40	6.34	5.78	5.43	5.20	5.04	4.63	4.46	4.37	4.32
	.99	604	31.3	13.31	9.23	7.62	6.80	6.31	5.99	5.77	5.22	5.00	4.89	4.81

Solution $h_{n_0}(k, P^*)$ of Equation (6.3.10) (concluded)

k	P^*	n_0												
		2	3	4	5	6	7	8	9	10	15	20	25	30
20	.75	22.72	6.21	4.39	3.77	3.46	3.28	3.16	3.08	3.02	2.85	2.78	2.74	2.72
	.80	29.22	7.08	4.85	4.12	3.76	3.56	3.42	3.32	3.26	3.07	2.99	2.95	2.92
	.85	39.98	8.29	5.46	4.56	4.13	3.89	3.73	3.62	3.54	3.33	3.24	3.19	3.16
	.90	61.4	10.25	6.35	5.18	4.64	4.34	4.14	4.01	3.91	3.66	3.56	3.50	3.46
	.95	125.3	14.52	8.05	6.28	5.51	5.08	4.82	4.64	4.51	4.18	4.04	3.97	3.92
	.975	252.9	20.43	10.09	7.49	6.40	5.83	5.48	5.24	5.07	4.66	4.49	4.40	4.34
	.99	635	32.1	13.52	9.34	7.69	6.86	6.36	6.03	5.80	5.25	5.03	4.91	4.84
21	.75	23.83	6.36	4.47	3.82	3.51	3.32	3.20	3.11	3.05	2.88	2.81	2.77	2.74
	.80	30.65	7.24	4.93	4.17	3.81	3.59	3.45	3.36	3.29	3.10	3.02	2.97	2.94
	.85	41.94	8.48	5.54	4.62	4.18	3.93	3.77	3.65	3.57	3.35	3.26	3.21	3.18
	.90	64.4	10.49	6.44	5.24	4.69	4.38	4.18	4.04	3.94	3.69	3.58	3.52	3.48
	.95	131.5	14.86	8.17	6.35	5.56	5.13	4.86	4.67	4.54	4.21	4.07	3.99	3.95
	.975	265.7	20.92	10.24	7.57	6.46	5.88	5.52	5.28	5.11	4.68	4.51	4.42	4.36
	.99	667	32.8	13.72	9.44	7.76	6.91	6.40	6.07	5.84	5.27	5.05	4.93	4.86
22	.75	24.94	6.50	4.54	3.88	3.55	3.36	3.23	3.15	3.08	2.91	2.83	2.79	2.76
	.80	32.08	7.40	5.01	4.23	3.85	3.63	3.49	3.39	3.32	3.12	3.04	2.99	2.96
	.85	43.90	8.67	5.63	4.67	4.22	3.97	3.80	3.69	3.60	3.38	3.29	3.24	3.20
	.90	67.4	10.72	6.54	5.30	4.74	4.42	4.22	4.08	3.98	3.71	3.60	3.54	3.51
	.95	137.7	15.19	8.29	6.42	5.61	5.17	4.89	4.71	4.57	4.23	4.09	4.02	3.97
	.975	278	21.39	10.39	7.65	6.52	5.92	5.56	5.31	5.14	4.71	4.53	4.44	4.38
	.99	700	33.6	13.93	9.54	7.83	6.96	6.44	6.11	5.87	5.30	5.07	4.95	4.88
23	.75	26.05	6.64	4.61	3.92	3.59	3.39	3.27	3.18	3.11	2.93	2.85	2.81	2.78
	.80	33.51	7.56	5.08	4.28	3.89	3.67	3.52	3.42	3.34	3.15	3.06	3.02	2.99
	.85	45.86	8.85	5.71	4.73	4.27	4.00	3.83	3.72	3.63	3.41	3.31	3.26	3.22
	.90	70.4	10.95	6.63	5.36	4.78	4.46	4.25	4.11	4.00	3.74	3.63	3.57	3.53
	.95	143.9	15.51	8.40	6.49	5.66	5.21	4.93	4.74	4.60	4.26	4.11	4.04	3.99
	.975	291	21.85	10.53	7.73	6.57	5.96	5.59	5.35	5.17	4.73	4.56	4.46	4.40
	.99	730	34.3	14.12	9.63	7.89	7.01	6.48	6.14	5.90	5.32	5.10	4.97	4.90
24	.75	27.16	6.78	4.68	3.97	3.63	3.43	3.30	3.20	3.14	2.95	2.88	2.83	2.81
	.80	34.93	7.72	5.15	4.33	3.93	3.70	3.55	3.45	3.37	3.17	3.08	3.04	3.01
	.85	47.82	9.03	5.78	4.78	4.31	4.04	3.87	3.75	3.66	3.43	3.33	3.28	3.24
	.90	73.5	11.17	6.72	5.42	4.82	4.49	4.28	4.14	4.03	3.76	3.65	3.59	3.55
	.95	150.1	15.83	8.51	6.55	5.71	5.25	4.96	4.77	4.63	4.28	4.14	4.06	4.01
	.975	303	22.30	10.67	7.80	6.63	6.01	5.63	5.38	5.20	4.76	4.58	4.48	4.42
	.99	762	35.1	14.31	9.73	7.95	7.05	6.52	6.18	5.93	5.37	5.12	4.99	4.92
25	.75	28.27	6.91	4.74	4.02	3.67	3.46	3.33	3.23	3.16	2.98	2.90	2.85	2.83
	.80	36.36	7.87	5.22	4.38	3.97	3.74	3.58	3.48	3.40	3.19	3.11	3.06	3.03
	.85	49.78	9.21	5.86	4.83	4.35	4.07	3.90	3.77	3.68	3.45	3.35	3.30	3.26
	.90	76.5	11.38	6.80	5.47	4.87	4.53	4.31	4.17	4.06	3.78	3.67	3.61	3.57
	.95	156.3	16.14	8.62	6.62	5.75	5.28	4.99	3.80	4.66	4.30	4.16	4.08	4.03
	.975	316	22.74	10.80	7.88	6.68	6.05	5.66	5.41	5.23	4.78	4.60	4.50	4.44
	.99	794	35.8	14.49	9.82	8.01	7.10	6.56	6.21	5.96	5.37	5.14	5.01	4.94

References and Author Index[†]

Adiri, I. and Avi-Itzhak, B. (1969). A time-sharing queue with a finite number of customers. *Journal of the Association for Computing Machinery*, 16, 315–323. [313]

Ahrens, J. H. and Dieter, U. (1974). *Non-Uniform Random Numbers*. Institute für Math. Statistik, Technische Hochscule in Graz, A 8010 Graz, Hamerlingg. 6, VI, Austria. [125]

Banks, J. and Carson, J. S. II (1984). *Discrete-Event System Simulation*. Prentice-Hall, Inc., Englewood Cliffs, New Jersey. [44]

Beckwith, N. B. and Dudewicz, E. J. (1996). A bivariate generalized lambda distribution (GLD-2) using Plackett's method of construction: distribution, examples and applications. *American Journal of Mathematical and Management Sciences*, 16, 333–393. [224, 228]

Bernhofen, L. T., Dudewicz, E. J., Levendovszky, J., and van der Meulen, E. C. (1996). Ranking of the best random number generators via entropy-uniformity theory. *American Journal of Mathematical and Management Sciences*, 16, 49–88. [137]

Bratley, P., Fox, B. L., and Schrage, L. E. (1983). *A Guide to Simulation*. Springer-Verlag, Inc., New York. [183]

Brody, T. A. (1984). A random-number generator. *Computer Physics Communications*, 34, 39–46. [127]

Browne, M. W. (1993). Coin-tossing computers found to show subtle bias. *The New York Times*, January 12, 1993. Reprinted in *Themes of the Times, Mathematics*, Spring 1994, 1. [89]

[†] The list in the brackets [ ] consists of page numbers where the reference either is cited or can be referred to for further details.

Casti, J. L. (1997). *Would-Be Worlds, How Simulation is Changing the Frontiers of Science*. John Wiley & Sons, Inc., New York [7]

Chen, H. J. and Mithongtae, J. (1986). Selection of the best exponential distribution with a preliminary test. *American Journal of Mathematical and Management Sciences*, 6, 219–249. [193]

Cooper, R. B. (1981). *Introduction to Queueing Theory*. North Holland, New York-Oxford. [44]

Dagpunar, J. (1988). *Principles of Random Variate Generation*. Oxford University Press, Oxford, England. [180]

Devroye, L. (1986). *Non-Uniform Random Variate Generation*. Springer-Verlag, Inc., New York. [181]

Dieter, U. (1991). Optimal acceptance-rejection methods for sampling from various distributions. *The Frontiers of Statistical Computation, Simulation, & Modeling*, Vol. I of the Proceedings of ICOSCO-I, Peter R. Nelson, Chief Editor, American Sciences Press, Inc., Columbus, Ohio, 113–136. [195]

Dudewicz, E. J. (1976). *Introduction to Statistics and Probability*. American Sciences Press, Inc., Columbus, Ohio. [205]

Dudewicz, E. J. (1983). Heteroscedasticity. *Encyclopedia of Statistical Sciences, Volume 3* (edited by N. L. Johnson, S. Kotz, and C. B. Read). John Wiley & Sons, Inc., New York, pp. 611–619. [300]

Dudewicz, E. J. (1992). The generalized bootstrap. *Bootstrapping and Related Techniques*, edited by K. H.Jöckel, G. Rothe, and W. Sendler, Vol. 376 of Lecture Notes in Economics and Mathematical Systems, Springer-Verlag, Berlin, 31–37. [321, 322]

Dudewicz, E. J. (1995). The heteroscedastic method: fifty+ years of progress 1945–2000. *MSI-2000: Multivariate Statistical Analysis in Honor of Professor Minoru Siotani on his 70th Birthday, Volume I* (edited by T. Hayakawa, M. Aoshima, and K. Shimizu). American Sciences Press, Inc., Columbus, Ohio, 179–197. [300]

Dudewicz, E. J. (editor) (1996). *Modern Digital Simulation Methodology, II: Univariate and Bivariate Distribution Fitting, Bootstrap Methods, & Applications to CensusPES and CensusPlus of the U.S. Bureau of the Census, Bootstrap Sample Size, and Biology and Environment Case Studies*. American Sciences Press, Inc., Columbus, Ohio. [7]

Dudewicz, E. J. (editor) (1997). *Modern Digital Simulation Methodology,*

III: Advances in Theory, Application, & Design–Electric Power Systems, Spare Parts Inventory, Purchase Interval and Incidence Modeling, Automobile Insurance Bonus-Malus Systems, Genetic Algorithms – DNA Sequence Assembly, Education, & Water Resources Case Studies. American Sciences Press, Inc., Columbus, Ohio. [7]

Dudewicz, E. J. and Bernhofen, L. T. (1998). A "perfect" random number generator? – TESTRAND Results on the Quality of Rey's Proposal "URN 39," draft in preparation. [132, 134]

Dudewicz, E. J., Chen, P., and Taneja, B. K. (1989). *Modern Elementary Probability and Statistics, with Statistical Programming in SAS, MINITAB, & BMDP.* American Sciences Press, Inc., Columbus, Ohio. [409]

Dudewicz, E. J. and Karian, Z. A. (1985). *Modern Design and Analysis of Discrete-Event Computer Simulations.* IEEE Computer Society Press, Washington, D.C. [7, 48, 92, 122, 126, 313]

Dudewicz, E. J. and Karian, Z. A. (1988). TESTRAND for the VAX-11 family of computers: a random number generation and testing library. *The American Statistician*, 42, 228. [97]

Dudewicz, E. J. and Karian, Z. A. (1996). The extended generalized lambda distribution (EGLD) system for fitting distributions to data with moments, II: tables, *American Journal of Mathematical and Management Sciences*, 16(3 & 4), 271–332. [216, 219, 220, 224]

Dudewicz, E. J., Karian, Z. A., and Marshall, R. J. III (1985). Random number generation of microcomputers. *Modeling and Simulation on Microcomputers: 1985* (edited by R. G. Lavery). The Society for Computer Simulation, La Jolla, California, 9–14. [17, 127, 131, 134, 143, 145]

Dudewicz, E. J., Levy, G. C., Lienhart, J., and Wehrli, F. W. (1989). Statistical analysis of magnetic resonance imaging data in the normal brain (data, screening, normality, discrimination, variability), & implications for expert statistical programming for ESS^{TM} (the Expert Statistical System). *American Journal of Mathematical and Management Sciences*, 9, 299–359. [226]

Dudewicz, E. J. and Mishra, S. N. (1988). *Modern Mathematical Statistics.* John Wiley & Sons, Inc., New York. [37, 104, 170, 171, 177, 178, 180, 182, 185, 191, 192, 205, 221, 233, 238, 298, 306, 309, 314, 323]

Dudewicz, E. J. and Ralley, T. G. (1981). *The Handbook of Random Num-*

ber Generation and Testing with TESTRAND Computer Code. American Sciences Press, Inc., Columbus, Ohio. [97, 101, 113, 115, 117, 119, 123, 126, 127, 134, 143, 148, 160]

Dudewicz, E. J. and Taneja, V. S. (1981). A multivariate solution of the multivariate ranking and selection problem. *Communications in Statistics*, A10, 1849–1868. [312]

Dudewicz, E. J. and van der Meulen, E. C. (1981). Entropy-based tests of uniformity. *Journal of the American Statistical Association*, 76, 967–974. [113, 114]

Dudewicz, E. J. and van der Meulen, E. C. (1984). On assessing the precision of simulation estimates of percentile points. *American Journal of Mathematical and Management Sciences*, 4, 335–343. [323]

Dudewicz, E. J., van der Meulen, E. C., SriRam, M. G., and Teoh, N. K. W. (1995). Entropy-based random number evaluation. *American Journal of Mathematical and Management Sciences*, 15, 115–153. [114, 136]

Edissonov, I. (1994). The new ARSTI optimization method: adaptive random search with translating intervals. *American Journal of Mathematical and Management Sciences*, 14, 143–166. [420]

Franta, W. R. and Maly, K. (1977). An efficient data structure for the simulation event set. *Communications of the ACM*, 20, 596–602. [50]

Gafarian, A. V. and Danesh-Ashtiani, M. (1981). Mean value estimation of a covariance stationary, continuous-time process, with simulation output applications. *American Journal of Mathematical and Management Sciences*, 1, 237–265. [324]

Gait, J. (1977). A new nonlinear pseudorandom number generator. *IEEE Transactions on Software Engineering*, 3, 359–363. [128]

Gentle, J. E. (1985). Monte Carlo methods. In *Encyclopedia of Statistical Sciences*, Vol. 5 (edited by S. Kotz, N. L. Johnson, and C. B. Read). John Wiley & Sons, Inc., New York, pp. 280–287. [1]

Ghoshdastidar, D. and Roy, M. K. (1975). A study on the evaluation of Shell's sorting technique. *The Computer Journal*, 18, 234–235. [49]

Gibbons, J. D. (1997). *Nonparametric Methods for Quantitative Analysis (Third Edition)*. American Sciences Press, Inc., Columbus, Ohio. [182]

Golden, B. L. (editor) (1993). *Vehicle Routing 2000: Advances in Time-*

Windows, Optimality, Fast Bounds, & Multi-Depot Routing. American Sciences Press, Inc., Columbus, Ohio. [420]

Golden, B. L. and Assad, A. A. (editors) (1986). *Vehicle Routing with Time-Window Constraints: Algorithmic Solutions.* American Sciences Press, Inc., Columbus, Ohio. [420]

Golden, B. L. and Eiselt, H. A. (editors) (1992). *Location Modeling in Practice: Applications (Site Location, Oil Field Generators, Emergency Facilities, Postal Boxes), Theory, & History.* American Sciences Press, Inc., Columbus, Ohio. [420]

Henriksen, J. O. and Crane, R. C. (1983). *GPSS/H User's Manual.* Wolverine Software Corporation, Annandale, Virginia. [128]

Hoaglin, D.C. (1976). Theoretical properties of congruential random-number generators: an empirical view. *Memorandum NS-340.* Department of Statistics, Harvard University, Cambridge, Massachusetts. [126]

Hogben, D., Peavy, S. T., and Varner, R. N. (1971). *OMNITAB II User's Reference Manual.* Technical Note 552, National Bureau of Standards, Washington, D.C. [125]

Hyakutake, H. (1988). A general treatment for selecting the best of several multivariate normal populations. *Sequential Analysis,* 7, 239–251. [312]

IBM (1970). System/360 scientific subroutine package, version III, programmer's manual, program number 360A-CM-03X. *Manual GH20-0205-4* (Fifth Edition). IBM Corporation, White Plains, New York. [125]

IBM (1971). *General Purpose Simulation System V, Introductory User's Manual.* IMB Corporation, White Plains, New York. [459]

Jennergren, L. P. (1984). Another method for random number generation on microcomputers. *Simulation,* 40, 79. [131]

Johnson, M. E. (editor) (1988). *Simulated Annealing (SA) & Optimization: Modern Algorithms With VLSI, Optimal Design, & Missile Defense Applications.* American Sciences Press, Inc., Columbus, Ohio. [419]

Karian, Z. A. (1985a). GPSS for microcomputers: a software review of GPSS/PC. *American Journal of Mathematical and Management Sciences,* 5, 93–101. [447]

Karian, Z. A. (1985b). GPSS/PC, Discrete-event simulation on the IBM-PC. *Byte,* 10, No. 10, 295–301. [447]

Karian, Z. A. (1987). PC-SIMSCRIPT II.5. *Byte*, 12, No. 8, 244–246. [447]

Karian, Z. A. and Dudewicz, E. J. (1986). Discrete-event simulation on microcomputers. *Modeling and Simulation on Microcomputers: 1986* (edited by C. C. Barnett), The Society for Computer Simulation, La Jolla, California, 146–150. [447]

Karian, Z. A. and Dudewicz, E. J. (1999). *Fitting Statistical Distributions to Data: The Generalized Lambda Distribution (GLD) and the Generalized Bootstrap (GB) Methods.* Forthcoming with CRC Press LLC, Boca Raton. [153, 224]

Karian, Z. A., Dudewicz, E. J., and McDonald, P.(1996). The extended generalized lambda distribution system for fitting distributions to data: history, completion of theory, tables, applications, the 'final word' on moment fits, *Communications in Statistics: Simulation and Computation*, 25(3), 611–642. [214, 221]

Kinderman, A. J., Monahan, J. F., and Ramage, J. G. (1977). Computer methods for sampling from Student's t distribution. *Mathematics of Computation*, 31, 1009–1018. [187]

Kinderman, A. J. and Monahan, J. F. (1980). New methods for generating Student's t and gamma variables. *Computing*, 23, 369–377. [187]

Knuth, D. E. (1973). *The Art of Computer Programming, Volume 3/Sorting and Searching.* Addison-Wesley Publishing Company, Inc., Reading, Massachusetts. [46]

Konishi, S. (1991). Normalizing transformations and bootstrap confidence intervals. *The Annals of Statistics*, 19, 2209–2225. [98]

Kruskal, J. B. (1969). Extremely portable random number generator. *Communications of the ACM*, 12, 93–94. [125]

Law, A. M. and Kelton, W. D. (1982). *Simulation Modeling and Analysis.* McGraw-Hill Book Company, New York. [301, 313]

Learmonth, G. P. and Lewis, P. A. W. (1973). Naval Postgraduate School random number generator package LLRANDOM. *Technical Report*, Naval Postgraduate School, Monterey, California. [123]

L'Ecuyer, P. (1987). A portable random number generator for 16-bit computers. *Modeling and Simulation on Microcomputers: 1987* (edited by P. Hogan). The Society for Computer Simulation, San Diego, California, 1987, 45–49. [129]

Lewis, P. A. W., Goodman, A. S., and Miller, J. M. (1969). A pseudo-random number generator for the System/360. *IBM Systems Journal*, 8, 136–146. [123]

Lewis, T. G. and Payne, W. H. (1973). Generalized feedback shift register pseudo-random number algorithm. *Journal of the ACM*, 20, 456–468. [124]

Lin, Y. (1997). Asymptotics of bootstrapping mean on some smoothed empirical distribution. *Statistics & Decisions*, 15, 301–306. [332]

Little, J. D. C. (1961). A proof for the queueing formula $L = \lambda w$. *Operations Research*, 16, 651–65. [43]

Lurie, D. and Mason, R. L. (1973). Empirical investigation of several techniques for computer generation of order statistics. *Communications in Statistics*, 2, 363–371. [124]

Mamrak, S. A. and Amer, P. D. (1979). *Computer Science & Technology: A Methodology for the Selection of Interactive Computer Services.* National Bureau of Standards Special Publication 500-44, U.S. Government Printing Office, Washington, D.C. [312]

Marsaglia, G. (1968). Random numbers fall mainly in the planes. *Proceedings of the National Academy of Sciences*, 61, 25–28. [116]

Marsaglia, G., Ananthanarayanan, K., and Paul, N. (1973). *How to Use the McGill Random Number Package "SUPER-DUPER."* Four page typed description, School of Computer Science, McGill University, Montreal, Quebec, Canada. [124]

Marsaglia, G. and Bray, T. A. (1968). One-line random number generators and their use in combinations. *Communications of the ACM*, 11, 757–759. [123, 124]

Mathews, I. L. (1986). The generation and testing of pseudo-random numbers on computers. *Honors Project*, Department of Mathematical Sciences, Denison University, Granville, Ohio. [128, 134]

Mathews, I. L., Karian, Z. A., and Dudewicz, E. J. (1987). The generation and testing of pseudo-random numbers with TESTRAND on VAX computers. *Modeling and Simulation on Microcomputers: 1987* (edited by P. Hogan). The Society for Computer Simulation, San Diego, California, 1987, 50–55. [128]

Mazumdar, M., Coit, D. E., and Shih, F.-R. (1999). An efficient Monte Carlo method for assessment of system reliability based on a Markov model.

American Journal of Mathematical and Management Sciences, 19, to appear. [325]

McQuay, W. K. (1973). Computer simulation methods for military operations research. *Technical Report AFAL-TR-73-341.* Air Force Avionics Laboratory, Air Force Systems Command, Wright-Patterson Air Force Base, Ohio, October 1973. [47]

Micceri, T. (1989). The unicorn, the normal curve, and other improbable creatures, *Psych. Bull.*, 105, 156–166. [221]

Minuteman Software. (1992). *GPSS/PC*™ *Reference Manual, General Purpose Simulation.* Minuteman Software, P. O. Box 131, Holly Springs, NC 27540. [448]

Miyazaki, H. (1987). Pseudorandom numbers generated by the doubly multiplicative congruential method. *New Perspectives in Theoretical and Applied Statistics* (edited by M. L. Puri, J. P. Vilaplana, and W. Wertz). John Wiley & Sons, Inc., New York, 319–325. [128]

Nelson, W. (1979). *How to Analyze Data With Simple Plots.* Volume 1 of The ASQC Basic References in Quality Control: Statistical Techniques (E. J. Dudewicz, Editor). American Society for Quality Control, Milwaukee, Wisconsin. [239]

Nie, N., Bent, D. H., and Hull, C. H. (1970). *SPSS—Statistical Package for the Social Sciences.* McGraw-Hill, Inc., New York. [125]

Öztürk, A. and Dudewicz, E. J. (1992). A new statistical goodness-of-fit test based on graphical representation. *Biometrical Journal*, 34, 403–427. [240]

Parsons, R. and Johnson, M. E. (1997). A case study in experimental design applied to genetic algorithms with applcations to DNA sequence assembly. *American Journal of Mathematical and Management Sciences*, 17, 369–396. [420]

Parzen, E. (1962). *Stochastic Processes.* Holden-Day, Inc., San Francisco, California. [169]

Payne, W. H., Rabung J. R., and Bogyo, J. P. (1969). Coding the Lehmer pseudo-random number generator. *Communications of the ACM*, 12, 85–86. [128]

Pearson, E. S. and Please, N. W. (1975). Relation Between the Shape of Population Distribution. *Biometrika*, 62, 223–241. [221]

Plackett, R. L. (1965). A class of bivariate distributions. *Journal of the American Statistical Association*, 60, 516–522. [225]

Ramberg, J. S., Dudewicz, E. J., Tadikamalla, P. R., and Mykytka, E. F. (1979). A probability distribution and its uses in fitting data. *Technometrics*, 21, 201–214. [216, 219, 220]

Rao, C. R. (1989). *Statistics and Truth, Putting Chance to Work*. First published by the Council of Scientific & Industrial Research, New Delhi, India. Reprinted and published in the U. S. A. by International Co-operative Publishing House, P. O. Box 245, Burtonsville, Maryland 20866–0245. [89]

Rey, W. J. J. (1990). Can I trust my random generator? A new test of uniformity. Abstract, International Conference on Bootstrapping and Related Techniques, University of Trier, Germany. [132]

Romeu, J. L. (1990). Development and evaluation of a general procedure for assessing multivariate normality. *Ph. D. Dissertation*, Syracuse University, Syracuse, New York. [240]

Ross, S. M. (1985). *Introduction to Probability Models*. Academic Press, Inc., New York. [44]

Schrage, L. (1979). A more portable FORTRAN random number generator. *ACM Transactions on Mathematical Software*, 5, 132–138. [128]

Schriber, T. J. (1974). *Simulation Using GPSS*. John Wiley & Sons, Inc., New York. [249]

Schriber, T. J. (1990). *An Introduction to Simulation Using GPSS/H*. John Wiley & Sons, Inc., New York. [295]

Schriber, T. J. and Andrews, R. W. (1984). ARMA-based confidence intervals for simulation output analysis. *American Journal of Mathematical and Management Sciences*, 4, 345–373. [324]

Seo, T. and Siotani, M. (1992). The multivariate Studentized range and its upper percentiles. *Journal of the Japan Statistical Society*, 22, 123–137. [89]

Sezgin, F. (1991). An improved combining method for random number generation. *The Frontiers of Statistical Computation, Simulation, and Modeling, Volume I of the Proceedings of ICOSCO-I (The First International Conference on Statistical Computing)* (P. R. Nelson, Chief Editor). American Sciences Press, Inc., Columbus, Ohio, 237–252. [128]

Shannon, R. E. (1975). *Systems Simulation, the Art and Science*. Prentice-

Hall, Inc., Englewood Cliffs, New Jersey. [304]

Shapiro, S. S. and Brain, C. W. (1982). Recommended distributional testing procedures. *American Journal of Mathematical and Management Sciences*, 2, 175–221. [239]

Shell, D. L. (1959). A high speed sorting procedure. *Communications of the ACM*, 2, 30–32. [49]

Stadlober, E. (1989). *Book Review* of "Principles of Random Variate Generation" by John Dagpunar. *Journal of Quality Technology*, Vol. 21, pp. 290–291. [181]

Sun, L. and Müller-Schwarze, D. (1995). Statistical resampling methods in biology: a case study of beaver dispersal patterns. *American Journal of Mathematical and Management Sciences*, 16, 463–502. [322]

Swain, C. G. and Swain, M. S. (1980). A uniform random number generator that is reproducible, hardware-independent, and fast. *Journal of Chemical Information and Computer Sciences*, 20, 56–58. [126]

Tadikamalla, P. R. (editor) (1984). *Modern Digital Simulation Methodology: Input, Modeling, and Output.* American Sciences Press, Inc., Columbus, Ohio. [180, 194]

Tadikamalla, P. R. and Johnson, M. E. (1981). A complete guide to gamma variate generation. *American Journal of Mathematical and Management Sciences*, 1, 213–236. [195]

Tadikamalla, P. R. and Johnson, M. E. (1982). Correction note. *American Journal of Mathematical and Management Sciences*, 2, 93. [195]

Taylor, H. M. and Karlin, S. (1984). *An Introduction to Stochastic Modeling.* Academic Press, Inc., New York. [44]

Thesen, A. (1985). An efficient generator of uniformly distributed random variables between zero and one. *Simulation*, 44, 17–20. [127]

Ulam, S. M. (1976). *Adventures of a Mathematician.* Scribner, New York. [1]

Ulrich, E. G. (1978). Event manipulation for discrete simulations requiring large numbers of events. *Communications of the ACM*, 21, 777–785. [49]

U.S. News and World Report (1997). Issue of 1/29/97. [1]

Vasicek, O. (1976). A test of normality based on sample entropy. *Journal of the Royal Statistical Society, Series B*, 38, 54–59. [114]

Vuillemin, J. (1978). A data structure for manipulating priority queues. *Communications of the ACM*, 21, 309–315. [50]

Walker, R. J. (1967). *An Instruction Manual for CUPL—The Cornell University Programming Language*. Cornell University, Ithaca, New York. [125]

West, W. W. III and Johnson, G. D. (1984). *SIMSCRIPT II.5 User's Manual, VAX/VMS*. C.A.C.I., 12011 San Vicente Boulevard, Los Angeles, California 90049. [128]

Whittlesey, J. R. B. (1968). A comparison of the correlational behavior of random number generators for the IBM 360. *Communications of the ACM*, 11, 643–644. [128]

Wilcox, R. R. (1990). Comparing the means of two independent groups, *Biometrical Journal*, 7, 771–780. [221]

Wilson, J. R. (1984). Variance reduction techniques for digital simulation. *American Journal of Mathematical and Management Sciences*, 4, 277–312. [325]

Zarling, R. L. (1971). Personal communication from Dr. Raymond L. Zarling. Marietta College, Marietta, Ohio. [125]

Zheng, X. (1994). Third-order correct bootstrap calibrated confidence bounds for nonparametric mean. *Mathematical Methods of Statistics*, 3, 62–75. [123]

Subject Index